Nano-Dimensional Zinc Oxide for Energy Harvesting, Sensing, and Environmental Remediation

Zinc oxide (ZnO) nanostructures have gained significant attention in recent years due to their applications in various fields. This book provides a comprehensive overview of the recent advances in ZnO-based nanostructures for energy harvesting, sensing, and environmental remediation applications. It covers advanced aspects of application-based research on ZnO nanostructures. The book also includes biosynthesized ZnO nanostructures and their role in targeted drug delivery.

Features:

- Explores various synthesis techniques for ZnO nanostructures, detailing their impact on material properties and applications in emerging technologies
- Discusses the role of ZnO-based nanostructures in solar cells, photocatalysis, and environmental remediation, including heavy metal detection and removal
- Highlights ZnO nanostructures' potential in electrochemical sensors for medical diagnostics, antimicrobial applications, and gas sensing technologies
- ZnO's piezoelectric properties, ultraviolet photodetectors, and its integration into next-generation electronic and optoelectronic devices are discussed in depth
- Provides insights into ZnO nanostructures role in plant growth enhancement, along with discussions on future research directions and technological advancements in the field

This is a reference book for academicians and researchers interested in the potential of nano-dimensional zinc oxide.

Nano-Dimensional Zinc Oxide for Energy Harvesting, Sensing, and Environmental Remediation

Edited by
Nupur Aggarwal
Naveen Kumar
Shivani Singla

CRC Press
Taylor & Francis Group
Boca Raton London New York

CRC Press is an imprint of the
Taylor & Francis Group, an **informa** business
A CHAPMAN & HALL BOOK

Designed cover image: ShutterStock

First edition published 2025
by CRC Press
2385 NW Executive Center Drive, Suite 320, Boca Raton FL 33431

and by CRC Press
4 Park Square, Milton Park, Abingdon, Oxon, OX14 4RN

CRC Press is an imprint of Taylor & Francis Group, LLC

ISBN: 978-1-032-73445-3 (hbk)
ISBN: 978-1-032-73580-1 (pbk)
ISBN: 978-1-003-46490-7 (ebk)

DOI: 10.1201/9781003464907

Typeset in Times
by SPi Technologies India Pvt Ltd (Straive)

Contents

Editors

Dr. Nupur Aggarwal is working as an associate professor at Chandigarh University, India, where she spearheads interdisciplinary research in materials science. Her work focuses on the synthesis of metal oxide nanoparticles, targeting applications in optoelectronics and spintronics. Over the past 9 years, Dr. Aggarwal has delved deeply into the field of dilute magnetic semiconductors, striving to unlock new potentials in electronic and spintronic devices by enhancing their magnetic and electrical properties. With more than 55 research publications in reputed SCI and SCOPUS-indexed journals, she plays a vital role as a reviewer and editor for esteemed journals, including the *American Journal of Nano Research and Applications (NANO)*, *PLOS ONE*, and *Frontiers in Nanotechnology*, where she helps shape the direction of future research in nanotechnology.

Dr. Naveen Kumar is an associate professor in the Galgotias Multidisciplinary Research & Development Cell (G-MRDC), Galgotias University, Greater Noida, India. He works on interdisciplinary research fields involving very recent and technologically significant strands: the synthesis of multiferroics for dielectric and ferroelectric studies. He has been working in the field of magnetic ferrites such as $BiFeO_3$, Fe_3O_4, and $Mg_{0.05}Zn_{0.05}Fe_2O_4$ for the past 8 years and has published more than 100 research articles in peer-reviewed SCI/SCOPUS-indexed journals and several research papers in international conferences. He is also an editor of reputed journals such as the *American Journal of Nano Research and Applications* (NANO), *Concepts in Magnetic Resonance Part A*, *PLOS ONE*, and *Synthesis, Characterization and Processing of New Materials for Innovative Applications*.

Dr. Shivani Singla is an assistant professor at Chandigarh University, India, and excels in interdisciplinary materials science, specializing in the synthesis and characterization of advanced glass and glass-ceramic materials for optics, photonics, and radiation shielding. Her research delves into nanoparticle interactions in glass matrices, enabling advancements in display technologies, cancer detection, and shielding applications. With over 20 publications in leading journals and 10+ presentations at global conferences, Dr. Singla's work has driven innovations in materials science, earning patents in optical materials and education technology.

Contributors

Nupur Aggarwal
Chandigarh University
Chandigarh, India

Sandeep Arya
University of Jammu
Jammu, India

Kanchan Bala
Government Mohindra College
Patiala, India

Indu Gupta
Chandigarh University
Chandigarh, India

Leena Jaswal
TRAMIET
Mandi, India

Rahul Jain
Chandigarh Group of Colleges
Punjab, India

Anu Kapoor
Chandigarh University
Ajitgarh, India

Naveen Kumar
Galgotias University
Greater Noida, India

Shubham Kumar Patial
Academy of Scientific and Innovative Research (AcSIR)
Ghaziabad, India
Materials Science and Sensor Applications, CSIR-Central Scientific Instruments Organisation
Chandigarh, India

and

School of Science
STEM College,
RMIT University
Melbourne, Victoria, Australia
Functional Materials and Microsystems
RMIT University
Melbourne, Victoria, Australia

Suresh Kumar
Maharishi Markandeshwar (Deemed to be University)
Haryana, India

Madan Lal
Graphic Era (Deemed to be University)
Dehradun, India

Mandeep Kaur
Chandigarh University
Chandigarh, India

Bastola Narayan
University of Bath
Bath, United Kingdom

Ranvir Singh Panwar
NMIMS (Deemed to be University)
Chandigarh, India

Payal Patial
Chandigarh University
Punjab, India

Anirudh Vashisht
Chandigarh University
Chandigarh, India

Deepika Sharma
IEC University
Himachal Pradesh, India

Deeksha Sharma
Chandigarh University
Chandigarh, India

Nandni Sharma
Chandigarh University
Chandigarh, India

Navdeep Sharma
Abhilashi University
Mandi, India

Prateeksha Sharma
Chandigarh University
Chandigarh, India

Anuj Sharma
Materials Science and Sensor Applications, CSIR-Central Scientific Instruments Organisation
Chandigarh, India

Gurbhej Singh
University Institute of Engineering
Chandigarh University
Chandigarh, India

Suman Singh
Academy of Scientific and Innovative Research (AcSIR)
Ghaziabad, India
Materials Science and Sensor Applications
CSIR—Central Scientific Instruments Organisation
Chandigarh, India

Shivani Singla
Chandigarh University
Chandigarh, India

Ashish Tiwari
University of Delaware
Newark, Delaware, United States

Neelam Vishwakarma
Academy of Scientific and Innovative Research (AcSIR)
Ghaziabad, India
Materials Science and Sensor Applications
CSIR—Central Scientific Instruments Organisation
Chandigarh, India

and

Functional Materials and Microsystems
RMIT University
Melbourne, Victoria, Australia

1 Synthesis Techniques for ZnO Nanostructures

Nupur Aggarwal, Deeksha Sharma, and Prateeksha Sharma

1.1 INTRODUCTION

Most of the divisions of current material sciences are popular and efficient, making it possible to manipulate at atomic scales and produce novel features, starting with nanoscale engineering and nanotechnology. Applying scientific ideas with nanoscale engineering solutions is beneficial in biomedical as well as biotechnology, benefiting human life in a wide range of areas [1]. The materials at the subatomic scale involve designs and characteristics that can be implemented to investigate new properties based on imaging, modeling, and other characterization techniques [2] Metal salt solutions and aqueous solutions of plant extract are employed during the biogenic preparation of metal oxide nanomaterials, primarily because the rate of reaction between the metal salts and plant extract is very high even at room temperature.

Biogenic or green route synthesis of metal oxide nanomaterials has an advantage since such synthesis does not involve toxic reagents, is easy to handle, no machinery is required, and uses water or ordinary solvents such as ethanol or ethyl acetate. These days, it is evident that many researchers are interested in the synthesis of metal oxide nanoparticles (MONPs) by the green route. The preparation of MONPs has received significant attention in the last decades due to the advantages, namely, the ability to produce on an industrial scale, low-cost synthesis, and compatibility of the manufactured MONPs with biological systems [3]. The biosynthesis of MONPs promises perspectives for straightforward synthesis with reduced environmental impacts.

1.2 ZINC OXIDE (ZnO)

ZnO is a dilute magnetic n-type semiconductor with high excitonic energy and a wide optical band gap, making it a promising candidate for next-generation applications across spintronics, optoelectronics, energy harvesting, and piezoelectric devices. Among various metal oxides like SnO_2, NiO, CuO, ZnO, and TiO_2 synthesized, ZnO stands out for its promising properties and relatively simple, cost-effective preparation process. Due to its large optical band gap ($E_g \approx 3.6$ eV) and high binding energy of around 60 meV, ZnO NPs have found applications in various industries, owing to their optical activity, catalytic properties, and antimicrobial potential [4]–[6].

DOI: 10.1201/9781003464907-1

Additionally, they have been used in the production of rubber, paint, water purification, protein adsorption, and dentistry applications. Piezoelectric and pyroelectric characteristics are present in ZnO NPs [7], which make it helpful in providing access to aquatic vegetation that is immune to all types of mechanical, chemical, and physical extinction methods. ZnO possesses various morphologies in the nano range, such as nanoribbons, nanoflowers, nanoparticles (NPs), nanorods, and nanowires [8].

1.2.1 Preparation of Zinc Oxide Nanoparticles

Conventional synthesis methods for nanostructures are shown in Figure 1.1. There are two prevalent approaches for fabricating NPs, namely:

1.2.1.1 Top-Down Approach

The top-down approach creates nanoscale structures by reducing the size of more extensive bulk materials. This technique involves breaking down a bulk material into smaller and smaller pieces using various mechanical and physical methods until the desired nanoscale dimensions are achieved. The goal is to produce NPs or nanostructures by starting with a macroscopic material and systematically reducing its dimensions.

In this approach, the size of the material is reduced mechanically as the bulk material is gradually broken down into nanoscale structures. Physical and chemical procedures are most typically used for size reduction. However, the method has flaws, such as surface flaws and increased porosity, which restricts its application [9]. NPs are made by breaking macroscopic systems to nanometers in the "top-down" synthesis.

1.2.1.2 Bottom-Up Approach

The bottom-up approach is a method for creating nanoscale structures by assembling them from atomic or molecular components. This technique builds NPs and nanostructures starting from individual atoms or molecules, which then self-organize into the desired forms under controlled conditions. The bottom-up approach produces highly uniform and precisely controlled nanostructures, often with fewer defects than the top-down approach.

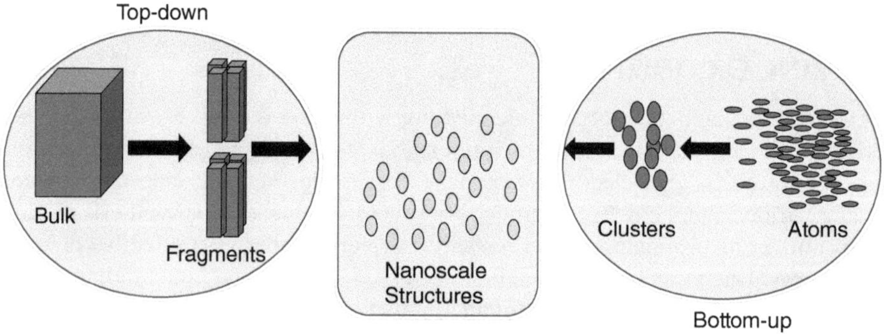

FIGURE 1.1 Conventional synthesis approaches for nanostructure formation.

The bottom-up technique is used to assemble NPs from atomic or molecular components. This approach is far more famous for making NPs since it progresses in a controlled way. In the bottom-up approach, we chemically added two or more precursors to produce nanostructures under controlled atmospheric conditions [10].

1.2.2 Various Methods of Synthesis

The nanomaterials can be prepared using physicochemical methods and biological techniques, as depicted in Figure 1.2.

(i) **Physical methods**: NPs, with their unique size-dependent properties, have revolutionized various fields, from electronics to medicine. Synthesizing NPs with precise control over size, shape, and composition is crucial for tailoring their properties for specific applications. In this chapter, we explore the fundamental principles and methodologies of NP synthesis using physical approaches.

Ball milling and laser ablation are widely used methods for ZnO NP synthesis. ZnO powders are mechanically ground into NPs in ball milling, often used for catalytic and antibacterial applications. In laser ablation, bulk ZnO targets are irradiated in water to produce high-purity NPs with controlled size and morphology. Additionally, physical vapor deposition (PVD) is employed to synthesize monodispersed ZnO NPs, particularly for optoelectronic and sensor applications.

Wirunchit et al. synthesized ZnO NPs for biological uses using a ball-milling method at temperatures between 400 and 900°C. The best results were at 800°C, producing small, round particles with strong antibacterial activity [11].

Chen et al. fabricated ZnO NPs using pulsed laser ablation of zinc powders in deionized water, resulting in irregular particles sized 10–50 nm with a wurtzite structure.

They showed optical absorption near 380 nm and emitted ultraviolet and visible light, indicating some defects [12].

Mohammed et al. made ZnO thin films on glass using the PVD method, resulting in 130 nm thick films with a pure wurtzite structure. They exhibited 90% transmittance in the visible range and an energy bandgap of about 3.24 eV [13].

(ii) **Chemical methods** for NP synthesis involve chemical reactions to induce the formation of NPs from precursor compounds. These methods offer versatility in controlling NP size, shape, composition, and surface properties.

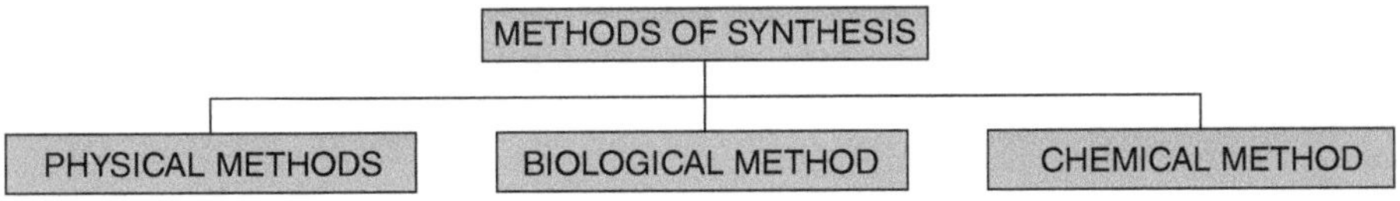

FIGURE 1.2 Depicting various methods of synthesis of nanoparticles (NPs).

For example, sol-gel synthesis uses zinc acetate hydrolysis to form a gel, which is calcined to produce ZnO NPs. In co-precipitation, zinc nitrate is mixed with sodium hydroxide to create ZnO NPs. Hydrothermal synthesis involves reacting metal precursors in a high-temperature, high-pressure solution, yielding well-crystallized ZnO NPs.

Arya et al. concluded that parameters like pH, additives, annealing temperature, and calcination significantly affect the morphology and optical properties of ZnO NPs synthesized via the sol-gel method, enabling their optimization for various applications [14].

Kotresh et al. synthesized ZnO NPs using co-precipitation at 50°C and 70°C with zinc nitrate hexahydrate and sodium hydroxide. Characterization revealed average crystalline sizes of 24 nm and 29 nm, respectively. The NPs exhibited broad absorption and narrow emission spectra, making them suitable for optoelectronic devices and biosensors [15].

Gerbreders et al. fabricated ZnO nanostructures using a nitrate-based precursor and hexamethylenetetramine (HMTA). The study examines how variations in reagent concentration, growth time, temperature, pH, and capping agents influence morphology, identifying optimal parameters for nine distinct shapes [16].

1.2.3 Hindrances of Utilizing Physical and Chemical Methods

Chemical and physical routes of production of NPs are relatively expensive. On the contrary, physical synthesis techniques necessitate complex instrumentation and expensive, high-power radiative heating equipment. The time duration, ablation period, and the nature of the laser light, i.e., wavelength, etc., all affect the synthesis of NPs using a laser ablation approach [17]. Alcohol, sodium citrate, and sodium borohydride stabilize and reduce carriers in the chemical method of NPs formation [18]. For the chemical synthesis of nanomaterials, synthetic and natural polymers such as cellulose, chitosan, natural rubber, and micelles are employed. Because these chemicals are non-biodegradable and hazardous, their use is restricted. According to a recent study, the hazardous nature of chemicals hurts the surface of NPs, causing contamination and restricting their use in medicinal applications [19]. Green synthesis or biosynthesis is recommended to eliminate all these problems and disadvantages of physicochemical synthesis procedures because it uses plant derivatives or microorganisms, which are cost-effective, environmentally benign, and non-toxic.

1.2.4 Biological Synthesis

Because it is straightforward, affordable, and primarily repeatable, the biosynthesis of NPs is advantageous over physicochemical methods. Microorganisms, such as bacteria and algae, are used to synthesize NPs, but the fabrication rate is prolonged, and it prepares NPs in a small range. Control over morphology is difficult for this method of synthesis. Also plant-based synthesis of NPs requires no sophisticated instrumentation, less energy and temperature, no use of hazardous chemicals, ease of handling, and it is cost-effective. Due to all these parameters, researchers are nowadays adopting this green synthesis route to fabricate nanomaterials, as it is

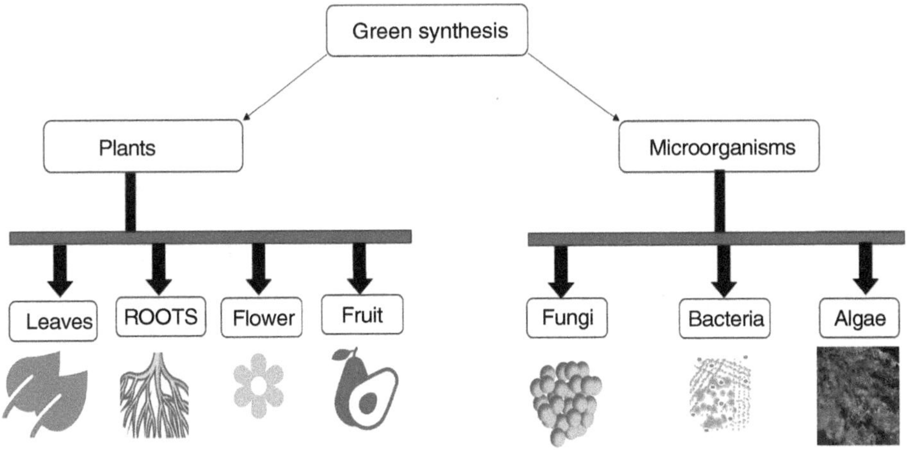

FIGURE 1.3 Different aspects of green route synthesis.

cost-effective, less contaminated, more efficient, and stable. Therefore, green route synthesis has become another safer approach for the fabrication of NPs. Synthesis via the green route involves different biological components, such as plants and micro-organisms. Researchers are optimizing different plant extracts and various parts of plants, such as fruits, flowers, roots, and leaves, which can synthesize nanomaterials, as shown in Figure 1.3.

For instance, the use of renewable resources, better solvents, and auxiliaries while making safer compounds are the key benefits of biological synthesis. The synthesis of nanomaterials at a large scale is still a challenge as it requires a sophisticated environment and controlled atmospheric conditions. However, its commercial use is expected to occur shortly as much research is ongoing for a better understanding of the chemical composition of a wide range of available flora and fauna, which can interact with metal ions to improve their characteristics [20, 21].

1.2.4.1 Plant-Mediated ZnO NPs

Despite using a physicochemical synthesis of nanostructures, the biological method is more acceptable because of its wide availability, good results, and eco-friendly nature. Most studies concentrated on environmentally friendly ways to create metal and oxide NPs. Plants may be used to quickly and cheaply develop safe NPs for human consumption.

The most evident technique for fast fabrication of ZnO NPs from plant derivatives involves thoroughly washing the plant component in regular water and sterilizing it with double-distilled water. The leaves and by-products are then dried and crushed in a mortar and pestle to get the dried powder and diluted later to prepare the extract. The process of creating NPs by biosynthesis utilizing different plant extracts is shown in Figure 1.4.

Zinc nitrate hexahydrate is a commonly used precursor to synthesize ZnO NPs using *Vitex negundo* plant extract. Biosynthesized ZnO NPs showed antimicrobial potential against *Staphylococcus aureus* and *Escherichia coli* [22].

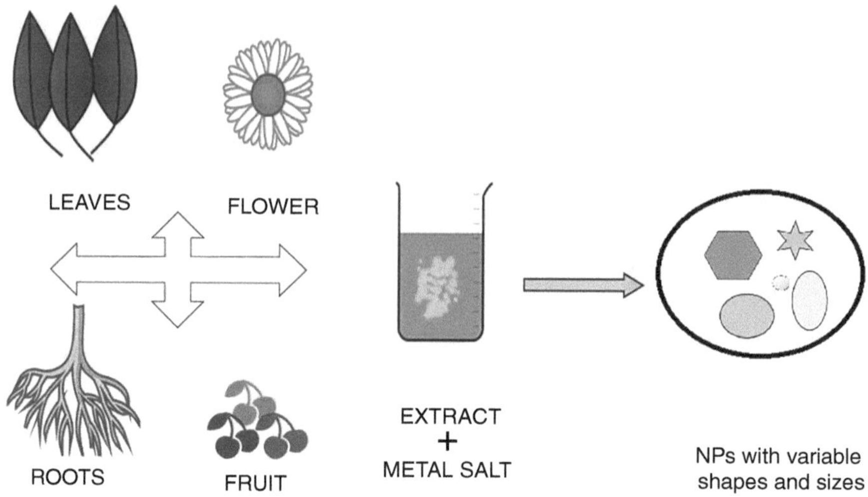

FIGURE 1.4 Biosynthesis of nanoparticles using different plant extracts.

Dobrucka et al. prepared ZnO NPs using *Trifolium pratense*; they discovered that the synthesized ZnO NPs were wurtzite in structure with sizes ranging from 60 to 70 nm [23]. The effectiveness of biosynthesized ZnO NPs synthesized from *Passiflora caerulea* leaf extract was treated against dandruff, microbes, and arthritis [24].

Santhosh Kumar [25] et al. tested them against a pathogenic culture isolated from the urine of a patient with a urinal infection. The findings demonstrated that the synthesized ZnO NPs had antibacterial properties that were effective against urinary tract infections.

Nava et al. [26] focused on the environmentally friendly, low-cost production of ZnO NPs made from camellia sinensis extract. Siripireddy et al. [27] used *Eucalyptus globulus* to investigate the effectiveness of ZnO NPs as effective photocatalysts for the removal of toxic pollutants in organic dyes, such as methylene blue and methyl orange, and the enhanced antioxidant activity of ZnO NPs.

The authors analyzed the impact of plant derivative and precursor zinc content, ambient temperature, and stirring duration. They concluded that employing the most miniature zinc precursor will yield better results [28].

Goswami et al. investigated how temperature variations between 20 and 100 °C alter the morphology and size of ZnO NPs. The reaction's temperature increased the particle size, which is consistent with research on the characteristics of ZnO made using the sol-gel synthesis method [29]. Due to the short reaction time, which slows grain expansion, quick synthesis appears to produce smaller particles.

According to a similar mechanism reported by Anbuvannan et al. [30], the antioxidants in the leaves of *Moringa oleifera* combined with the Zn^{2+} ions result in ZnO NPs on calcination. This study used the FTIR (Fourier Transform Infrared Spectroscopy) technique to detect phytochemicals in the extract and ZnO NPs produced after treatment at different temperatures (100 °C and 500 °C). ZnO NPs calcined at 100 °C showed high-frequency Zn-O stretching bend in the FTIR spectrum that may indicate

TABLE 1.1
Synthesis of ZnO Nanoparticles Using Various Plant Extracts for Different Applications

Plant Extract	Morphological Analysis	Applications	References
Tecoma castanifolia	Spherical/70–75 nm	Anticancer	[32]
Vitis vinifera	Spherical/40–60 nm	Antimicrobial	[33]
Monsonia burkeana	Hexagonal/5–15 nm	Antibacterial	[34]
Pongamia pinnata	Spherical/~100 nm	Antibacterial	[35]
Costus pictus D. Don	Hexagonal and spherical/40–80 nm	Antimicrobial	[36]
Mangifera indica	Spherical and hexagonal/45–60 nm	Anticancer	[37]
Ocimum basilicum	Irregular particle shapes/10–25 nm	Antibacterial	[38]
Dysphania ambrosioides	Quasi-spherical and hexagonal shapes/7–130 nm	Antibacterial	[39]
Cassia fistula	Spherical shape/3–68 nm	Antibacterial	[40]
Eclipta prostrata	Hexagonal/17–30 nm	Antioxidant	[41]
Sargassum sp.	Irregular particle shapes/80–100 nm	Antibacterial, anti-inflammatory	[42]

the formation of the ZnO compound [31]. Table 1.1 lists the applications of different plant extracts in synthesizing different NPs and highlights those applications in various sectors.

1.2.4.2 Microbial Synthesis

Microbial synthesis of ZnO NPs is an innovative and environmentally friendly approach that uses microorganisms such as bacteria, fungi, and algae to produce NPs. This method leverages the natural metabolic processes of these organisms to reduce zinc salts into ZnO NPs, often under mild conditions. Microorganisms are cultured in nutrient media to grow and increase, after which their biomass or cell-free extracts are harvested. These extracts contain enzymes, proteins, and metabolites that act as reducing and stabilizing agents. When introducing a zinc salt solution, these biological molecules facilitate the reduction of zinc ions (Zn^{2+}) to ZnO, forming NPs. The NPs are then isolated through centrifugation and purified through rinsing with distilled water and ethanol. This green synthesis method is advantageous due to its eco-friendliness, cost-effectiveness, and ability to produce biocompatible NPs. However, it may face challenges such as variability in NP properties and scalability issues. The synthesis of NPs mediated by microorganisms is depicted in Figure 1.5, which also highlights the microbial pathways involved in the manufacture of nanomaterials through environmentally friendly biosynthetic processes.

The growing field of nanoscience has seen a great inclination toward bio and green synthesis of nanomaterials using microorganisms. These methods offer environmental friendliness, cost-effectiveness, and energy efficiency. Various organisms have been utilized for nanomaterial synthesis, leading to diverse applications across

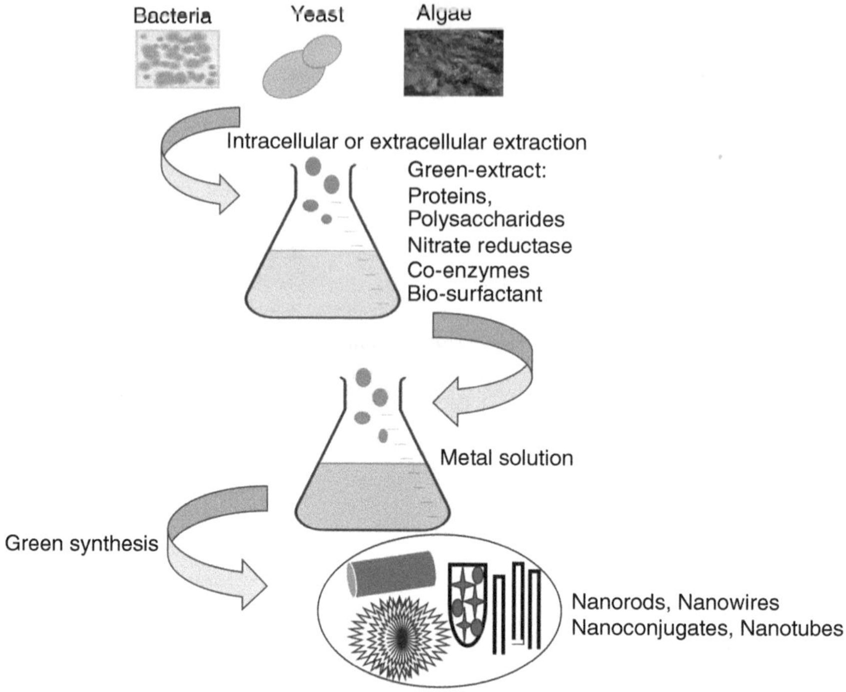

FIGURE 1.5 Synthesis of nanoparticles by microorganisms.

agriculture, bioremediation, the food industry, biotechnology, and biomedicine. Continued advancements in bioproduction technologies promise to expand the use of nanomaterials, offering potential benefits to society [43]. The manufacture of different NPs utilizing different microbes is summarized in Table 1.2, along with some of the potential uses for them in various sectors.

1.2.4.2.1 Algal Synthesis

Algae can produce ZnO NPs through a sustainable and environmentally friendly process called algal synthesis. Algae, both microalgae and macroalgae, are cultured in nutrient-rich media under controlled light and temperature conditions to promote growth and biomass production. After that, the algal biomass is collected and treated to create an extract that includes bioactive substances like proteins, polysaccharides, and antioxidants. These compounds serve as natural reductants and stabilizers. In this synthesis process, the Zn^{2+} ions are reduced to ZnO via the bioactive compounds found in the algal extracts combined with an aqueous solution containing a zinc salt, such as zinc nitrate or zinc acetate, and the mixture eventually leads to the production of ZnO NPs. This is a green method and also economically efficient since algae are an abundant and renewable resource. In addition, ZnO NPs generated in this way commonly show biocompatibility and can therefore be applied to a wide range of biomedical and environmental uses. However, practical-scale applications may still face some drawbacks, such as variation in composition with algal extracts and the

TABLE 1.2
Synthesis of ZnO Nanoparticles Using Various Microorganisms for Different Applications

Microbes	Morphological Analysis	Applications	References
Arthrospira platensis	Spherical/30–55 nm	Anticancer	[44]
Cystoseira crinita	Multilayered rectangular shape/23–200 nm	Water purification	[45]
Chlorella vulgaris	Rod-shaped/length and width of 633.77 nm and 75.61 nm, respectively	Biomedical textiles	[46]
Nostoc sp.	Spherical/~28 nm	Antioxidant activity	[47]
Spirulina platensis	Spherical, cube, and nanorod shapes/120–180 nm	Antimicrobial coatings	[48]
Evolvulus alsinoides	Spherical/~100 nm	Biomedical	[49]
Datura metel (*D. Metel*)	Spherical/23.67 nm	Antibacterial	[50]
Ziziphus spina-christi	Spherical	Antibacterial	[51]
Kombucha SCOBY	Spherical/23 nm	Antibacterial	[52]
Pseudomonas aeruginosa	Hexagonal/44 nm	Antibacterial	[53]
Arthrospira platensis	Spherical/30–55 nm	Biomedical	[54]
Cystoseira crinita	Multilayered rectangular shape/23–200 nm	Antioxidant	[55]

difficulty in scale-up production. The general mechanism behind the creation of NPs mediated by algae is depicted in Figure 1.6.

Algae-mediated green synthesis of MONPs is regarded as an eco-friendly and cost-effective bioremediation technique. The synthesis of NPs using algae includes simplicity, scalability, and fast production. The method is slowly gaining popularity as an authentic solution for searching for pollutants from water resources such as

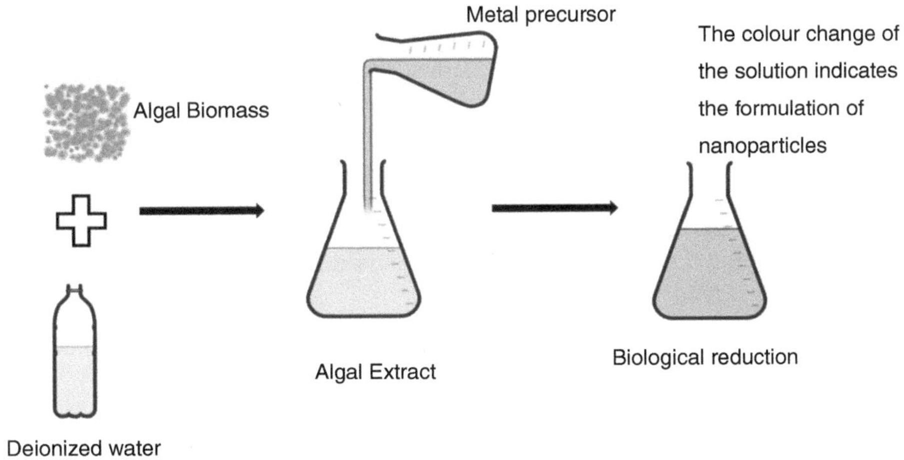

FIGURE 1.6 General mechanism for algal-mediated synthesis.

TABLE 1.3
Synthesis of ZnO Nanoparticles Using Various Algae Extracts for Different Applications

Algae Extract	Morphological Analysis	Applications	References
Marine brown algae (Cystoseira)	Spherical/>100 nm	Dye-sensitized solar cells	[57]
Halimeda opuntia	Hexagonal/33 nm	Antibacterial agent in aquaculture	[58]
Sargassum algae	Spherical/30–50 nm	Antidiabetic agent	[59]
Sargassum wightii ethanol	Spherical, and hexagonal/ 20–31 nm	Anticancer	[60]
Marine algae	Spherical/6.3–15.7 nm, Rod-shaped/242–862 nm	Bioaccumulation	[61]
Tetraselmis India	Spherical/20–40 nm	Antioxidant	[62]
Sargassum *vulgare* (SVE)	Spherical/50–150 nm	Antifungal	[63]
Gracilaria edulis (GE)	Rod shaped/66 - 95 nm	anticancer	[64]
Spirulina platensis	Nanorods with spherical and cubic shapes/120–180 nm	Antioxidant	[65]
Sargassum natans	Laminar particles (flakes)/length 300 nm and thickness 20 nm	Anti-inflammatory	[66]
Phaeodactylum tricornutum	Spherical/39–51 nm	Food preservation	[67]

well, seawater, industrial wastewater, and all other kinds of water [56]. Table 1.3 illustrates how different algae extracts are used to synthesize different NPs, demonstrating their usefulness in a range of fields.

1.3 CONCLUSION

The production of MONPs utilizing plant derivatives is very demanding as it is cost-effective and accessible from any toxic product; hence, it is an environment-friendly synthesis route. This review describes different biological components and methodologies for fabricating ZnO NPs and their application in other fields, such as improving their antibacterial, anticancer, and photocatalytic activities. The biological synthesis of ZnO NPs is a chemical-free, cost-effective, environment-benign, and safe approach, enhancing its application area. Though there are numerous forms of plant derivatives and different synthesis routes using this biological component, this review has inferred that no universal route for plant extraction leads to specific outcomes for certain predefined applications.

REFERENCES

1. Mabrouk, M., Das, D. B., Salem, Z. A., & Beherei, H. H. (2021). Nanomaterials for biomedical applications: production, characterizations, recent trends and difficulties. *Molecules*, *26*(4), 1077.

2. Kalinin, S. V., Dyck, O., Balke, N., Neumayer, S., Tsai, W. Y., Vasudevan, R.,... & Strelcov, E. (2019). Toward electrochemical studies on the nanometer and atomic scales: Progress, challenges, and opportunities. *ACS Nano*, *13*(9), 9735–9780.
3. Annu, A. A., & Ahmed, S. (2018). Green synthesis of metal, metal oxide nanoparticles, and their various applications. *Handbook of Ecomaterials*, *2018*, 1–45.
4. Rai, R. S., & Bajpai, V. (2021). Recent advances in ZnO nanostructures and their future perspective. *Advanced Nano Research*, *11*(1), 37–54.
5. Chavali, M. S., & Nikolova, M. P. (2019). Metal oxide nanoparticles and their applications in nanotechnology. *SN Applied Sciences*, *1*(6), 607.
6. Mohammed, M., Adam, T., Ihmedee, F., Mohammed, A., AL_Essa, S. A. K., & Betar, B. (2024). Zinc oxide nanoparticles: A comprehensive review on its synthesis, anticancer and drug delivery applications. *Al-Rafidain Journal of Engineering Sciences*, *2*(2), 185–202.
7. Moarref, Z., Fahim, N. H., & Montazer, M. (2023). Zinc Oxide Piezoelectric Nanogenerators for Low-frequency Applications. *Nanotechnology in Electronics: Materials, Properties, Devices*, 63–95.
8. Ahmad, A. A., Alsaad, A. M., Aljarrah, I. A., Al-Bataineh, Q. M., & Telfah, A. D. (2022). Optical, electronic, and structural properties of different nanostructured ZnO morphologies. *The European Physical Journal Plus*, *137*(6), 752.
9. El-Khawaga, A. M., Zidan, A., & Abd El-Mageed, A. I. (2023). Preparation methods of different nanomaterials for various potential applications: A review. *Journal of Molecular Structure*, *1281*, 135148.
10. Abid, N., Khan, A. M., Shujait, S., Chaudhary, K., Ikram, M., Imran, M.,... & Maqbool, M. (2022). Synthesis of nanomaterials using various top-down and bottom-up approaches, influencing factors, advantages, and disadvantages: A review. *Advances in Colloid and Interface Science*, *300*, 102597.
11. Wirunchit, S., Gansa, P., & Koetniyom, W. (2021). Synthesis of ZnO nanoparticles by Ball-milling process for biological applications. *Materials Today: Proceedings*, *47*, 3554–3559.
12. Chen, W., Yao, C., Gan, J., Jiang, K., Hu, Z., Lin, J.,... & Wu, J. (2020). ZnO colloids and ZnO nanoparticles synthesized by pulsed laser ablation of zinc powders in water. *Materials Science in Semiconductor Processing*, *109*, 104918.
13. Mohammed, R., Ahmed, S., Abdulrahman, A., & Hamad, S. (2020). Synthesis and characterizations of ZnO thin films grown by physical vapor deposition technique. *Journal of Applied Science and Technology Trends*, *1*(2), 135–139.
14. Arya, S., Mahajan, P., Mahajan, S., Khosla, A., Datt, R., Gupta, V.,... & Oruganti, S. K. (2021). influence of processing parameters to control morphology and optical properties of Sol-Gel synthesized ZnO nanoparticles. *ECS Journal of Solid State Science and Technology*, *10*(2), 023002.
15. Kotresh, M. G., Patil, M. K., & Inamdar, S. R. (2021). Reaction temperature based synthesis of ZnO nanoparticles using co-precipitation method: Detailed structural and optical characterization. *Optik*, *243*, 167506.
16. Gerbreders, V., Krasovska, M., Sledevskis, E., Gerbreders, A., Mihailova, I., Tamanis, E., & Ogurcovs, A. (2020). Hydrothermal synthesis of ZnO nanostructures with controllable morphology change. *CrystEngComm*, *22*(8), 1346–1358.
17. Wang, Z.L., 2008. Splendid one-dimensional nanostructures of zinc oxide: a new nanomaterial family for nanotechnology. *ACS Nano*, *2*(10), pp.1987–1992.
18. Rajan, A., Cherian, E. and Baskar, G., 2016. Biosynthesis of zinc oxide nanoparticles using Aspergillus fumigatus JCF and its antibacterial activity. *International Journal of Modern Science and Technology*, *1*(2), pp.52–57.

19. Aggarwal, N., Mehta, P., Bhambra, J., Sharma, A., Ali, D., Sharma, N.K., Sharma, S., Kumar, G. and Kumar, N., 2023. Biosynthesis of Ocimum basilicum leaf extract mediated zinc oxide nanoparticles and their cytotoxic activity. *Biomass Conversion and Biorefinery*, *13*(9), pp.1–8.
20. Gad El-Rab, S.M., Abo-Amer, A.E. and Asiri, A.M., 2020. Biogenic synthesis of ZnO nanoparticles and its potential use as antimicrobial agent against multidrug-resistant pathogens. *Current Microbiology*, *77*, pp.1767–1779.
21. Kalpana, V.N., Payel, C. and Rajeswari, V.D., 2017. Lagenaria siceraria aided green synthesis of ZnO NPs: anti-dandruff, Antimicrobial and Anti-arthritic activity. *Research Journal of Chemistry and Environment*, *21*(11), pp.14–19.
22. Santhosh Kumar, J., Kumar, S.V. and Rajeshkumar, S., 2017. Synthesis of zinc oxide nanoparticles using plant leaf extract against urinary tract infection pathogen. *Resource-Efficient Technologies*, *3*(4), pp.459–465.
23. Dobrucka, R. and Długaszewska, J., 2016. Biosynthesis and antibacterial activity of ZnO nanoparticles using Trifolium pratense flower extract. *Saudi Journal of Biological Sciences*, *23*(4), pp.517–523.
24. Youssef, F.S., Ismail, S.H., Fouad, O.A. and Mohamed, G.G., 2024. Green Synthesis and Biomedical Applications of Zinc Oxide Nanoparticles. *Review. Egyptian Journal of Veterinary Sciences*, *55*(1), pp.287–311.
25. Santhosh Kumar, J., Kumar, S.V. and Rajeshkumar, S., 2017. Synthesis of zinc oxide nanoparticles using plant leaf extract against urinary tract infection pathogen. *Resource-Efficient Technologies*, *3*(4), pp.459–465.
26. Nava, O.J., Luque, P.A., Gómez-Gutiérrez, C.M., Vilchis-Nestor, A.R., Castro-Beltrán, A., Mota-González, M.L. and Olivas, A., 2017. Influence of Camellia sinensis extract on Zinc Oxide nanoparticle green synthesis. *Journal of Molecular Structure*, *1134*, pp.121–125.
27. Siripireddy, B. and Mandal, B.K., 2017. Facile green synthesis of zinc oxide nanoparticles by Eucalyptus globulus and their photocatalytic and antioxidant activity. *Advanced Powder Technology*, *28*(3), pp.785–797.
28. Bandeira, M., Giovanela, M., Roesch-Ely, M., Devine, D.M. and da Silva Crespo, J., 2020. Green synthesis of zinc oxide nanoparticles: A review of the synthesis methodology and mechanism of formation. *Sustainable Chemistry and Pharmacy*, *15*, p.100223.
29. Goswami, M., Adhikary, N.C. and Bhattacharjee, S., 2018. Effect of annealing temperatures on the structural and optical properties of zinc oxide nanoparticles prepared by chemical precipitation method. *Optik*, *158*, pp.1006–1015.
30. Irfan, M., Munir, H. and Ismail, H., 2021. Moringa oleifera gum based silver and zinc oxide nanoparticles: green synthesis, characterization and their antibacterial potential against MRSA. *Biomaterials Research*, *25*(1), p.17.
31. M. Anbuvannan, M. Ramesh, G. Viruthagiri, N. Shanmugam, N. Kannadasan, Synthesis, characterization and photocatalytic activity of ZnO nanoparticles prepared by biological method, Spectrochim. *Spectrochimica Acta Part A*. *143* (2015) 304–308, doi:10.1016/j.saa.2015.01.124
32. Sharmila, G., Thirumarimurugan, M. and Muthukumaran, C., 2019. Green synthesis of ZnO nanoparticles using Tecoma castanifolia leaf extract: Characterization and evaluation of its antioxidant, bactericidal and anticancer activities. *Microchemical Journal*, *145*, pp.578–587.
33. Hussein, B.Y. and Mohammed, A.M., 2021. Green synthesis of ZnO nanoparticles in grape extract: Their application as anticancer and antibacterial. *Materials Today: Proceedings*, *42*, pp.A18–A26.

34. Ngoepe, N.M., Mbita, Z., Mathipa, M., Mketo, N., Ntsendwana, B. and Hintsho-Mbita, N.C., 2018. Biogenic synthesis of ZnO nanoparticles using Monsonia burkeana for use in photocatalytic, antibacterial and anticancer applications. *Ceramics International*, *44*(14), pp.16999–17006.
35. Sundrarajan, M., Ambika, S. and Bharathi, K., 2015. Plant-extract mediated synthesis of ZnO nanoparticles using Pongamia pinnata and their activity against pathogenic bacteria. *Advanced Powder Technology*, *26*(5), pp.1294–1299.
36. Suresh, Joghee, Ganeshan Pradheesh, Vincent Alexramani, Mahalingam Sundrarajan, and Sun Ig Hong. "Green synthesis and characterization of zinc oxide nanoparticle using insulin plant (Costus pictus D. Don) and investigation of its antimicrobial as well as anticancer activities." *Advances in Natural Sciences: Nanoscience and Nanotechnology 9*, no. 1 (2018): 015008.
37. Rajeshkumar, S., Kumar, S.V., Ramaiah, A., Agarwal, H., Lakshmi, T. and Roopan, S.M., 2018. Biosynthesis of zinc oxide nanoparticles using Mangifera indica leaves and evaluation of their antioxidant and cytotoxic properties in lung cancer (A549) cells. *Enzyme and Microbial Technology*, *117*, pp.91–95. http://dx.doi.org/ 10.1016/j.arabic.2014. 11 015
38. Abdelsattar, A.S., Farouk, W.M., Gouda, S.M., Safwat, A., Hakim, T.A. and El-Shibiny, A., 2022. Utilization of Ocimum basilicum extracts for zinc oxide nanoparticles synthesis and their antibacterial activity after a novel combination with phage. *Materials Letters*, *309*, p.131344.
39. Álvarez-Chimal, R., García-Pérez, V. I., Álvarez-Pérez, M. A., Tavera-Hernández, R., Reyes-Carmona, L., Martínez-Hernández, M., & Arenas-Alatorre, J. Á. (2022). Influence of the particle size on the antibacterial activity of green synthesized zinc oxide nanoparticles using Dysphania ambrosioides extract, supported by molecular docking analysis. *Arabian Journal of Chemistry*, *15*(6), 103804.
40. Naseer, M., Aslam, U., Khalid, B., & Chen, B. (2020). Green route to synthesize Zinc Oxide Nanoparticles using leaf extracts of Cassia fistula and Melia azedarach and their antibacterial potential. *Scientific Reports*, *10*(1), 9055.
41. Tran, X. T., Bien, T. T. L., Van Tran, T., & Nguyen, T. T. T. (2024). Biosynthesis of ZnO nanoparticles using aqueous extracts of Eclipta prostrata and Piper longum: characterization and assessment of their antioxidant, antibacterial, and photocatalytic properties. *Nanoscale Advances.*, *6*(19), 4885–4899.
42. Lopez-Miranda, J. L., Molina, G. A., González-Reyna, M. A., España-Sánchez, B. L., Esparza, R., Silva, R., & Estévez, M. (2023). Antibacterial and anti-inflammatory properties of ZnO nanoparticles synthesized by a green method using sargassum extracts. *International Journal of Molecular Sciences*, *24*(2), 1474.
43. Tauseef, A., Hisam, F., Hussain, T., Caruso, A., Hussain, K., Châtel, A., & Chénais, B. (2023). Nanomicrobiology: emerging trends in microbial synthesis of nanomaterials and their applications. *Journal of Cluster Science*, *34*(2), 639–664.
44. El-Belely, E. F., Farag, M. M., Said, H. A., Amin, A. S., Azab, E., Gobouri, A. A., & Fouda, A. (2021). Green synthesis of zinc oxide nanoparticles (ZnO-NPs) using Arthrospira platensis (Class: Cyanophyceae) and evaluation of their biomedical activities. *Nanomaterials*, *11*(1), 95.
45. Elrefaey, A. A. K., El-Gamal, A. D., Hamed, S. M., & El-Belely, E. F. (2022). Algae-mediated biosynthesis of zinc oxide nanoparticles from Cystoseira crinite (Fucales; Sargassaceae) and it's antimicrobial and antioxidant activities. *Egyptian Journal of Chemistry*, *65*(4), 231–240.
46. Morowvat, M. H., Kazemi, K., Jaberi, M. A., Amini, A., & Gholami, A. (2023). Biosynthesis and antimicrobial evaluation of zinc oxide nanoparticles using Chlorella vulgaris biomass against multidrug-resistant pathogens. *Materials*, *16*(2), 842.

47. Minhas, L. A., Mumtaz, A. S., Kaleem, M., Farraj, D. A., Kamal, K., Minhas, M. A. H.,… & Mahmoud, R. M. (2023). Green synthesis of zinc oxide nanoparticles using Nostoc sp. and their multiple biomedical properties. *Catalysts*, *13*(3), 549.
48. Rilda, Y., Sari, L. M., Armaini, A., Imelda, I., & Pardi, H. (2024). Characterization and evaluation of the antibacterial activity of facile biosynthesis Zinc oxide nanorods using Spirulina platensis. *Separation Science and Technology*, *59*(4), 686–696.
49. Yadav, A., Jangid, N. K., & Khan, A. U. (2024). Biogenic synthesis of ZnO nanoparticles from Evolvulus alsinoides plant extract. *Journal of Umm Al-Qura University for Applied Sciences*, *10*(1), 51–57.
50. Akila, K., Thambidurai, S., Suresh, N., & Prabu, K. M. (2024). Photocatalytic and bactericidal activity of zinc oxide nanoparticles biosynthesized with Datura metel leaf extract. *Ionics*, *30*(6), 3637–3649.
51. Shnawa, B. H., Jalil, P. J., Al-Ezzi, A., Mhamedsharif, R. M., Mohammed, D. A., Biro, D. M., & Ahmed, M. H. (2024). Evaluation of antimicrobial and antioxidant activity of zinc oxide nanoparticles biosynthesized with Ziziphus spina-christi leaf extracts. *Journal of Environmental Science and Health, Part C*, *42*(2), 93–108.
52. El-Fallal, A. A., Elfayoumy, R. A., & El-Zahed, M. M. (2023). Antibacterial activity of biosynthesized zinc oxide nanoparticles using Kombucha extract. *SN Applied Sciences*, *5*(12), 332.
53. Jayabalan, J., Mani, G., Krishnan, N., Pernabas, J., Devadoss, J. M., & Jang, H. T. (2019). Green biogenic synthesis of zinc oxide nanoparticles using Pseudomonas putida culture and its in vitro antibacterial and anti-biofilm activity. *Biocatalysis and Agricultural Biotechnology*, *21*, 101327.
54. El-Belely, E. F., Farag, M. M., Said, H. A., Amin, A. S., Azab, E., Gobouri, A. A., & Fouda, A. (2021). Green synthesis of zinc oxide nanoparticles (ZnO-NPs) using Arthrospira platensis (Class: Cyanophyceae) and evaluation of their biomedical activities. *Nanomaterials*, *11*(1), 95.
55. Elrefaey, A. A. K., El-Gamal, A. D., Hamed, S. M., & El-Belely, E. F. (2022). Algae-mediated biosynthesis of zinc oxide nanoparticles from Cystoseira crinite (Fucales; Sargassaceae) and it's antimicrobial and antioxidant activities. *Egyptian Journal of Chemistry*, *65*(4), 231–240.
56. Alprol, A. E., Mansour, A. T., Abdelwahab, A. M., & Ashour, M. (2023). Advances in green synthesis of metal oxide nanoparticles by marine algae for wastewater treatment by adsorption and photocatalysis techniques. *Catalysts*, *13*(5), 888.
57. Jalali, T., Ghanavati, F., & Osfouri, S. (2024). Green synthesis of ZnO nanoparticles using marine brown algae (Cystoseira) extract comprising sol-gel, and combustion techniques based on dye-sensitized solar cells application. *International Journal of Modern Physics B*, *38*(14), 2450178.
58. Anand, K. V., Mahalakshmi, D., Selvan, S. M., Ravi, M., Kannan, M., Govindaraju, K., & Shalan, A. E. (2024). Biomass extract of green macroalga Halimeda opuntia assisted ZnO nanoparticles: Preparation, physicochemical characterization, and antibacterial activity against Vibrio harveyi. *Biomass Conversion and Biorefinery*, *14*(2), 2225–2233.
59. Rudayni, H. A., Rabie, A. M., Aladwani, M., Alneghery, L. M., Abu-Taweel, G. M., Al Zoubi, W.,… & Bellucci, S. (2023). Biological activities of sargassum algae mediated ZnO and Co-doped ZnO nanoparticles as enhanced antioxidant and anti-diabetic agents. *Molecules*, *28*(9), 3692.
60. Sundaresan, U., & Kasi, G. (2023). Synthesis of ZnO nanoparticles using Sargassum wightii ethanol extract and their antibacterial and anticancer applications. *Biomass Conversion and Biorefinery*, *14*(20), 26173–26191.

61. Peng, X., Palma, S., Fisher, N. S., & Wong, S. S. (2011). Effect of morphology of ZnO nanostructures on their toxicity to marine algae. *Aquatic Toxicology*, *102*(3–4), 186–196.
62. Thirumoorthy, G. S., Balasubramaniam, O., Kumaresan, P., Muthusamy, P., & Subramani, K. (2021). Tetraselmis indica mediated green synthesis of zinc oxide (ZnO) nanoparticles and evaluating its antibacterial, antioxidant, and hemolytic activity. *BioNanoScience*, *11*(1), 172–181.
63. Karkhane, M., Lashgarian, H. E., Mirzaei, S. Z., Ghaffarizadeh, A., Sepahvand, A., & Marzban, A. (2020). Antifungal, antioxidant and photocatalytic activities of zinc nanoparticles synthesized by Sargassum vulgare extract. *Biocatalysis and Agricultural Biotechnology*, *29*, 101791.
64. Priyadharshini, R. I., Prasannaraj, G., Geetha, N., & Venkatachalam, P. (2014). Microwave-mediated extracellular synthesis of metallic silver and zinc oxide nanoparticles using macro-algae (Gracilaria edulis) extracts and its anticancer activity against human PC3 cell lines. *Applied biochemistry and biotechnology*, *174*, 2777–2790.
65. Rilda, Y., Sari, L. M., Armaini, A., Imelda, I., & Pardi, H. (2024). Characterization and evaluation of the anti-bacterial activity of facile biosynthesis Zinc oxide nanorods using Spirulina platensis. *Separation Science and Technology*, *59*(4), 686–696.
66. López-Miranda, J. L., Mares-Briones, F., Molina, G. A., González-Reyna, M. A., Velázquez-Hernández, I., España-Sánchez, B. L.,… & Estévez, M. (2023). Sargassum natans I algae: an alternative for a greener approach for the synthesis of ZnO nanostructures with biological and environmental applications. *Marine Drugs*, *21*(5), 297.
67. Ouzakar, S., Senhaji, N. S., Saidi, M. Z., El Hadri, M., El Baaboua, A., El Harsal, A., & Abrini, J. (2023). Antibacterial and antifungal activity of zinc oxide nanoparticles produced by Phaeodactylum tricornutum culture supernatants and their potential application to extend the shelf life of sweet cherry (Prunus avium L.). *Biocatalysis and Agricultural Biotechnology*, *49*, 102666.

2 Solar Cell Properties of ZnO-Based Nanostructures

Indu Gupta, Rahul Jain, and Nandni Sharma

2.1 INTRODUCTION

Fossil fuel combustion for energy that drives the economy of the 21st century is ravaging climatic conditions. A transition from fossil fuels to renewable energy needs to take place to maintain the existence of our society. The change is necessary due to two main reasons; one is the limited and exhaustible amount of natural resources that cannot be replaced immediately, and another is the greenhouse effect due to the CO_2 emissions derived from burning fossil fuels. Therefore, addressing these pressing energy and environmental problems arising from traditional energy usage is a duty of our human beings. Photovoltaics (PVs) may serve as a potential low-carbon energy source via the CO_2 emission free power generation. Single-crystalline silicon (Si) p-n junction, for example, belonging to 1-Gen PVs, have been the mainstream in PV market in the past 20 years until now, with conversion efficiencies exceeding 25% [1].

To overcome these problems, the second-generation PV was introduced, relying on thin-film technology with thinner absorbing layers made from $Cu(In,Ga)Se_2$ (CIGS), poly- or amorphous Si, etc. Eventually, these thin-film PVs are targeted to provide cost-competitiveness, coming from simplified production methods [2, 3]. However, the limited thickness of the active layers leads to lower light absorption as well as an increased nonradiative recombination center density, which makes their efficiency (η) less than that of single-crystalline Si cells [7–10]. Many approaches to PVs have been reported to achieve high performance while being affordable, and among them, the third-generation PV based on semiconductor nanostructures (NSs) have gained attention. These nanostructured PV, comprising nanowires (NWs), nanotubes (NTs), and nanoparticles (NPs), provide another angle to increase η with a lower total cost outlay [4, 5]. To this end, researchers have been working hard to probe the advanced features of these nanostructured PV. In conclusion, first-generation PV, in particular the monocrystalline Si ones, are facing increased competition from the second- and third-generation applications that are not only cheaper but also more efficient.

In particular, zinc oxide (ZO) NSs have attracted significant interest for use in PVs, owing to their unique properties like wide surface area, high electron mobility,

DOI: 10.1201/9781003464907-2

and tunable bandgap. ZO NSs can be utilized in solar cell applications in several ways:

1. As an electron transport layer (ETL): ZO NSs can be utilized as an ETL in PVs. This layer assists in transporting the generated electrons of the solar cell to the electrode, which improves the efficiency of the solar cell. NSs of ZO possess superior electron mobility, which ensures that the electrons are well transported.
2. As a photoanode: Photoanodes of DSSCs can also be made with ZO NSs. The photoanode can then absorb the incoming photon energy to generate e^- and transfer them to the electrode through the ETL. Thus, the large surface area of the ZO NS would cause more incident photons to be absorbed, and it is anticipated to increase the solar cell's efficiency.
3. As a buffer layer: ZO-nanostructured films can be used as an intermediate layer between the components of a thin-film solar cell. This enables one to improve the interface between the electrode and the active layer, thus having an increased solar cell's efficiency.
4. As a light-trapping layer: ZO-nanostructured films can be utilized as a layer for trapping light in thin-film solar cells. As a light-trapping layer, the structure will help improve photon absorption, thus increasing the solar cell's efficiency.

ZO NSs have diverse prospective applications in PV as shown in Figure 2.1. Prominent characteristics of ZO that make them great options for solar cell applications include:

1. High electron mobility: ZO is characterized by high electron mobility. Thus, it is a good choice for the solar cell's ETL. Because of the high electron mobility, the e^- produced in the solar cell easily transport to the electrode, thus resulting in improved efficiency of the solar cell.

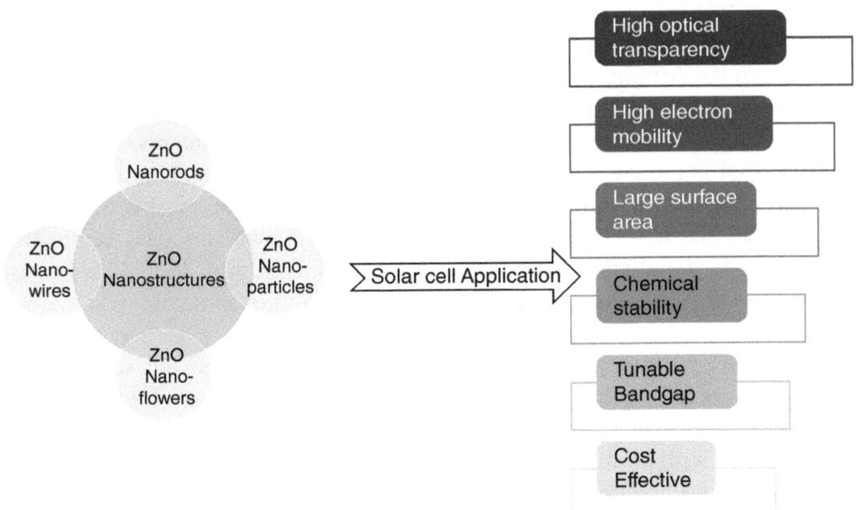

FIGURE 2.1 Solar cell properties of ZO nanostructures.

2. Large surface area: Large surface area-to-volume ratios are a characteristic of nanostructured ZO that makes it perfect for use as an electrode in PV. A larger surface area means that more photons are absorbed, thus increasing the solar cell's efficiency.
3. Tunable bandgap: The bandgap of ZO can be engineered using the dimensions and configuration of the NSs. This, in turn, allows the developing solar energy cells to have a maximum efficiency specifically designed and optimized for their absorption characteristics.
4. Optical transparency: ZO NSs are capable of optical transparency, particularly in the electromagnetic spectrum's visible range. The transparent conductive oxides based on ZO NSs enable the construction of transparent electrodes in PV, leading to both light transmission into the active layer and efficient charge extraction.
5. Chemical stability: The standard ambient conditions make ZO chemically stable, which means that it has a very long lifespan in PV. It is comparatively inert to oxidation and moisture, as well as many other weather elements, ensuring the viability of the solar cell devices.
6. Low cost and abundance: Among various semiconductor materials like Si or gallium arsenide, ZO is a cheap and abundantly available material. The cost factor involved makes ZO NSs very promising in terms of volume production as well as commercialization of PV.

Due to high electron mobility, high conductivity, ample electron affinity, and stability, the binary compound semiconductor ZO materials belonging to groups II–VI have also been considered for PV applications. ZO has a very broad E_g, spanning from 3.1 to 3.3 eV, and absorbs exclusively ultraviolet light [6]. ZO has a variety of different nanomorphologies, including nanospheres (NSps), NWs, nanorods (NRs), nanoflowers (NFs), NTs, nanocrystals (NCs), and three-dimensional (3D) NSs with core-shell arrangements. These unique characteristics have led to the widespread utilization of ZO in various fields. Applications range from sensors, surface coatings, and porous ceramics to photodetectors, nanopiezoelectric devices, supercapacitors, and even PV. Multiple methods are available to synthesize ZO NSs, employing different approaches such as biological, physical, or chemical pathways. Some of these techniques include green synthesis, utilizing microorganisms, hydrothermal approaches, sol-gel procedures, electrochemistry, inkjet printing, atomic layer deposition, and sputtering techniques. ZO can also be used with substances that have smaller energy gaps, such as organic polymers, dye sensitizers, and semiconductors with smaller band gaps, to expand the visible spectrum of its light absorption. Because of several benefits, ZO distinguishes apart from other semiconductor nanostructured PV.

The most recent developments in the use of ZO in third-generation solar cell technology are thoroughly examined in this chapter. It extensively explores various PV architectures, such as sensitized PV and heterojunction PV, with a specific focus on their integration with promising perovskite materials. This chapter concludes by outlining future trends and potential developments in the field of PV utilizing ZO NSs. This review article aims to shed light on this important, fascinating, and rapidly growing field, providing valuable insights into the current state and future directions of PV based on ZO NSs.

2.2 APPLICATION OF ZO NSs FOR PHOTOVOLTAICS

2.2.1 Sensitized Photovoltaics (SPVs)

The first introduction and concept of SPV was by O'Regan and Grätzel in 1991. Since then, it has gained a lot of attention on a feasible front that can be explored as an alternative to traditional Si PVs. These PVs promise an easy and economical way of manufacturing, and in addition, they exhibit long-term stability while having energy conversion efficiencies greater than 10% [7]. In DSSCs, ruthenium (Ru)-based dyes are used as photosensitizers, which adsorb on mesoporous metal oxide photoelectrodes having large surface areas and will help them effectively capture solar energy. The dye injects e- into the metal oxide film's conduction band when it is optically stimulated. From there, it diffuses through the semiconducting layer to the back contact. The oxidized dye is replenished at the same time that redox couplings in the solution diffuse and reduce at the counter electrode. Because typical dyes like N3 and N719 are expensive, scientists are looking at less expensive alternatives such as semiconductor quantum dots (QDs) with small band gaps, whose optical characteristics may be altered to match the solar spectrum. Impact ionization enables QDSCs to generate multiple e^--h pairs per photon, resulting in quantum yields exceeding 1 [8–10]. However, early studies on QDSCs demonstrated performance levels below expectations, with efficiencies below 1% in two-electrode configurations [11, 12]. It was not until Diguna et al. reported an η of 2.7% in 2007 that the development of QDSCs saw a resurgence [13]. Various morphologies of the photoanode, such as the inverse opal structure, have also been investigated to enhance light-harvesting η by facilitating better dye penetration and photon confinement nearby the photonic E_g wavelength, leading to substantial enhancements in dye absorption [14]. In the subsequent sections, we will explore recent advancements in DSSCs, PSSCs, and QDSCs, with a particular focus on the utilization of different ZO-nanostructured materials in these solar cell technologies.

2.2.2 Dye-Sensitized PV (DSPVs)

Matsumura et al. used ZO porous disks sensitized with rose Bengal under monochromatic light at 562 nm in 1980 to obtain an energy conversion η of 2.5% [15]. Subsequently, it was reported that the efficiencies of nanoporous ZnO thin films could achieve up to 2% when exposed to 56 mW/cm^2 of Ru-complex dye and 2.5% when exposed to 99 mW/cm^2 of mercurochrome sensitizer.[16, 17]. The ZO surface was bound to the mercurochrome dye via a carboxylate linkage similar to that of TiO_2. However, because of the Ru-based dye's adsorption behavior, ZO did not perform as well as TiO_2. During dye adsorption, it was found that agglomerates of Zn^{2+} and Ru dye molecules filled the ZO electrode's nanopores, forming an insulating barrier that impeded the dye's ability to efficiently inject electrons into the semiconductor [18, 19]. ZO is more acidic than TiO_2, as evidenced by the fact that its isoelectric point is roughly –9 as opposed to –6 for TiO_2. As a result, when acidic dyes are present, ZO's chemical stability is weakened [20]. As a result, when acidic dyes are present, ZO's chemical stability is weakened [20]. ZO surfaces may get etched during dye adsorption as a result of the Ru dye solution's relatively acidic pH due to the deprotonation of carboxyl groups in the solution. The acidic carboxylic groups dissolve ZO at the crystal surface, preventing efficient electron injection through the formation of Zn2+/dye complexes [20]. It has been attempted to enhance the performance of ZO electrodes sensitized with Ru bipyridyl complexes. To stop dye aggregation and the formation of Zn2+/dye complexes, one technique was to add basic KOH to the dye solution. Moreover, excluding organic additives and applying pressure during ZO film preparation helped enhance interfacial kinetics, resulting in an η of up to 5% [21, 22]. The ZO electrode's photo-electrochemical characteristics are greatly influenced by the electrode morphology, which includes the size, shape, and porosity of the particles. For example, a 150 nm thick ZO film composed of 8 nm spherical particles exhibited efficient electron transport and minimized recombination losses in the nanostructured electrode [23]. Conventional NP films used in DSPVs provide more surface area for effective light-harvesting dye molecule absorption. However, the sluggish trap-limited diffusion mechanism, particularly evident at higher wavelengths, hinders electron transport within the NP film, thus compromising device η [24]. To overcome this limitation, various oriented 1D NSs like NWs, NRs, and NTs have been explored as potential solutions to enhance diffusion of e^- in the anode. NWs, with aspect ratios (length/diameter) reaching up to 1000, and NRs, typically less than 10 depending on synthesis methods and conditions, emerged as promising candidates. In 2005, a seeded growth technique was used to construct ZO nanowire photoanodes with array lengths of 20–25 μm and surface areas up to one-fifth of an NP film [25]. Initially, the thin layer of ZO QDs was coated to the fluorine-doped tin oxide (FTO) substrate by a dip coating method. This layer was roughly 10–15 nm thick. These nuclei then facilitated nanowire generation through the thermal decomposition of a zinc complex. These NW-based DSPVs demonstrated an η of 1.5% under one sun's illumination after being sensitized with Ru dye [26]. In subsequent studies, femtosecond transient absorption spectroscopy was employed to assess e^- injection from photoexcited Ru dyes into NWs and NPs. Compared to NPs, which exhibited a tri-exponential response, NWs displayed bi-exponential kinetics with faster electron injection. Further advancements involved the application of core-shell structures to ZO NW photoanodes by coating the NWs with a thin shell of amorphous aluminum oxide or anatase titanium dioxide through atomic layer deposition [27]. By acting as a tunnel barrier, an extremely thin alumina shell decreased recombination and increased the open-circuit voltage (V_{oc}). However, electron injection was hampered by growing shell thickness. Conversely, titania shells increased the fill factor (FF) and V_{oc} while decreasing recombination rates [28]. It was found that the alignment of ZO NRs affected their PV effectiveness in addition to their size. When sensitized with N179 dye, vertically aligned ZO NRs exhibited modest power conversion efficiency (PEC); hydrothermally grown and vapor-deposited ZO NRs, both approximately 3.5 μm in length, achieved efficiencies of 0.22% and 0.09%, respectively. The addition of Gold (Au) NPs to the ZO NRs' surface significantly increased efficiency by creating a Schottky barrier that prevents electrons from returning to the N719 dye and electrolyte. This led to a 1.2% improvement in η [29]. Additionally, ZO NRs with aluminum (Al) doping performed better, increasing η from 0.05% to 1.34%. Trivalent Al^{3+} ions were added to the divalent Zn^{2+} ion sites, which improved e^- concentration and electrical conductivity and made it easier for electrons to migrate into the Al-doped ZO conduction band. A ZO NT array was created by chemically etching the core area of the ZO NRs that were electrochemically deposited [30].

Employing a 5.1 μm-long ZO NT photoanode sensitized with N719 dye resulted in an η of 1.18%. On the contrary, under Air Mass (AM) 1.5 illumination, ZO NT photoanodes templated by anodic aluminum oxide showed a η of 1.6%. A remarkable photovoltage of 739 mV and an FF of 0.64 demonstrate that a direct charge collection path over a thickness of tens of micrometers was made possible by combining atomic layer deposition to coat nanotube pores [31]. However, total η was hindered by low photocurrent due to insufficient light harvesting caused by slight roughness factor and reflectivity/scattering of the photoanode (light from the counter electrode side). Increasing the surface area was identified as crucial for further enhancing PEC. Selecting the right dye is pivotal for enhancing the PEC of ZO-based DSPVs. An appropriate dye increases electron injection η, depending on the semiconductor's density of electron-accepting states (DOS), energy level alignment between the dye and semiconductor, electronic coupling, and dye lifetime [32]. Selecting a dye group that matches the semiconductor's conduction and valence band energy levels—which correspond to the dye's lowest unoccupied molecular orbital (LUMO) and highest occupied molecular orbital (HOMO) levels, respectively—is essential for the electronic coupling between the dye and semiconductor. Furthermore, charge transfer is facilitated by a longer dye lifetime in the excited state. Due to its alkaline nature and vulnerability to low pH settings, which can result in the loss of Zn2+ ions from the ZO lattice, ZO is incompatible with Ru-based dyes, unlike TiO_2 [33]. Due to ZO film release, decreased concentration of free electrons, increased recombination, and the conversion of the porous film into a denser one, an improper immersion period of the ZO film in Ru dye can lower DSPV η. Recent research has focused on metal-free dyes like indoline, heptamethine-cyanine, carbazole, anthocyanin, and xanthene dyes, showing compatibility with ZO. Some studies have achieved efficiencies exceeding 5% for flexible ZO-based DSPVs using these dyes. In a recent study by Selopal et al., efficient metal-free organic sensitizers named B18, BTD-R, and CPTD-R were developed for ZO-based DSPVs [34]. Among these dyes, B18 exhibited superior PV properties compared to others when used in hierarchically structured ZO and commercial TiO_2. These results indicated better electron-injection potential, improved light harvesting, and slight shift in the dye-sensitive photoanode conduction band for the device with B18 dye. In conclusion, selecting a compatible dye in ZO-based DSPVs is essential for optimizing device performance and achieving higher η. To enhance electron transport within the ZO photoanode and consequently improve cell η, one strategy is to introduce dopants into the ZO film. Dopants are essential for filling the semiconductor's holes, which increases the recombination of injected electrons. Various materials have been investigated as dopants for ZO films, encompassing iodine, boron, silver, lithium, lanthanum, strontium, indium, and neodymium.

For instance, Zhao et al. reported the use of indoline D205 and N719 as sensitizers in dye-sensitized PVs to investigate iodine-doped ZO-based photoanodes. When iodine doping was introduced, cells' efficiency was noticeably higher than when iodine was absent. The DSPVs based on D205-I-ZO and N719-I-ZO demonstrated η improvements of 20.3% and 17.9%, respectively [35]. Using N3 dye as a sensitizer, Mahmood and Park reported on cells made with boron-doped ZO nanosheets. Boron doping led to a remarkable enhancement in PCE, reaching 6.75%, compared to undoped ZO nanosheets (2.62%) and films consisting solely of nanosized crystallites (3%) [36]. According to Lanjewar et al., doping ZO with silver significantly reduced the band gap, allowing for better absorption of visible light. When 10.3 wt % Ag doping was added, the band gap decreased from 3.28 eV for pure ZO film to 2.65 eV for Ag: ZO, increasing the PCE from 0.55% to 2.02%. [37]. Aksoy et al. recently conducted a study where lithium was incorporated into ZO powder using the N719 dye for dye-sensitized PVs. When lithium was added, there is a formation of NF morphology, and cell η increased to 1.23%, attributed to morphology changes and improved crystallinity in the Li-doped ZO-based DSPVs [38]. In the same way, Goel et al.'s study showed that doping ZO with lanthanum in nanopowder PVs produced better results than pure ZO-based cells. After lanthanum doping, light-harvesting η improved from 0.20% to 0.36% [39, 40]. Additionally, Chava et al. investigated indium-doped ZO films, achieving a maximum photo-conversion η of 2.7% with 0.2 dopant concentration [41]. These films comprised aggregated spheres of nanocrystallites and nanosized crystallites, attributing the enhanced performance to strong light scattering phenomena and pore size of the aggregates, facilitating dye infiltration and electrolyte diffusion.

2.2.3 Two-Dimensional (2D) DSPVs

Typically, mesoporous network films made by sintering metal oxide NPs like ZO and TiO_2 makeup DSPVs [42–45]. These films are used to construct a 2D planar photoanode. Parallel to the working electrode is the counter electrode, which is usually a glass substrate covered with platinum. To facilitate the absorption of light-harvesting dye molecules and electronic conduction, the mesoporous layer is placed on a glass substrate that has a transparent conducting oxide (TCO). When the dye is exposed to sunlight, photogenerated electrons from the dye molecules are pumped into the nanocrystalline sheets' conduction band, oxidizing the dye. Iodide/triiodide (I^-/I^{3-}) redox system-containing electrolyte undergoes electron transfer to return the dye to its initial state. The ability of the photoanodes to minimize electron-hole recombination rates, transport electrons efficiently, and absorb light is what essentially determines how well DSPVs operate. All of these elements work together to increase the DSPVs' overall η. 1D single-crystalline NSs, like ZO, have recently generated a lot of interest because of their special structures and beneficial qualities. These NSs are a viable substitute for the mesoporous nanocrystalline structures frequently found in DSPV systems because they provide a direct channel for effective electron collecting. Like TiO_2, ZO is a wide E_g semiconductor that has higher electron mobility, which helps in DSPVs by promoting electron transport and reducing recombination losses. Researchers Baxter and Aydil first reported the development and functioning of a ZO nanowire-based DSPV with internal quantum efficiencies (IQEs) of 70% and energy conversion efficiencies of 0.5% [46]. Nevertheless, by expanding the nanowire surface area, which raises current densities and efficiencies, the η of these cells can be further increased. Dense and aligned ZO nanowire arrays were introduced by Law et al. [24] and synthesized by a seeded growth technique in DSPVs. In comparison to NP films, these NWs' high aspect ratios and densities allowed for increased electron diffusivity and diffusion coefficients. Even at 25 μm thicknesses, the electron transport η of nanowire films was nearly identical to that of TiO_2 films, leading to significantly greater current densities than ZO NP cells. One way to address the shortcomings of 1D NS arrays, such as insufficient interior surface areas, inadequate dye loading, and inefficient light harvesting, is to create hierarchical ZO NSs with multi-scale topologies. For instance, when compared to NR arrays of comparable size and density, hierarchical ZO NF films have shown improved overall efficiencies. Other strategies have demonstrated increased electron transport and dye adsorption, resulting in improved DSPV performance. One such strategy is the use of hybrid photoanodes, which combine ZO NWs and NPs. One of the challenges in attaining greater conversion η and photocurrent density in DSPVs is the recombination rate. Semiconducting core-shell nanomaterials have been used to lessen this [47]. For example, by suppressing recombination sites and lowering recombination through decreased visible absorption, ZO core-ZnS shell NWs demonstrated a notable increase in DSPV η [25]. Furthermore, ZO nanowire arrays with insulating Al_2O_3 or anatase TiO_2 shells deposited on them showed enhanced V_{oc} and FF, which increased total η. Other 1D ZO NSs such as NTs and nanotips have also been studied by DSPVs due to their huge specific aspect ratios and direct electron transport paths [48, 49]. Space constraints prevent a detailed discussion of the impact of ZO NS growth settings, electrolyte species, and dye molecule types on DSPV performance, despite their considerable research.

2.2.4 Three-Dimensional (3D) DSPVs

There are restrictions on the surface area, dye loading capacity, and adaptability for remote operation of DSPVs built on 2D planar substrates. Furthermore, to limit the growth in surface area by minimizing losses such as e-h recombination, it is imperative to maintain a substantially smaller e^- transport distance (d) than the e^- diffusion length (L_n).

To overcome these constraints, Weintraub et al. introduced a revolutionary 3D optical fiber/nanowire DSPV structure [50]. In this innovative design, ZO NWs developed perpendicular to the surface of an optical fiber when light was irradiated along the fiber's axial direction. Because of the way the dye molecules and light were arranged inside the fiber, internal reflections were used to keep the electron transit distance short. Up to a six-fold increase in DSPV η was achieved when internal axial lighting was used instead of external light illumination in conventional 2D configurations. The absolute full-sun η at 3.3% was 120 times more than the highest value reported for ZO NWs made on flat surfaces. Even though the hybrid 3D structure of optical fiber and nanowire showed improved performance, the planar electrode's unique geometry made scaling up for large-scale applications difficult. The same research team overcame the issue by using a quartz slide in place of the optical fiber in a planar waveguide-nanowire integrated 3D DSPV [51]. ZO nanowire arrays that were aligned and growing perpendicular to both surfaces were positioned between quartz slides in this configuration. The quartz slide guided the propagation of light by acting as a planar waveguide. Every time light reached the waveguide-nanowire interface, photons were absorbed by dye molecules and linked into the ZO NWs. When compared to planar illumination, this resulted in a significant η enhancement of an average of 5.8 factors. ZO NWs in this planar waveguide design obtained up to 2.4% full solar η. By functioning inside the unit's volume as opposed to just on its surface, which is illuminated by light, the 3D design idea completely transforms conventional PV systems. This creates opportunities for use in isolated areas, including subterranean or deep below, where light can reach an external surface while the solar cell is hidden somewhere else. In particular, the planar waveguide type provides a practical method for creating large-scale, highly efficient 3D PV systems [51].

2.2.5 Quantum Dot-Sensitized PV (QDSPVs)

There are two primary reasons why ZO-based DSPVs have a lower conversion η than DSSCs that use TiO_2 NPs. First, excessive Zn^{2+}/dye agglomerates are formed when protons from the dyes dissolve Zn atoms at the ZO surface, making ZO unstable in acidic dyes [52]. Second, because molecular relaxation and ultrafast electron injection compete with one another, the electron-injection kinetics from dyes to ZO are sluggish [53]. To address these issues, a novel class of QDSPVs has been developed using semiconductor QDs as photosensitizers [54]. Compared to dyes, QDs have advantages including improved solar spectrum matching because of their tunable absorption spectra depending on the particle size. Additionally, QDs can produce many e-h pairs for every photon, potentially boosting the device's η. Therefore, it is anticipated that QDSPVs will have a power conversion η higher than the 31% Shockley-Queisser limit [55]. Leschkies et al. used QDs and the enhanced electron transport of 1D ZO NSs to develop ZO nanowire-based CdSe QDSPVs [56]. As the photoanode, they employed a ZO nanowire array that was grown on a SnO2/glass substrate doped with F and coated with CdSe QDs using bifunctional linker molecules. The space between the electrodes of the photocathode, which was an F-doped SnO2/glass substrate coated with Pt, was filled with a liquid electrolyte that contained the I^{3-}/I^- redox pair. The creation of a type II heterojunction between ZO and CdSe QDs allowed for efficient electron transport from the photoexcited QDs to ZO. The cells showed excellent IQEs of 50–60%, FF of 0.4%, V_{oc} of 0.5–0.6 V, and short-circuit current densities (J_{sc}) of 1 to 2 mA/cm^2. Nevertheless, QDSPVs still have relatively poor efficiencies, which are frequently caused by carrier recombination at the semiconductor surface due to sluggish e^- transport and e^- buildup in the semiconductor layer. ZO NRs grown on graphene were used by Chen et al. to demonstrate a novel architecture for QDSPVs with the goal of increasing electron transit [57]. Lowering the barrier between ZO and the FTO electrode and aligning the Fermi levels of both graphene and ZO allowed for effective electron transport from CdSe QDs to the graphene-ZO NS. This QDSPV had a high FF of 61.8% and a J_{sc} of 4.65 mA/cm^2, which is among the highest FF values for ZO NR-based devices. QDSSCs have also made use of metal NPs in addition to semiconductor QDs. For instance, Chen et al. described gold NP-sensitized ZO NRs for Schottky barrier PV, whereby surface plasmon resonance was utilized to give improved optical absorption in the visible light range [58]. When a sensitizing dye (N719) was added, the dye-free Schottky barrier cell's FF and η increased to 0.63 V and 1.2% from 0.50 and 0.7%, respectively. Two basic strategies for optimizing QDSPVs have been investigated. First, it has been suggested to switch from liquid electrolytes to other hole transport media. Leschkies et al. created QDSPVs without the need of liquid electrolytes by introducing colloidal PbSe QDs into ZO nanowire arrays [59]. Photocurrent production was made possible by this design, which enabled excitons formed within the QD film to dissociate at heterointerfaces or when subjected to a high electric field. Second, research has been done on covering the QD surface with a protective layer. CdS, ZnS, and ZnSe are the examples of wide band gap materials that have been used as shells to boost photoluminescence quantum yield by passivating surface nonradiative recombination sites. Kuo et al. improved J_{sc} by 88% in QDSPVs by covering Zn-doped CuInS2:ZnS QDs with a thin ZnS shell [60]. Charge transport η was enhanced, and stimulated electron recombination was successfully avoided. While perfect 1D ZO-nanostructured QDSPVs are still unattainable, electrolyte-type optimization can increase η and maximal durability. Light harvesting η can be increased by co-sensitizing different-sized QDs, and another interesting way to improve it is to create low-cost counter electrodes with strong electrocatalytic activity. To sum up, QDSPVs have the potential to be more efficient than conventional DSPVs. Researchers are advancing in their efforts to enhance the η of QDSPVs by employing QDs as photosensitizers and refining electron transport and recombination mechanisms.

2.3 CONCLUSIONS

This chapter includes the utilization of ZO as an active ingredient in PV technologies such as DSPVs, QDPVs, and PSPVs. PVs based on ZO NS are yet undergoing developments despite the emerging technologies because of their possibility for cost-effective and reproducible PV devices. Performance demands and toxicity issues, however, have to be resolved. To optimize η for ZO PVs, current ongoing research being pursued primarily centers on the two fronts. The first is the invention of incredibly simple synthesis techniques for 1D ZO NSs, which have excellent conducting properties. Second, this involves improved synthesis techniques, yielding NSs for better performance in PV applications. However, research is being done to find other dye molecules that will solve ZO's stability issues in the electrolyte. The right identification of dye molecules will help alleviate the stability challenges concerning ZO in the electrolyte. In addition to this, the development of higher power conversion efficiency and longer device life under several environments is a promise for these advances. Setting up a cost-effective and reproducible method of preparing ZO and then environmentally sustainable methods will give it buoyancy in the market. Having the ability to make solar cell devices customizable in terms such as flexibility, durability, and washability is also critical for its commercial deployment.

REFERENCES

1. Zhao, J., Wang, A., & Green, M. A. (1999). 24·5% Efficiency silicon PERT cells on MCZ substrates and 24·7% efficiency PERL cells on FZ substrates. *Progress in Photovoltaics: Research and Applications*, *7*(6), 471–474.
2. Green, M. A. (2006). Consolidation of thin-film photovoltaic technology: the coming decade of opportunity. *Progress in Photovoltaics: Research and Applications*, *14*(5), 383–392.
3. Shah, A. V., Schade, H., Vanecek, M., Meier, J., Vallat-Sauvain, E., Wyrsch, N.,... & Bailat, J. (2004). Thin-film silicon solar cell technology. *Progress in Photovoltaics: Research and Applications*, *12*(2–3), 113–142.
4. Hochbaum, A. I., & Yang, P. (2010). Semiconductor nanowires for energy conversion. *Chemical Reviews*, *110*(1), 527–546.
5. Nozik, A. J. (2010). Nanoscience and nanostructures for photovoltaics and solar fuels. *Nano Letters*, *10*(8), 2735–2741.
6. Senger, R. T., & Bajaj, K. K. (2003). Optical properties of confined polaronic excitons in spherical ionic quantum dots. *Physical Review B*, *68*(4), 045313.
7. Grätzel, M. (2001). Photoelectrochemical cells. *Nature*, *414*(6861), 338–344.
8. Nozik, A. J. (2002). Quantum dot PV. *Physica E: Low-dimensional Systems and Nanostructures*, *14*(1–2), 115–120.
9. Schaller, R. D., & Klimov, V. I. (2004). High efficiency carrier multiplication in PbSe nanocrystals: implications for solar energy conversion. *Physical Review Letters*, *92*(18), 186601.
10. Robel, I., Subramanian, V., Kuno, M., & Kamat, P. V. (2006). Quantum dot PV. Harvesting light energy with CdSe nanocrystals molecularly linked to mesoscopic TiO2 films. *Journal of the American Chemical Society*, *128*(7), 2385–2393.
11. Plass, R., Pelet, S., Krueger, J., Grätzel, M., & Bach, U. (2002). Quantum dot sensitization of organic– inorganic hybrid PV. *The Journal of Physical Chemistry B*, *106*(31), 7578–7580.
12. Niitsoo, O., Sarkar, S. K., Pejoux, C., Rühle, S., Cahen, D., & Hodes, G. (2006). Chemical bath deposited CdS/CdSe-sensitized porous TiO_2 PV. *Journal of Photochemistry and Photobiology A: Chemistry*, *181*(2–3), 306–313.
13. Diguna, L. J., Shen, Q., Kobayashi, J., & Toyoda, T. (2007). High efficiency of CdSe quantum-dot-sensitized TiO_2 inverse opal PV. *Applied Physics Letters*, *91*(2), 023116.
14. L. J. Diguna, M. Murakami, A. Sato, Y. Kumagai, T. Ishihara, N. Kobayashi, Q. Shen and T. Toyoda, *Japanese Journal of Applied Physics*., 2006, 45, 5563.
15. Matsumura, M., Matsudaira, S., Tsubomura, H., Takata, M., & Yanagida, H. (1980). Dye sensitization and surface structures of semiconductor electrodes. *Industrial & Engineering Chemistry Product Research and Development*, *19*(3), 415–421.
16. Rensmo, H., Keis, K., Lindström, H., Södergren, S., Solbrand, A., Hagfeldt, A.,... & Muhammed, M. (1997). High light-to-energy conversion efficiencies for PV based on nanostructured ZnO electrodes. *The Journal of Physical Chemistry B*, *101*(14), 2598–2601.
17. Hara, K., Horiguchi, T., Kinoshita, T., Sayama, K., Sugihara, H., & Arakawa, H. (2000). Highly efficient photon-to-electron conversion with mercurochrome-sensitized nanoporous oxide semiconductor PV. *Solar Energy Materials and PV*, *64*(2), 115–134.
18. Horiuchi, H., Katoh, R., Hara, K., Yanagida, M., Murata, S., Arakawa, H., & Tachiya, M. (2003). Electron injection efficiency from excited N^3 into nanocrystalline ZnO films: effect of (N^{3-} Zn^{2+}) aggregate formation. *The Journal of Physical Chemistry B*, *107*(11), 2570–2574.

19. Keis, K., Lindgren, J., Lindquist, S. E., & Hagfeldt, A. (2000). Studies of the adsorption process of Ru complexes in nanoporous ZnO electrodes. *Langmuir*, *16*(10), 4688–4694.
20. Parks, G. A. (1965). The isoelectric points of solid oxides, solid hydroxides, and aqueous hydroxo complex systems. *Chemical Reviews*, *65*(2), 177–198.
21. Keis, K., Magnusson, E., Lindström, H., Lindquist, S. E., & Hagfeldt, A. (2002). A 5% efficient photoelectrochemical solar cell based on nanostructured ZnO electrodes. *Solar Energy Materials and PV*, *73*(1), 51–58.
22. Keis, K., Bauer, C., Boschloo, G., & Hagfeldt, A. (2002). K. Westermark, H. Rensmo, H. Siegbahn. *Journal of Photochemistry and Photobiology A*, *148*, 57–64.
23. Keis, K., Vayssieres, L., Rensmo, H., Lindquist, S. E., & Hagfeldt, A. (2001). Photoelectrochemical properties of nano-to microstructured ZnO electrodes. *Journal of the Electrochemical Society*, *148*(2), A149.
24. Law, M., & Green, L. E. (2005). J. c. Johnson, R. Saykally and P. Yang. *Nature Mater*, *4*, 455–459.
25. Law, M., Greene, L. E., Radenovic, A., Kuykendall, T., Liphardt, J., & Yang, P. (2006). ZnO– Al_2O_3 and ZnO– TiO2 core– shell nanowire dye-sensitized PV. *The Journal of Physical Chemistry B*, *110*(45), 22652–22663.
26. Zhao, Q., Xie, T., Peng, L., Lin, Y., Wang, P., Peng, L., & Wang, D. (2007). Size-and orientation-dependent photovoltaic properties of ZnO nanorods. *The Journal of Physical Chemistry C*, *111*(45), 17136–17145.
27. Wibowo, A., Marsudi, M. A., Amal, M. I., Ananda, M. B., Stephanie, R., Ardy, H., & Diguna, L. J. (2020). ZnO nanostructured materials for emerging solar cell applications. *RSC Advances*, *10*(70), 42838–42859.
28. Chen, Z. H., Tang, Y. B., Liu, C. P., Leung, Y. H., Yuan, G. D., Chen, L. M.,… & Lee, S. T. (2009). Vertically aligned ZnO nanorod arrays sensitized with gold nanoparticles for Schottky barrier photovoltaic cells. *The Journal of Physical Chemistry C*, *113*(30), 13433–13437.
29. Yun, S., Lee, J., Chung, J., & Lim, S. (2010). Improvement of ZnO nanorod-based dye-sensitized solar cell efficiency by Al-doping. *Journal of Physics and Chemistry of Solids*, *71*(12), 1724–1731.
30. Han, J., Fan, F., Xu, C., Lin, S., Wei, M., Duan, X., & Wang, Z. L. (2010). ZnO nanotube-based dye-sensitized solar cell and its application in self-powered devices. *Nanotechnology*, *21*(40), 405203.
31. Martinson, A. B., Elam, J. W., Hupp, J. T., & Pellin, M. J. (2007). ZnO nanotube based dye-sensitized PV. *Nano Letters*, *7*(8), 2183–2187.
32. Vittal, R., & Ho, K. C. (2017). Zinc oxide based dye-sensitized PV: A review. *Renewable and Sustainable Energy Reviews*, *70*, 920–935.
33. Pourjafari, D., & Oskam, G. (2019). ZnO-based dye-sensitized PV. In *Nanomaterials for Solar Cell Applications* (pp. 145–204). Elsevier.
34. Selopal, G. S., Wu, H. P., Lu, J., Chang, Y. C., Wang, M., Vomiero, A.,… & Diau, E. W. G. (2016). Metal-free organic dyes for TiO2 and ZnO dye-sensitized PV. *Scientific Reports*, *6*(1), 1–12.
35. Zhao, J. X., Zheng, Y. Z., Lu, X. H., Chen, J. F., Tao, X., & Zhou, W. (2013). Characterizing the Role of Iodine Doping in Improving Photovoltaic Performance of Dye-Sensitized Hierarchically Structured ZnO PV. *ChemPhysChem*, *14*(9), 1977–1984.
36. Mahmood, K., & Park, S. B. (2013). Highly efficient dye-sensitized solar cell with an electrostatic spray deposited upright-standing boron-doped ZnO (BZO) nanoporous nanosheet-based photoanode. *Journal of Materials Chemistry A*, *1*(15), 4826–4835.

37. Lanjewar, M., & Gohel, J. V. (2017). Enhanced performance of Ag-doped ZnO and pure ZnO thin films DSSCs prepared by sol-gel spin coating. *Inorganic and Nano-Metal Chemistry*, *47*(7), 1090–1096.
38. Aksoy, S., Polat, O., Gorgun, K., Caglar, Y., & Caglar, M. (2020). Li doped ZnO based DSSC: Characterization and preparation of nanopowders and electrical performance of its DSSC. *Physica E: Low-dimensional Systems and Nanostructures*, *121*, 114127.
39. Goel, S., Sinha, N., Yadav, H., Joseph, A. J., & Kumar, B. (2017). Experimental investigation on the structural, dielectric, ferroelectric and piezoelectric properties of La doped ZnO nanoparticles and their application in dye-sensitized PV. *Physica E: Low-dimensional Systems and Nanostructures*, *91*, 72–81.
40. Rajan, A. K., & Cindrella, L. (2019). Ameliorating the photovoltaic conversion efficiency of ZnO nanorod based dye-sensitized PV by strontium doping. *Superlattices and Microstructures*, *128*, 14–22.
41. Chava, R. K., & Kang, M. (2017). Improving the photovoltaic conversion efficiency of ZnO based dye sensitized PV by indium doping. *Journal of Alloys and Compounds*, *692*, 67–76.
42. Saboor, A., Shah, S. M., & Hussain, H. (2019). Band gap tuning and applications of ZnO nanorods in hybrid solar cell: Ag-doped verses Nd-doped ZnO nanorods. *Materials Science in Semiconductor Processing*, *93*, 215–225.
43. Zhang, Q., Dandeneau, C. S., Zhou, X., & Cao, G. (2009). ZnO nanostructures for dye-sensitized PV. *Advanced Materials*, *21*(41), 4087–4108.
44. Hagfeldt, A., Boschloo, G., Sun, L., Kloo, L., & Pettersson, H. (2010). Dye-sensitized PV. *Chemical Reviews*, *110*(11), 6595–6663.
45. Zhang, Q., & Cao, G. (2011). Nanostructured photoelectrodes for dye-sensitized PV. *Nano Today*, *6*(1), 91–109.
46. Li, L., Zhai, T., Bando, Y., & Golberg, D. (2012). Recent progress of one-dimensional ZnO nanostructured PV. *Nano Energy*, *1*(1), 91–106.
47. Chung, J., Myoung, J., Oh, J., & Lim, S. (2010). Synthesis of a ZnS shell on the ZnO nanowire and its effect on the nanowire-based dye-sensitized PV. *The Journal of Physical Chemistry C*, *114*(49), 21360–21365.
48. Martinson, A. B., Elam, J. W., Hupp, J. T., & Pellin, M. J. (2007). ZnO nanotube based dye-sensitized PV. *Nano Letters*, *7*(8), 2183–2187.
49. Yang, Z., Xu, T., Ito, Y., Welp, U., & Kwok, W. K. (2009). Enhanced electron transport in dye-sensitized PV using short ZnO nanotips on a rough metal anode. *The Journal of Physical Chemistry C*, *113*(47), 20521–20526.
50. Weintraub, B., Wei, Y. and Wang, Z.L., (2009). Optical fiber/nanowire hybrid structures for efficient three-dimensional dye-sensitized PV. *Angewandte Chemie International Edition*, *48*(47), 8981–8985.
51. Wei, Y., Xu, C., Xu, S., Li, C., Wu, W., & Wang, Z. L. (2010). Planar waveguide– nanowire integrated three-dimensional dye-sensitized PV. *Nano Letters*, *10*(6), 2092–2096.
52. Horiuchi, H., Katoh, R., Hara, K., Yanagida, M., Murata, S., Arakawa, H., & Tachiya, M. (2003). Electron injection efficiency from excited N^3 into nanocrystalline ZnO films: effect of (N^{3-} Zn^{2+}) aggregate formation. *The Journal of Physical Chemistry B*, *107*(11), 2570–2574.
53. Asbury, J. B., Anderson, N. A., Hao, E., Ai, X., & Lian, T. (2003). Parameters affecting electron injection dynamics from ruthenium dyes to titanium dioxide nanocrystalline thin film. *The Journal of Physical Chemistry B*, *107*(30), 7376–7386.
54. Yang, Z., Chen, C. Y., Roy, P., & Chang, H. T. (2011). Quantum dot-sensitized PV incorporating nanomaterials. *Chemical Communications*, *47*(34), 9561–9571.

55. Shockley, W., & Queisser, H. J. (1961). Detailed Balance Limit of Efficiency of pn Junction PV,. *Journal of Applied Physics*, *32*, 510–519.
56. Leschkies, K.S., Divakar, R., Basu, J., Enache-Pommer, E., Boercker, J.E., Carter, C.B., Kortshagen, U.R., Norris, D.J. and Aydil, E.S., (2007). Photosensitization of ZnO nanowires with CdSe quantum dots for photovoltaic devices. *Nano Letters*, *7*(6), 1793–1798.
57. Chen, J., Li, C., Eda, G., Zhang, Y., Lei, W., Chhowalla, M., Milne, W.I. & Deng, W.Q., (2011). Incorporation of graphene in quantum dot sensitized PV based on ZnO nanorods. *Chemical Communications*, *47*(21), 6084–6086.
58. Leschkies, K.S., Divakar, R., Basu, J., Enache-Pommer, E., Boercker, J.E., Carter, C.B., Kortshagen, U.R., Norris, D.J. and Aydil, E.S., (2007). Photosensitization of ZnO nanowires with CdSe quantum dots for photovoltaic devices. *Nano Letters*, *7*(6), 1793–1798.
59. Leschkies, K.S., Jacobs, A.G., Norris, D.J. and Aydil, E.S., (2009). Nanowire-quantum-dot PV and the influence of nanowire length on the charge collection efficiency. *Applied Physics Letters*, *95*(19), 193103(1-3).
60. Kuo, K.T., Liu, D.M., Chen, S.Y. and Lin, C.C., (2009). Core-shell $CuInS_2$/ZnS quantum dots assembled on short ZnO nanowires with enhanced photo-conversion efficiency. *Journal of Materials Chemistry*, *19*(37), 6780–6788.

3 Environmental Applications of ZnO Nanostructures

Navdeep Sharma, Leena Jaswal, Madan Lal, Sandeep Arya, and Ashish Tiwari

3.1 INTRODUCTION

The 21st century has witnessed a remarkable revolution in various fields, including catalysis, energy storage, agriculture, biosensors, and biomedical applications, thanks to nanotechnology for providing significant methods to synthesize nanoscale materials. These applications rely on a diverse range of materials like metals, metal oxides, and allotropes of carbon. In the realm of nanotechnology research, the control of nanostructured material morphology has emerged as a key focus, as it directly influences their properties and applications, driving further advancements [1–4]. Among semiconducting metal oxides like tin oxide (SnO_2), titanium oxide (TiO_2), and ZnO, which offer unique physicochemical properties based on size and shape, ZnO emerges as a prominent II–VI semiconductor, with zinc belonging to group II and oxygen to group VI of the periodic table. It possesses a direct bandgap of approximately 3.37 eV at room temperature and exhibits excellent stability in both chemical and thermal environments. ZnO is a versatile n-type semiconducting material that finds applications in a broad spectrum of fields, such as solar cells, batteries, supercapacitors (SCs), luminescent materials, photocatalysis, piezo-electrics, biomedical sensors, biosensors, and optoelectronics [5–7]. Additionally, ZnO exhibits remarkable properties such as high electron mobility, transparency, a large exciton binding energy (~60 meV), strong luminescence at room temperature, thermal and mechanical stability, a broad radiation absorption range, and photostability.

Technological advancements have facilitated the design of microstructured and nanostructured ZnO with tailored size, structure, porosity, and morphology. This progress has significantly enhanced the performance and functionality of ZnO nanostructures across diverse applications [8–10]. Recently, nanostructures of ZnO, such as nanowires, nanorods, nanoneedles, nanorings, nanospheres, and nanotubes, have garnered significant interest. These nanostructures are renowned for their exceptional properties, including chiefly large specific area, nontoxicity, high sensitivity, good compatibility, and high isoelectric point. Moreover, they exhibit higher dissolution rates and increased solubility compared to their macroscopic counterparts [11–14].

DOI: 10.1201/9781003464907-3

The distinct morphological characteristics of ZnO contribute to its versatility and broad-ranging applications in the fields of environment, biology, energy, and beyond.

The synthesis and fabrication processes are very crucial and have a significant role in shaping the morphology of ZnO. Parameters such as surfactant, temperature, concentration, and time are crucial in various synthetic methods. The literature documents several techniques for synthesizing well-structured ZnO, including chemical routes as well as green synthesis routes for the preparation of nanomaterials such as nanoparticles (NPs), nanopowder, composites, and films. These methods encompass the sol-gel process, wet chemical techniques, hydrothermal method, precipitation, solvothermal process, and microwave techniques [15–17]. Alternatively, green synthesis represents an effective and environmentally friendly approach to producing ZnO nanomaterials. By utilizing naturally occurring molecules, biogenic synthesis enables the creation of unique NPs while minimizing the use of harmful reagents and maximizing process efficiency, aligning with the principles of green chemistry [18]. Biogenic synthesis of NPs reduces environmental and health impacts compared to chemical methods, paving the way for diverse applications. Valuable resources for this approach include mainly plant extracts, different plant parts, and microorganisms [19, 20].

This chapter comprises comprehensive study of the extensive environmental applications of ZnO, shedding light on its remarkable properties. Our investigation delves deeply into the profound contributions made by various ZnO nanostructures, with a specific emphasis on their integration within environmental contexts. Figure 3.1 provides a

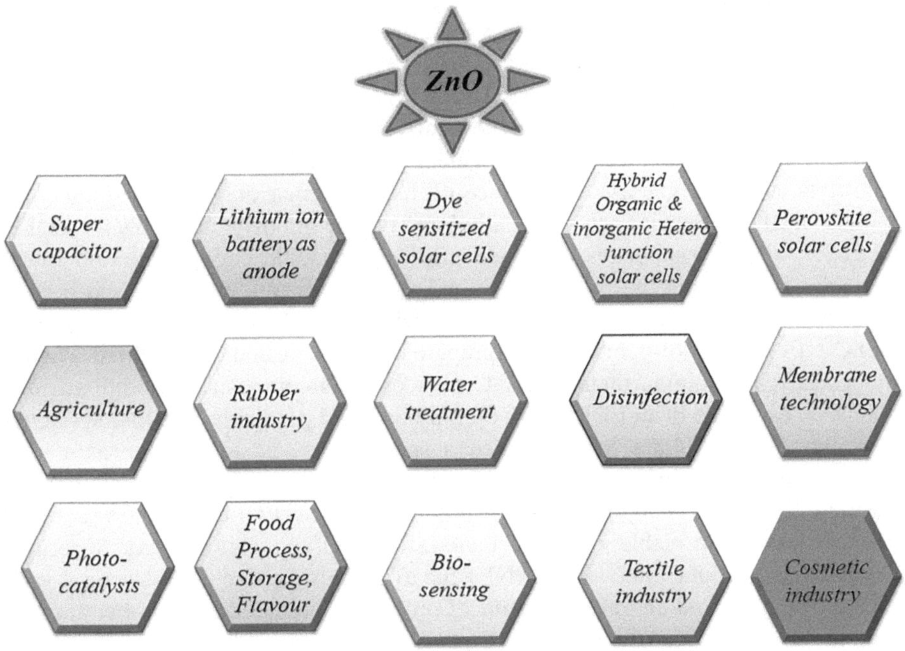

FIGURE 3.1 Representation of diverse environmental applications of ZnO nanostructures.

schematic representation of ZnO, serving as a visual aid for the diverse environmental applications discussed throughout this chapter. Additionally, this chapter incorporates illustrative diagrams to enhance the comprehension of these applications. This chapter offers a wide-ranging overview of the underlying principles of ZnO and its versatile applications within the environmental sphere and invaluable insights to readers interested in exploring sustainable solutions and emerging trends within this field.

3.2 RENEWABLE ENERGY APPLICATIONS

ZnO nanostructures possess distinctive properties that render them highly promising nanomaterials having a wide range of renewable energy applications, like SCs, energy storage systems, and solar cells. This section focuses on the specific environmental applications of ZnO nanostructures in the realm of renewable energy and explores the recent advancements in this field.

3.2.1 Super Capacitor (SC)

ZnO-based materials exhibit unique properties that make them highly promising for energy harvesting and storage applications. With their tunable bandgap, varied morphologies, and sizes, these materials are well-suited for use as electrodes in SCs. The high surface area of nanostructured ZnO enables more efficient charge storage, while its distinctive morphology enhances electrical conductivity. In the current scenario, research is concentrated on developing innovative synthesis methods and controlling the properties of ZnO-based materials for targeted applications [21]. SCs can be classified into three categories based on their charge storage mechanisms: (i) electrochemical double-layer capacitors (EDLCs), (ii) pseudocapacitors, and (iii) hybrid capacitors. EDLCs store charge electrostatically without any charge transfer between the electrode and electrolyte, relying on electrodes with high surface areas and thin electrolyte layers to increase capacitance [22]. Selvakumar et al. reported the combination of ZnO NPs with activated carbon to create composite materials for EDLC electrodes. These composites exhibited improved specific capacitance and lower resistance, showcasing the potential of ZnO for enhancing their utility and performance for EDLCs and other energy storage devices [23]. Anuj reported the synthesis of ZnO NPs via the sol-gel method to prepare a ZnO-activated charcoal (AC) nanocomposite by combining them with AC powder. This composite material was utilized as an electrode in an EDLC cell, employing a 6M potassium hydroxide (KOH) electrolyte. The resultant SC displayed a specific capacitance of approximately 118.122 mF cm^{-2}, showing negligible variation at elevated scan rates and current densities. These results underscore the significant potential of the ZnO-AC (1:1 ratio) nanocomposite as a viable electrode material for SC applications [24]. Shaarawy et al. demonstrated the effect of ZnO NPs due to the particle size with structural variation on the electrical and electrochemical properties prepared by soft-wet and dry methods. ZnO samples exhibited supercapacitive behavior, with capacitance ranging from 77 to 330 Fg^{-1}, depending on the morphology. Experimental data revealed that nanowires due to their nanocrystalline size and large surface area exhibit the highest capacitance and cycling stability [25]. Purushothaman et al. reported the synthesis of ZnO NPs with an average size (~19 nm) via spray pyrolysis method without a capping agent. Results showed that ZnO electrodes exhibited EDLC behavior with a specific capacitance of 365 F/g in a 2M KOH electrolyte at a scan rate of 5 mV/s as confirmed by using cyclic voltammetry technique [26]. Moreover, Bishwakarma et al. produced zinc oxide (ZnO) NPs using an electrochemical discharge machining process and tested them as potential electrodes for SCs. The results showed that electrodes have a specific capacitance of approximately 708.75 Fg^{-1} at a current density of approximately 1 A/g, further its capacitance after 1000 charging and discharging cycles retained about 90.42% [27].

In summary, the results from various studies suggest that ZnO-based materials hold significant potential in the field of SC applications.

3.2.2 Lithium-Ion Battery as Anode

The rapid industrial growth and increased consumption of fossil fuels have caused severe environmental problems. To tackle this issue, the development of secondary batteries has emerged as a promising and environmentally friendly solution for energy storage [28, 29]. Among rechargeable batteries, lithium-ion batteries (LIBs) have gained significant consideration because they provide a best substitute for fossil fuels and have made remarkable advancements in powering mobile devices and commercial vehicles like Tesla [30]. Controlling the morphology of anode materials is crucial for achieving optimal electrochemical performance; in this concern, ZnO NPs become a reliable and cost-effective choice for LIBs. Further, ZnO nanomaterials exhibit a high lithium storage capacity of approximately 987 $mAhg^{-1}$, along with good chemical stability and processability [31]. Kundu et al. synthesized ZnO nanorods with chemical methods and expose their performance as an anode in a LIB. The results showed that the nanorod anode exhibited high initial discharge capacity and maintained good reversible capacity throughout cycling, in comparison to other ZnO morphologies. The small size of the ZnO nanorods facilitated faster alloying and de-alloying and caused reduced volume changes during the cycling process [32]. Further, Xiao et al. demonstrated in detail the impact of structural patterns and crystallinity levels on the performance of nanostructured ZnO as an anode material in LIBs. The study revealed that ZnO microspheres calcined at approximately 600°C exhibited excellent performance as an anode material, demonstrating a specific capacity at 1st cycle of around 1328.2 $mAhg^{-1}$ and 662.8 mA h g^{-1} (50th cycle) at 0.1 C within the potential range of 0.05–3.00 V [33]. Furthermore, Laurenti et al. investigated ZnO nanostructures synthesized via chemical vapor deposition as LIB anodes. Results analysis exposed that the high surface area and effective electrolyte wetting facilitated excellent electrical transport and electrochemical behavior, along with a stable specific capacity (~70 μAh cm^{-2}) even under extended cycling rates and high coulombic efficiency. Notably, these findings were achieved without the use of binders or conducting additives, highlighting the effectiveness of ZnO nanostructures for LIB applications [34]. To improve the performance of nanostructured ZnO as a LIB anode, Xiao et al. coated it with polydopamine-derived carbon and controlled crystallization through subsequent heat treatment. The coated ZnO

NPs exhibited reversible capacities of approximately 1330.9 mAhg^{-1} (1st cycle) and 527.0 mAhg^{-1} (100th cycle), respectively. The heat treatment improved crystallization, and the nanostructure accommodated volume changes during cycling, resulting in enhanced capacity and retention compared to uncoated ZnO microspheres [35]. Moreover, Cauda et al. demonstrated the effective functioning of ZnO NPs in functioning as LIB anode for extended cycling capacity and used in other energy-related applications. Their findings highlighted the potential of flower-like ZnO nanostructures as stable and long-lasting anodic materials for future energy storage and conversion devices [36].

3.2.3 Dye-Sensitized Solar Cells: A Green Approach

ZnO NPs possess valuable properties that make them suitable as semiconductor materials in Dye-sensitized solar cells (DSSCs). With its high electron mobility, wide bandgap, and excellent solar-spectrum absorption capability, ZnO NPs can be utilized as an alternative or co-sensitizer to TiO_2 in DSSCs. Ongoing research is dedicated to optimizing the role of ZnO NPs in this field [37]. Rathnasamy et al. implemented an eco-friendly approach by synthesizing ZnO NPs from Carica papaya leaf extract. They demonstrated the potential use of these NPs as photoanodes in DSSCs, achieving a respectable energy conversion efficiency of approximately 1.6% [38]. Additionally, Sharma et al. explored the green synthesis route of ZnO NPs using Calotropis gigantea leaf extract. This research highlights the significance of using natural resources to produce sustainable and eco-friendly nanomaterials. Although the achieved energy conversion efficiency of 0.66% is relatively low, this research lays the foundation for future optimization and improvement of photoanode performance in DSSCs [39]. Shashanka et al. reported the successful preparation of ZnO NPs via a green synthesis route using Tilia tomentosa leaf extract, with the resulting material being calcined at around 400°C. The resulting NPs exhibited a bandgap of approximately 3.55 eV. These semiconducting NPs were utilized to construct a DSSC, which demonstrated an efficiency of 1.97% at 100mW/cm^2 due to increased dye molecule absorption on the ZnO surface. Hence, the role of green-synthesized ZnO NPs in DSSCs offers a promising and straightforward approach to sustainable solar energy production [40].

In a study by Alanazi et al., it was discovered that DSSCs can convert visible light into electricity using natural dyes extracted from green and red cabbage leaves. Under optimal conditions, red cabbage extract outperformed green cabbage extract, with a maximum power (P_m) of 1.36 mW/cm^2, a fill factor (FF) of 92.34%, and an efficiency (η) of 0.161%, compared to P_m = 0.349 mW/cm^2, FF = 44.19%, and η = 0.095% for green cabbage. These natural dyes hold potential for future use as sensitizers in solar cell research and development [41]. Kumar et al. exposed the biological processing of rod-shaped ZnO NPs synthesized from Colocasia esculenta tuber extract. These NPs were exposed to microwave irradiation and annealed at around 400 degrees. When utilized as a photoanode in DSSCs, they achieved an efficiency range of 1.29% to 1.60% [42]. Furthermore, Wong et al. emphasized the importance of balancing several factors such as electron lifetime, electron transport, dye adsorption, and electron injection to optimize DSSC performance. Their study found that orange-red emitting defects positively affected the short circuit current density. However, these defects also led to a reduction in the open-circuit voltage and FF. Additionally, a high concentration of nonradiative defects was shown to hinder photovoltaic performance and encourage NP aggregation [43].

3.2.4 Hybrid Organic and Inorganic Heterojunction Solar Cells (HSCs)

ZnO NPs have shown great promise for solar cells because of their excellent conductivity, stability, and cost-effectiveness. These nanostructured ZnO materials have been developed and utilized as photoanodes on conducting substrates [44]. In the context of hybrid solar cells (HSCs), ZnO NPs can serve as an inorganic semiconductor alongside an organic conjugated polymer, which facilitate as the light harvester and electron donor. This combination of organic and inorganic materials in HSCs offers flexibility and enables low-cost fabrication [45]. Furthermore, ZnO is commonly employed as an electron acceptor in HSCs, replacing the organic semiconductor used in fully organic solar cells (OSCs). This is attributed to the higher electron mobility and superior physiochemical stability of ZnO NPs compared to most available n-type organic semiconductors [46]. In 2004, Beek et al. reported the synthesis of nanocrystalline ZnO NPs with an average diameter of ~5nm via hydrolysis method involving the condensation of zinc acetate dihydrate with KOH in methanol and shown their applicability in the HSCs. The organic p-type semiconductor poly[2-methoxy-5-(3′,7′-dimethyloctyloxy)-1,4-phenylenevinylene] (MDMO-PPV) was utilized to create bulk heterojunction HSCs. Under an intensity equivalent to ~0.71 sun, the configuration achieved power conversion efficiency (PCE) of approximately 1.6%. The PCE remained moderately stable at a higher intensity equivalent to ~1.7 sun, with a value of 1.4% [47]. However, the inclusion of hydrophilic presynthesized ZnO NPs in the polymer blend was found to impede crystallization and decrease hole mobility. The nonuniform distribution of ZnO particles, particularly at the PEDOT/PSS interface, formed a polymer layer that obstructed excitons, resulting in decreased internal quantum efficiency and charge generation [48]. Therefore, the inefficiency of electron transport in ZnO-based HSCs can be attributed to the nonuniform dispersal of particles during the manufacturing process and the absence of a well-blended organic-inorganic interface, ultimately leading to decreased device efficiency [49]. Another approach to produce ZnO-based HSCs involves using diethylzinc as a precursor, dissolved in an organic solvent and mixed with the polymer to create a thin film. On exposure to moisture, the mixture undergoes a hydrolysis reaction, resulting in well-dispersed ZnO NPs throughout the polymer film. Subsequent thermal annealing near the polymer glass transition temperature (T_g) can enhance polymer crystallinity, thereby improving hole mobility and packing [50–52]. Notably, ZnO was found to crystallize at a lower temperature of 110°C, leading to more evenly distributed ZnO NPs in the film when employing a precursor-based method with post-fabrication thermal annealing. As a result, the performance of nc-ZnO:P3HT solar cells was significantly improved, increasing their PCE by 0.5% to approximately 1.4% [53]. Additionally, integrating the 3D morphology of ZnO NPs into P3HT using the precursor method resulted in a notable improvement in the PCE of solar cells. Achieving an active layer thickness of approximately 167 nm, a PCE of 2% was attained. Additionally, it was noted that thinner cells exhibited lower performance, potentially attributed to coarse phase separation, which led to the absence of a continuous pathway [54]. To further improve the efficiency of ZnO-based solar cells, several advancements can be made. These involve exploring new synthesis methods for ZnO NPs, optimizing the interface between organic and inorganic components, incorporating novel materials, and developing scalable fabrication techniques.

3.2.5 Perovskite Solar Cells (PSCs)

ZnO NPs are utilized in PSCs due to their advantageous electronic properties, transparency, and ease of synthesis. These NPs play a crucial role in enhancing charge transport, preventing perovskite layer degradation, and acting as a scaffold for uniform film growth. The working principle of PSCs, illustrated in Figure 3.2, is similar to that of DSSCs. It involves two main steps: i) Light absorption and charge generation: When the perovskite layer is illuminated, electron-hole pairs are generated. ii) Charge separation and collection: The electrons migrate to the conduction band of the semiconductor, while the holes move to the valence band of the hole transport material (HTM). This separation leads to the generation of an electric current [55, 56]. The pioneering work of Kumar et al. introduced ZnO NPs as an electron transport layer (ETL) in PSCs. Their findings revealed that ZnO NPs served as a dual-function layer, acting as both a compact layer to block holes and an electron-transporting layer with a mesoscopic nanorod structure. Compared to devices with only compacted ZnO NPs layers, planar devices incorporating these layers achieved a significantly higher conversion efficiency of approximately 8.90% (compared to the typical 2.62% for compacted ZnO NPs devices) [57]. Perovskites possess the crystal structure ABX3, where A and B represent cations with different ionic radii, and X

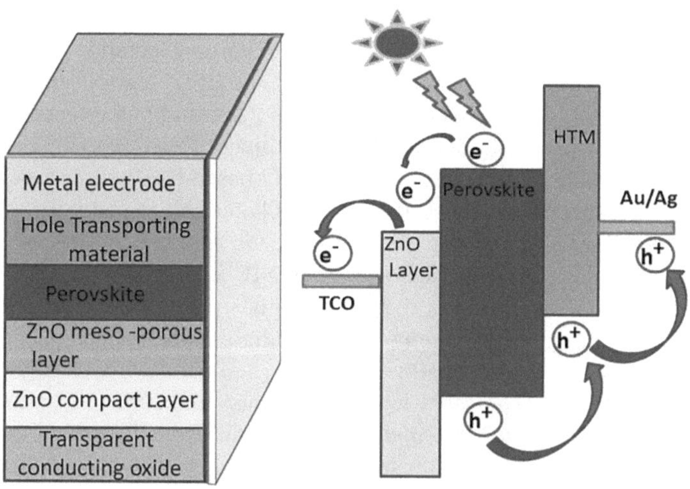

FIGURE 3.2 Schematic representation of the perovskite solar cell mechanism.

denotes an anion. This unique structure makes perovskites highly versatile and holds great promise for various applications [58]. Bi et al. employed a ZnO nanorod array as the ETL in PSCs. Their study demonstrated the effectiveness of ZnO nanorods in PSCs, highlighting their excellent long-term stability and suggesting that perovskite materials are well-suited for sensitizing ZnO nanorods [59].

Various morphological forms of ZnO, such as NPs, nanorods, nanowires, nanocolumns, and nanowalls, have been employed in PSCs. These different nanostructures possess varying electronic transmission capacities and surface areas, which influence factors like perovskite layer morphology, loading, the quality of the ZnO-perovskite interface, and overall perovskite material quality. Therefore, the utilization of different ZnO morphologies allows for manipulation of the interface in PSCs. Additionally, it has been observed that charge recombination at the ZnO-perovskite interface can limit cell performance. To mitigate this, doping processes have played a significant role in modifying the electronic properties and morphology of ZnO. This approach enhances the proportion of free charges and, consequently, the conductivity in solar cells [60, 61]. Doping in ZnO can occur through the replacement of either the Zn^{2+} cation or the O^{2-} anion. Cationic dopants commonly involve metals, while anionic dopants consist of nonmetals. Doping has the ability to shift the Fermi level toward the conduction band, resulting in improved conductivity and a decrease in the work function of ZnO [62]. Dong et al. (2014) demonstrated the fabrication of highly efficient PSCs using Al-doped ZnO (Al-ZnO) nanorods. Introducing interfacial modification with Al-ZnO led to significant improvements in tuning the energy band structure, reducing charge recombination and enhancing the open-circuit voltage and PCE. The average PCE increased from approximately 8.50% to 10.07%, with a maximum efficiency of around 10.7% [63]. Zuo et al. reported a remarkable enhancement in substrate miscibility between sol-gel ZnO ETLs and perovskite by depositing a self-assembling monolayer of 3-amino propanoic acid (C3-SAM). This led to the formation of a crystalline structure in the perovskite film through the conversion of the amino group to ammonium via hydrogen ion exchange. Importantly, the insertion of the C3-SAM layer between the ZnO and perovskite layers significantly improved the PCE of $CH_3NH_3PbI_3$ PSCs, with the maximum efficiency increasing from 11.96% to 15.67% [64]. Liu et al. investigated the effect of surfactants such as polyethyleneimine (PEI), polyvinyl pyridine (PVP), and ethanolamine on ZnO-based devices with compact layers prepared using solution processing and electron beam deposition. The results showed that ethyleneimine treatment was the most effective in improving photovoltaic performance by passivating surface defects [65].

3.2.6 Agriculture

Agriculture plays a crucial role in the economies of developing countries, but it faces numerous challenges such as climate change, urbanization, and increasing food demand. To promote sustainability in agriculture, various techniques have been employed, including precision farming, agroforestry, and conservation agriculture. These techniques aim to optimize resource utilization, reduce environmental impact, and foster economic development [66]. In recent years, nanotechnology has emerged as a valuable tool in agriculture for enhancing crop yields and improving crop quality through applications in seed germination and plant growth. Different types of NPs, such as carbon nanotubes (CNTs), silica (SiO_2), ZnO, TiO_2, and gold (Au), have been utilized in these applications. Recent studies have demonstrated positive effects on seed germination in various crops, including tomatoes, soybeans, wheat, rice, maize, pearl millet, and barley. However, the influence of these NPs on seed germination is dependent on genetic traits, environmental factors, and soil parameters [67–70]. Nanotechnology has transformed agriculture and food production by offering an efficient means of delivering pesticides and fertilizers with high specificity. NPs and nanocapsules enable targeted delivery, reducing collateral damage and facilitating controlled release. Nanosensors aid in determining the precise amount of essential farm inputs such as fertilizers and pesticides, as well as detecting pesticide residue, soil nutrients, and soil moisture. Nanofertilizers are rapidly absorbed by plants, while nanoencapsulated slow-release fertilizers reduce consumption and minimize environmental pollution. ZnO NPs have demonstrated the potential to enhance the growth and yield of food crops. Researchers have treated peanut seeds with varying concentrations of ZnO NPs and found that a concentration of approximately 1000 ppm with a mean particle size of 25 nm resulted in improved seed germination, seedling vigor, and overall plant growth. ZnO NPs have also been effective in promoting stem and root growth in peanut plants [71]. Colloidal solutions of ZnO NPs are utilized as nanofertilizers in agriculture, providing plant nutrients and improving soil health without the harmful effects associated with chemical fertilizers. One of the key advantages of nanofertilizers is that they can be used in smaller amounts, with 40–50 kg being sufficient for an adult tree compared to the 150 kg needed for traditional fertilizers. Nanopowders have also shown effectiveness as both fertilizers and pesticides [72, 73]. Seed germination depends on multiple factors, including antioxidant enzyme activities, water utilization rates, and oxygen utilization rates. When wheat was treated with ZnO NPs, a higher germination rate was observed as compared to the control group [74, 75]. Batsmanova et al. demonstrated that treating wheat seeds with metal NPs led to an average yield increase of 20–25% in resulting plants [76]. In a study by Mahajan et al., the growth of mung seedlings was stimulated up to a concentration of 20 ppm of ZnO, but growth was retarded beyond that concentration. Gram seedlings showed exciting growth at an optimum ZnO concentration of approximately 1 ppm. However, exceeding the limit resulted in retarded growth due to the accumulation of ZnO NPs by the roots [77]. Prasad et al. reported the significant impact of nanosized ZnO concentration on its benefits for agriculture. An optimal concentration of 1000 ppm of ZnO NPs improved the germination, root growth, shoot growth, dry weight, and pod yield of treated peanut seeds [78]. Burman et al. observed that a concentration of 1.5 ppm of ZnO NPs led to improved shoot dry weight and antioxidant activity in chickpea seedlings [79]. Ramesh et al. conducted a study demonstrating that treating cluster bean seedlings with ZnO NPs at a concentration of approximately 10 ppm enhanced nutrient growth and aided in plant growth [80]. Additionally, Zhang et al. reported the potential of ZnO NPs as a promising micronutrient fertilizer for rice crops. When ZnO NPs were utilized at the basal stage, they resulted in enhanced rice yield, improved quality, increased Zn content, and improvements in dry matter accumulation, leaf area index, SPAD value, and photosynthetic potential of rice crops [81]. Chandio et al. investigated the effect of HA-ZnO NPs (NPs combined with humic acid) on wheat seed germination and growth. The results indicated that using appropriate concentrations of HA-ZnO NPs as a co-fertilizer effectively improved seed germination and seedling growth. The most favorable outcome was observed at a concentration of 20 ppm HA-ZnO NPs, which had the highest positive effect on seedling growth compared to the control group [82]. Yan et al. reported the significant effect of ZnO NPs on rice seedling growth, biochemical reactions, and arsenic uptake. Their findings suggested that ZnO NPs can be used as a fertilizer to support plant growth and reduce the accumulation of arsenic in rice. Based on plant growth and arsenic accumulation, a concentration of approximately 100 mg/L of ZnO NPs was found to be optimal [83]. Additionally, Taheri et al. demonstrated that the use of ZnO NPs increased shoot dry matter and leaf area indexes by approximately 63.8% and 69.7%, respectively. These findings indicate that zinc NPs can enhance corn growth and yield in mineral-poor soils [84]. ZnO nanostructures have shown promising results in improving crop growth and productivity by enhancing zinc uptake and nutrient absorption. However, it is important to carefully consider the environmental impacts and conduct further research on potential risks to ensure their safe and responsible use in agriculture.

3.2.7 Rubber Industry

ZnO plays a crucial role in the rubber industry, enabling the production of high-quality rubber products that can withstand various applications. The rubber industry is a major consumer of ZnO, with global production reaching approximately 105 tons per year. This essential ingredient is used in the production of cross-linked rubber products, contributing to their durability and performance [85]. To enhance the thermal conductivity of silicone rubber, various fillers such as metal oxides, metal powders, and inorganic particles like Al_2O_3, MgO, Al_2N_3, SiO_2, and ZnO are incorporated. Among these fillers, nanosized ZnO particles have shown remarkable thermal conductivity even at low filling levels. However, they tend to aggregate within the polymer matrix due to weak particle surface interaction. To overcome this challenge, surface modification techniques and optimized processing methods can be employed to prevent aggregation and improve dispersion. This ensures improved thermal conductivity in silicone rubber while preserving its electrical resistance properties [86]. One effective technique to enhance ZnO NPs compatibility is the use of a coupling agent like vinyltriethoxysilane. Yuan et al. demonstrated the successful modification of ZnO NP surfaces, leading to enhanced dispersion and improved thermal conductivity in silicone rubber nanocomposites. The modified NPs formed a cross-linking structure with the silicone rubber matrix, resulting in improved mechanical properties [87]. Proper dispersion of ZnO is crucial for achieving desired rubber properties. While high-intensity mechanical mixing is commonly used, the addition of a fatty acid coating, such as propionic or stearic acid, can make the surface of ZnO hydrophobic and improve dispersion times [88]. Moreover, ZnO is extensively used as a cross-linking agent in the production of carboxylated elastomers, which exhibit exceptional mechanical properties such as high tensile strength, tear resistance, hardness, and hysteresis. The improved mechanical performance of these elastomers can be attributed to their ability to undergo stress relaxation through the slippage of elastomer chains on the surface of ionic clusters, as well as the reformation of ionic bonds during external deformation. Additionally, ionic elastomers possess thermoplastic characteristics, allowing them to be processed in a molten state similar to thermoplastic polymers [89]. While the use of ZnO cross-linking in carboxylate elastomers offers many benefits, it also has some drawbacks, including scorchiness, poor flexibility, and high compression set. Zinc peroxide or zinc peroxide/ZnO systems can address the scorchiness issue but require longer vulcanization times. Despite these challenges, ZnO is commonly used due to its ability to react with carboxylic groups, forming carboxylic salts. The surface properties of ZnO influence its cross-linking activity and the interphase size with elastomer chains [90]. The use of ZnO NPs as cross-linking agents in carboxylated nitrile elastomer formulations provides several advantages. It improves mechanical properties, increases crosslink density, and reduces the overall consumption of ZnO. The unique three-dimensional snowflake morphology of the NPs, with a surface area of approximately 24 m^2/g, is crucial for their effectiveness. This approach promotes environmental sustainability by addressing concerns about excessive ZnO usage, minimizing ecological risks, and fostering a more sustainable future. Additionally, ZnO NPs reduce agglomeration within the elastomer matrix and exhibit heat shrinkability in the vulcanizates [91]. Sabura et al. conducted successful synthesis of nanosized ZnO and utilized it as a curing agent in neoprene rubber. They discovered that the optimal dosage of nano-ZnO was lower than that of commercial ZnO, while maintaining comparable rubber properties. The NPs ranged in the particle size from approximately 15 to 30 nm, with surface areas ranging from about 12 to 30 m^2/g. This research demonstrates the potential of nano-ZnO as a cost-effective and efficient alternative in rubber curing applications [92]. In another study, Bindu et al. demonstrated that ZnO NPs can be effectively used as a replacement for conventional micro-ZnO in the vulcanization of natural rubber (NR). The vulcanizate containing nano-ZnO exhibited better properties and required a lower optimum dosage of 0.5 parts per hundred rubber (phr) compared to 5 phr of micro-sized ZnO. This substitution can lead to cost reduction and environmental benefits due to the reduced amount of ZnO in rubber compounds [93]. Furthermore, nanosized ZnO particles have a higher specific surface area (30–70 m^2/g) compared to traditional ZnO (6.0 m^2/g), providing more available zinc. The replacement of micron-sized ZnO with nano-ZnO improves rubber properties, including abrasion resistance and tear strength [93, 94]. ZnO is considered the best activator for sulfur vulcanization in rubber manufacturing, as it enhances curing speed and cross-linking density by promoting the formation of short sulfide cross-links [95]. Additionally, Xuan et al. reported that ZnO NPs capped with octylamine are effective activators for sulfur vulcanization in NR. These NPs reduce the required amount of ZnO compared to commercial systems while improving curing efficiency and mechanical performance in rubber composites. They significantly reduce ZnO levels under industrial vulcanization conditions [96]. Furthermore, Sahoo et al. found that the incorporation of ZnO NPs in carboxylated nitrile rubber enhances the state of cure, maximum torque, and mechanical properties. The vulcanizates showed a 120% increase in tensile strength, a 20% increase in elongation at the break, and a 30% increase in modulus at 300% elongation compared to those cured with rubber grade ZnO.

The optimum cure time and scorches time also slightly decreased with the addition of ZnO NPs [97]. The future prospects of ZnO in the rubber industry involve the development of eco-friendly alternatives to achieve sustainable and efficient rubber production with minimal environmental impact.

3.2.8 Water Treatment

Clean water is vital for human health, but its scarcity is becoming a pressing issue due to industrialization, urbanization, and climate change. Contaminated freshwater sources pose a significant risk, leading to waterborne diseases such as typhoid fever, dysentery, cholera, and diarrhea, which cause countless deaths worldwide [98, 99]. Addressing the urgent need for water purification is crucial for ensuring a healthy and sustainable future. Water purification aims to eliminate harmful chemicals, contaminants, and solids from water to make it safe for consumption. Traditional methods like filtration, sedimentation, distillation, and chlorination have limitations and can contribute to antibiotic resistance. Therefore, researchers are actively seeking alternative, effective, and sustainable purification methods [100, 101]. Among these alternatives, ZnO NPs have gained increasing attention as a promising solution for water purification due to their unique physicochemical properties [102]. These NPs possess a high surface area, reactivity, and distinctive electronic and optical properties, making them efficient in various water treatment applications [103]. ZnO NPs can act as highly effective photocatalysts, disinfectants, adsorbents, and sensors for the removal of various contaminants from water [104]. They can effectively eliminate both organic and inorganic pollutants and improve the filtration capabilities of water treatment membranes [105]. The scarcity and contamination of water are growing concerns, and ZnO NPs offer great potential as an efficient and cost-effective solution for water purification. They exhibit unique properties such as high thermal and chemical stability, unsaturated surfaces, and effective adsorption behavior toward organic and inorganic pollutants in water. Moreover, they possess a higher surface area and oxidizing power, including efficient photon absorption, compared to other semiconductors. Additionally, they are nontoxic, readily available, and cost-effective, making them an attractive solution for wastewater purification [106]. Scientists have been studying the potential of ZnO NPs in removing heavy metals from industrial wastewater. Spherical ZnO NPs have been found to efficiently remove lead (Pb^{2+}), cadmium (Cd^{2+}), and mercury (Hg^{2+}) from water solutions [107, 108]. Furthermore, ZnO NPs have been demonstrated as effective adsorbents for removing Cd^{2+}, Cu^{2+}, Ni^{2+}, and Pb^{2+} from water. X-ray diffraction (XRD) analysis revealed a crystallite size of approximately 16.70 nm, while scanning electron microscopy (SEM) confirmed their rod-like shape with a size of around 25 nm. Brunauer–Emmett–Teller (BET) analysis indicated a specific surface area of approximately 31.20 m^2/g with a total pore volume of approximately 12.09 $cm^3\ g^{-1}$ and a pore size of approximately 15.81 nm. The results showed that the adsorption of heavy metal ions increased as the pH of the solution became more alkaline. The pH of the solution plays a crucial role in the adsorption of heavy metals by nanoadsorbents. Under acidic conditions, hydrogen ions compete strongly with heavy metal ions in the solution [109, 110]. Xie et al. observed that increasing the absorbent dose (from 0.5 to 5 g/l) resulted in higher removal efficiency of Cd^{2+}, Cu^{2+},

Ni^{2+}, and Pb^{2+}. Pb^{2+} exhibited the highest removal percentage of approximately 81.5% at a dosage of approximately 5 mg/g, which was attributed to the increased binding sites with higher nanoabsorbent doses. The maximum removal efficiency for all four ions was achieved within approximately 180 minutes, with no significant increase observed beyond this time frame. The maximum uptake of metal ions was reported to be approximately 114.5 mg/g [111]. Salmani et al. conducted a study demonstrating that ZnO NPs exhibited a remarkable removal efficiency of approximately 89.6% for Cd^{2+} ions, following the pseudo-second-order and Langmuir isotherm models [112]. The efficiency of removal was found to be sensitive to variations in pH and ionic strength, with an improvement in Cd^{2+} ion removal observed when the solution pH was increased from 4 to 7. The maximum removal occurred within a shorter contact time, and various factors such as solution pH, contact time, metal concentration, and temperature played crucial roles in achieving efficient removal using ZnO NPs as adsorbents. ZnO NPs have proven to be effective adsorbents for removing heavy metal ions and dyes from water due to their high porous structure and large surface area, making them highly promising materials for water purification [113]. These NPs provide numerous active sites for adsorption, successfully eliminating various pollutants from water, including heavy metals, organic compounds, and dyes [114]. The adsorption process of pollutants onto ZnO NPs is influenced by factors such as particle size, concentration, pH, and the type of pollutant. This process is driven by electrostatic and van der Waals interactions between the surface of the NPs and the molecules of the pollutants. To enhance the adsorption capacity of ZnO NPs, their surface properties can be modified through methods such as doping, surface functionalization, and altering the surface charge [115].

3.2.9 Disinfection

Developing an effective antibacterial agent to combat waterborne pathogens is a challenging task. However, nanotechnology provides new opportunities for water treatment without the formation of disinfection by-products. Nanomaterials exhibit enhanced properties at the nanoscale due to their increased surface area. Among various inorganic materials such as silver (Ag) and TiO_2, ZnO NPs stand out as a particularly promising antibacterial agent. They possess strong activity even at low concentrations, remain stable under harsh processing conditions, and are safe for human and animal use. Additionally, ZnO offers superior durability, selectivity, and heat resistance compared to organic materials [116] [117]. ZnO NPs exhibit effective antimicrobial properties due to their ability to release oxygen species from their surface, leading to oxidative stress that damages bacterial DNA, cell membranes, and proteins. This surface activity causes decomposition of the cell wall, leakage of cellular contents, and ultimately, cell death [116]. Figure 3.3 schematically represents the disinfection mechanisms involved with ZnO nanostructures, aiding in understanding their practicality [116]. Furthermore, Heinlaan et al. suggested that the damage to bacterial cells caused by ZnO NPs is not a result of their entry into the cell, but rather the interaction between the bacterial cell and the NPs. This interaction alters the microenvironment within the contact area, resulting in cell damage [117]. Brayner et al. demonstrated and reinforced that ZnO NPs can damage and disrupt bacterial cell walls due to the abrasive nature of ZnO, increasing membrane permeability

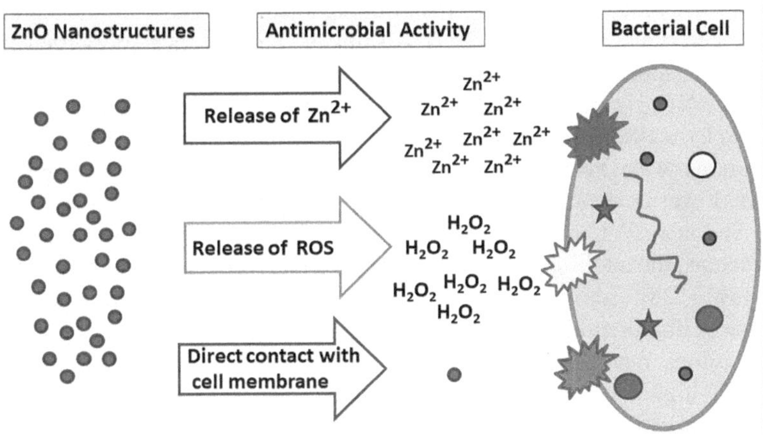

FIGURE 3.3 Schematic illustration of the ZnO disinfection mechanism.

and enabling the internalization of NPs into bacterial cells [118]. Another potential mechanism for the antibacterial activity of ZnO NPs involves the release of Zn^{2+} ions, which can damage the cell membrane and penetrate the intracellular contents. Li et al. reported that the toxicity of nano-ZnO against *Escherichia coli* was primarily attributed to the released Zn^{2+} ions [119]. Moreover, Shinde et al. found that the antimicrobial activity of ZnO is primarily dependent on NP size rather than the release of Zn^{2+} ions or oxygen. The study indicated that the release of Zn^{2+} ions and reactive oxygen species (ROS) was low, suggesting that particulate ZnO is responsible for the antimicrobial activity [120]. Zhang et al. demonstrated the crucial role of hydrogen peroxide (H_2O_2) in the disinfection properties of ZnO nanofluids. The results showed that smaller ZnO particles/aggregates exhibited better disinfection activity, and H_2O_2 was responsible for the disinfection activity of larger particles/aggregates against *Escherichia coli* bacteria [121].

3.2.10 ZnO Nanostructured Membrane Technology

In recent decades, membrane technology has become increasingly popular for water treatment, desalination, and wastewater management, with polymeric membranes being widely utilized. Membrane filtration is a highly effective technique that offers several advantages in water and wastewater treatment, including ease of operation, high productivity, selective separation, efficient contaminant removal, and the absence of harmful by-products [122, 123]. Membranes function by selectively separating materials, allowing permeate to pass through while retaining unwanted substances as retentate on their surface [124]. Two main types of membranes are used in water treatment: inorganic and polymeric. Inorganic membranes exhibit better resistivity and mechanical strength; whereas, polymeric membranes are more adaptable, cost-effective, and allow for controllable pore distribution [125, 126]. The synthesis of ZnO NPs by Awasthi et al. demonstrated their effectiveness against the formation of Bacillus subtilis biofilms, likely due to the induction of oxidative stress. Incorporating ZnO NPs into polymer membranes enhances their hydrophilicity, mechanical strength, and acts as a cross-linking agent, emphasizing the significance of ZnO in membrane technology [127]. Alsalhy et al. investigated polyvinyl chloride (PVC)/ZnO membranes with varying concentrations of ZnO NPs. The addition of 0.1 g of ZnO NPs significantly increased the mean roughness and pure water permeability (PWP) while maintaining a narrow pore size distribution. However, higher concentrations of ZnO NPs (0.3–0.4 g) led to decreased PWP due to NP aggregation. Therefore, lower concentrations are preferred to prevent aggregation and minimize fouling. The neat PVC membrane exhibited a PWP value of approximately 38.88 (L/h); whereas, the addition of 0.1 g of ZnO NPs increased the PWP value to around 122.22 (L/h) [128]. Additionally, the shape of ZnO nanomaterials affects membrane performance. Rajabi et al. reported that using nanorods instead of spherical NPs in polyethersulfone membranes resulted in a smaller water contact angle, potentially due to their higher migration toward the membrane surface [129]. Furthermore, Malini et al. investigated the antibacterial activity of chitosan/nano-ZnO membranes using the disk diffusion method. The results indicated that the chitosan membrane alone exhibited low antibacterial activity; whereas, the addition of nano-ZnO significantly enhanced the antibacterial properties of the composite membrane [130]. As a result, the composite membrane can serve as a multifunctional material for water treatment by simultaneously removing both bacteria and pollutants.

3.2.11 Photo Catalysts

Photocatalysis involves the use of a catalyst that generates an electron-hole pair when exposed to light, leading to reduction or oxidation reactions on the catalyst's surface. This process is depicted in Figure 3.4. Organic pollutants can be directly oxidized by

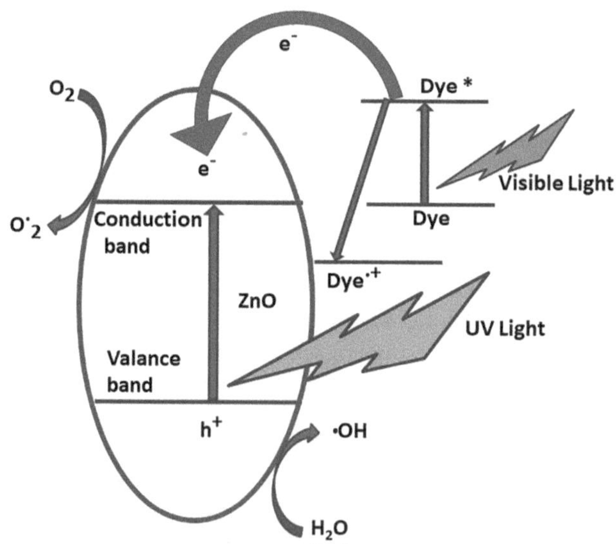

FIGURE 3.4 Schematic representation of oxygen generation in ZnO photocatalysis.

reacting with the photo-generated electron or indirectly through reactive groups in the solution, such as the hydroxyl radical (·OH) [131]. ZnO is a potent photocatalyst that produces ROS on light exposure, effectively decomposing pollutants into harmless substances. It is increasingly recognized as an environmentally friendly solution for remediation purposes [104, 132]. ZnO possesses photocatalytic properties similar to those of TiO_2 due to its bandgap energy of approximately 3.2 eV. However, ZnO's photocatalytic activity under ultraviolet (UV) light is susceptible to photocorrosion, which diminishes its efficiency in aqueous solutions. Therefore, while ZnO shows promising photocatalytic potential, concerns regarding its activity variability throughout the process persist [133]. Despite the issue of photocorrosion, several studies have demonstrated that ZnO can exhibit higher photocatalytic efficiency than TiO_2 in the degradation of specific dyes, even in aqueous solutions [134]. ZnO with a small particle size in the nanometer range demonstrates improved stability due to enhanced crystallinity and reduced defects. Additionally, incorporating other components can enhance ZnO's photocatalytic properties, enabling it to operate across a broader visible spectrum range [135]. Mousa et al. synthesized a TiO_2/ZnO heterojunction composite with enhanced antibacterial activity for dye photodegradation and wastewater treatment. The composite structure efficiently degraded dyes under direct sunlight within 120 minutes, attributed to the presence of rutile and anatase structures of TiO_2 and ZnO, a lower band gap, and the formation of a semiconductor-semiconductor heterojunction [136].

Chimupala et al. conducted a synthesis of controlled-size ZnO NPs using Tween80 as a surfactant. These NPs exhibited robust photocatalytic activity in the degradation of methylene blue dye. Interestingly, the application of mechanical force significantly enhanced the efficiency of dye degradation. The findings suggest that combining vibrational energy and solar harvesting could serve as a promising strategy for the development of effective photocatalysts in wastewater treatment [137]. In a separate study, Ranjitha et al. successfully synthesized ZnO nanostructures using a chemical additive-free hot water treatment method. These nanostructures demonstrated remarkable effectiveness in degrading methylene blue present in wastewater. This research highlights the potential of the approach for scalable, cost-effective, and environmentally friendly production of ZnO nanostructures for wastewater treatment [138].

3.2.12 Food Processing, Storage, Packaging, Flavor, and Quality Control

ZnO NPs have garnered significant attention in the field of food processing and storage due to their antimicrobial properties and stability under harsh conditions. They are considered safe for human consumption and recognized as such by the United States Food and Drug Administration (FDA) [139, 140]. Researchers aim to enhance the food industry by improving shelf life, controlling food-borne pathogens, ensuring safety, and optimizing the nutritional value, activity, and quality of food products [141]. Concerns have been raised regarding the harmful impact of pathogens on stored grains, which poses risks to human health. The antimicrobial properties of ZnO NPs play a crucial role in regulating important aspects of food materials such as moisture content, absorption, monolayer formation, solubility, and physical attributes [142]. To achieve extended shelf life and safe storage, reducing contamination is of utmost importance. Understanding the effects of storage conditions on microbial populations in minimally processed foods is crucial. ZnO NPs play a pivotal role in preserving freshness and nutritional value, with the potential to provide zinc supplements in functional foods [143, 144]. Galstyan et al. demonstrated that ZnO NPs possess oxygen-scavenging capabilities, antimicrobial activity, and temperature tolerance. They can be applied in films and coatings to create effective barriers against oxygen, carbon dioxide, and flavor compounds. These properties enable the development of nanosensors and intelligent packaging systems for advanced food quality control, process monitoring, authenticity verification, and freshness determination [145]. Chaudhary et al. reported that ZnO NPs exhibit distinctive properties, including a high specific surface area, biocompatibility, stability, electron communication, and nontoxicity. These characteristics make them ideal for portable, real-time monitoring systems for food quality analysis and process monitoring [146]. Meng et al. demonstrated the role of ZnO NPs in detecting naturally formed compounds such as trimethylamine (TMA), ethanol, gaseous amines, and NH_3 during the biodegradation of fish and animal products. This detection capability enables the assessment of food product freshness [147]. Moreover, studies have revealed that ZnO NPs have been utilized as antibacterial preservatives and anti-odor agents, effectively extending the shelf life of food materials [148]. From a safety standpoint, Kim et al. emphasized that the incorporation of ZnO NPs in food packaging should comply with the maximum migration limits set by regulatory authorities. For instance, the National Institute of Health (NIH) imposes a maximum daily consumption limit of approximately 40 mg/day for zinc in food contact materials [149].

3.2.13 Biosensing

Biosensors find extensive applications in various industries such as healthcare, chemical/biological analysis, the food industry, and environmental monitoring. These biosensors can be categorized based on their detection principles, including photometric, calorimetric, electrochemical, piezoelectric, and other types [150]. ZnO nanomaterials possess several desirable characteristics for biosensing applications. These include a high isoelectric point, high catalytic efficiency, and strong adsorption capability, making them suitable for the adsorption of specific proteins through electrostatic interactions [151]. The prominent properties of ZnO NPs, such as high surface area, biocompatibility, stability, low toxicity, and high electron transfer capability, enable their use in biosensing applications. They are employed for immobilizing biomolecules in biosensors and detecting analytes such as glucose, phenol, H_2O_2, cholesterol, and urea. ZnO NPs serve as sensors for gases and biochemicals in biomedical applications. The pore properties of ZnO play a vital role in gas sensors by enabling optimal adsorption. ZnO nanowires and NPs exhibit high sensitivity and selectivity, accurately detecting ethanol and acetone at a rapid pace [152]. The field of biosensor technology has been elevated by advancements in engineering and nanotechnology, enabling miniaturization and multi-functionality. The integration of sensing biomolecules onto a solid support, known as a matrix, is an essential part of biosensors. A suitable matrix enhances signal transduction, facilitates biomolecule

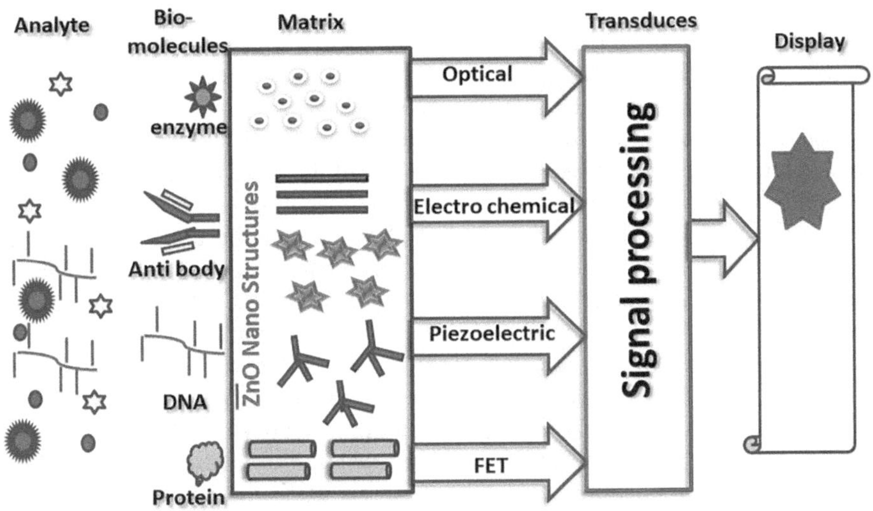

FIGURE 3.5 Schematic representation of biosensor organization.

immobilization, and preserves or enhances their activity. The physicochemical properties of the matrix determine the immobilization method and the operational stability of the biosensor, as depicted in Figure 3.5 [153]. Nanoporous and nanostructured ZnO films offer cost-effective fabrication methods and the versatility to create various nanostructures, making them a promising and affordable platform for biosensor development [154]. Fu et al. reviewed the field of biosensors and found that ZnO-based configurations have been developed using different platforms, including field-effect transistors (FETs), ion-sensitive field-effect transistors (ISFETs), optical devices, piezoelectric devices, and electrochemical transducers. Piezoelectric devices such as surface acoustic wave (SAW) sensors, quartz crystal microbalance (QCM), and film bulk acoustic resonator (FBAR), incorporating ZnO as the sensing material, have been employed for biosensing applications [155].

ZnO nanostructures are primarily utilized in electrochemical transduction for biosensing applications. However, recent studies have explored optical transduction and protein modification of FETs. FETs, particularly ISFETs, offer the advantages of miniaturization and multi-analyte sensing on microchips, potentially replacing electrochemical detection in miniaturized biosensors. Ali et al. demonstrated the wireless remote monitoring of glucose using ZnO nanowire arrays and existing general packet radio services (GPRS)/global system for mobile communication (GSM) networks. These advancements open up possibilities for the use of ZnO-based nanosensors in monitoring health parameters outside of conventional laboratories [156].

3.2.14 Textile Industry

Water-repellent, self-cleaning, and UV-blocking fibers offer significant advantages to both the military and business sectors. The defense textile industry faces challenges in improving functionalities, preserving fabric quality, and managing cross-infection risks. Therefore, the textile industry holds immense potential for commercializing nanotechnological products, and ongoing research is dedicated to developing fabrics with these desirable properties [157–159]. Nanostructured materials have shown to enhance textile properties and find expanding applications in the textile sector due to their increased surface area and affinity with fabrics [160]. Extensive research has demonstrated that ZnO improves antibacterial and self-cleaning properties in fabrics while enhancing UV absorption. The use of nanosized ZnO coatings in textiles is bio-compatible and offers superior air permeability and UV-blocking capabilities compared to bulk materials [161, 162]. Rajendran et al. coated nano-ZnO on cotton fabrics, resulting in antimicrobial properties. Nano-ZnO-treated fabrics exhibited stronger antibacterial effects than bulk-treated and untreated fabrics, particularly against S. aureus bacteria [163]. Mao et al. highlighted the attractiveness of ZnO nanostructures as UV-protective textile coatings. Various methods have been explored, including the hydrothermal growth of ZnO NPs on SiO_2-coated cotton fabric, which yielded outstanding UV-blocking properties [164]. Bashari et al. demonstrated the water-repellent and antibacterial properties of carnauba wax NPs and ZnO NPs on cotton, cotton/nylon6, and nylon6 fabrics using the layer-by-layer (LBL) self-assembly method. The study emphasizes the potential of carnauba wax in creating environmentally friendly water-repellent fabrics, with the added benefit of effective antibacterial effects through this innovative finishing method [165]. Shaban et al. showed that sol-gel coating cotton fibers with nano-ZnO created fabrics with antibacterial and superhydrophobic surfaces. The cotton fabric was cleaned with acetone and ethanol through ultrasonication at 80°C for 10 minutes. Zinc acetate dehydrate and methoxy ethanol served as precursors, while monoethanolamine acted as the stabilizer. The pH of the solution was adjusted using sodium hydroxide or acetic acid. Magnesium acetate tetrahydrate was used as the dopant material. The prepared ZnO NPs were spin-coated onto the fabric using the sol-gel method at 1100 rpm for 60 seconds [166]. ZnO nanostructures in the textile industry offer potential environmental benefits. However, considering the lifecycle impact of textiles, responsible manufacturing and waste management practices are crucial for environmental sustainability. Table 3.1 shows the linkage of ZnO properties with the textile industry.

3.2.15 Cosmetic Industry

ZnO nanostructures in cosmetics provide various benefits such as UV protection, antimicrobial effects, stability, and safety. These nanostructures enhance the efficacy

TABLE 3.1
Properties of ZnO and Its Role in Transforming Textile Sustainability

Over All Description of ZnO Nanostructures in Textile
Antibacterial : Improves antibacterial properties of fabrics, inhibiting microbial growth
Self-cleaning: Enhances self-cleaning capabilities, facilitating easy dirt and stain removal
UV-blocking: Provides effective protection against harmful UV radiation
Water repellency: Creates water-repellent properties, preventing water absorption in fabrics
Superhydrophobic: Generates superhydrophobic surfaces, causing water to bead and roll off fabrics

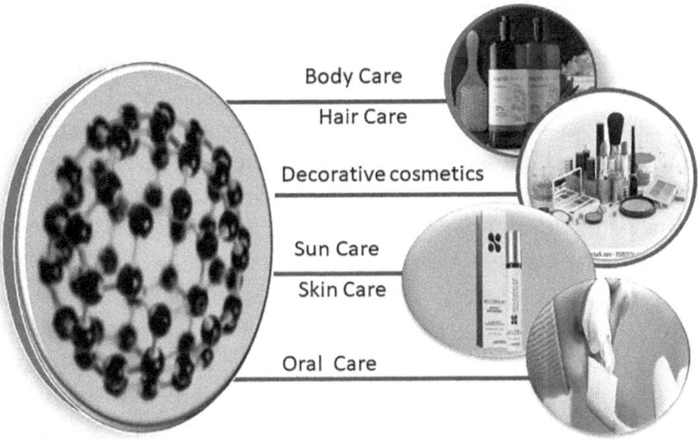

FIGURE 3.6 Cosmetic applications of ZnO nanostructure.

of cosmetic products, offering protection against harmful UV rays and promoting skin well-being. Additionally, ZnO exhibits antibacterial and antifungal properties, making it valuable for medical disinfectants and dermatological applications [167, 168]. ZnO NPs effectively shield against both UV-A and UV-B radiation, preventing sunburn, skin damage, and reducing the risk of skin cancer. They also contribute to enhancing the sun protection factor (SPF) of sunscreen products by absorbing UV radiation [169]. Reinosa et al. demonstrated that combining nanosized TiO_2 with ZnO particles in a composite formulation resulted in a higher SPF compared to using TiO_2 or ZnO alone [170]. Furthermore, ZnO and TiO_2 NPs offer superior UV protection compared to Avobenzone. They possess nongreasy, odorless, and colorless properties, effectively blocking UV radiation without leaving a visible residue. These advantages make them popular choices in sunscreens, contributing to even coverage and a smooth appearance in cosmetic creams [167]. Figure 3.6 showcases the diverse uses of ZnO NPs in cosmetics, while Table 3.2 highlights their usefulness in various applications.

TABLE 3.2
Efficacy of ZnO Nanostructures in Cosmetics

ZnO Nanostructured Usefulness in Cosmetic
i) Prolonging cosmetics shelf life through advanced preservation and stability
ii) Maximizing surface area to expose enhanced product performance and efficiency
iii) Empowering solubility enhancement for optimal product formulation and performance
iv) Catalyzing rapid absorption for enhanced skin penetration and benefits
v) Elevating cosmetic product texture with its transformative role
vi) Unleashing potent environmental benefits in minimal quantities/tiny doses with ZnO
vii) Advancing environmental benefits: ZnO action on the top layer of the skin

3.3 CONCLUSION AND FUTURE SCOPE

In conclusion, ZnO exhibits exceptional properties that play a crucial role in driving eco-friendly solutions across various sectors. It finds applications in renewable energy generation, water treatment, the rubber industry, cosmetics, biosensing, agriculture, and the food industry, highlighting its diverse environmental uses. ZnO enables efficient conversion of solar energy, contributes to the effective purification of water, enhances the durability of rubber products, provides UV protection in cosmetics, facilitates accurate detection of pollutants, promotes sustainable agriculture practices, and ensures food safety. Through its versatile applications, ZnO actively supports environmental conservation efforts, reduces dependence on fossil fuels, mitigates pollution, and contributes to a cleaner and more sustainable future.

The future scope of ZnO nanostructures in environmental applications lies in advancing research through novel synthesis methods, performance optimization, and a deeper understanding of underlying mechanisms. To further enhance efficiency and broaden its applicability across environmental sectors, the integration of ZnO with other materials and technologies is essential. It is crucial to prioritize sustainability by emphasizing environmentally friendly practices and adopting green synthesis approaches in future endeavors. Collaborative interdisciplinary efforts and advancements in characterizing the behavior of ZnO will drive innovation and sustainable solutions, paving the way toward a greener future.

REFERENCES

1. Nasrollahzadeh, M., Sajadi, S. M., Sajjadi, M., & Issaabadi, Z. (2019). An introduction to nanotechnology. In *Interface science and technology* (Vol. *28*, pp. 1–27). Elsevier.
2. Saravanakumar, B., & Kim, S. J. (2014). Growth of 2D ZnO nanowall for energy harvesting application. *The Journal of Physical Chemistry C*, *118*(17), 8831–8836.
3. Dimkpa, C. O., McLean, J. E., Britt, D. W., & Anderson, A. J. (2012). Bioactivity and biomodification of Ag, ZnO, and CuO nanoparticles with relevance to plant performance in agriculture. *Industrial Biotechnology*, *8*(6), 344–357.
4. Nasir, S., Hussein, M. Z., Zainal, Z., & Yusof, N. A. (2018). Carbon-based nanomaterials/allotropes: A glimpse of their synthesis, properties and some applications. *Materials*, *11*(2), 295.
5. Kumari, M. A., Devi, L. G., Maia, G., Chen, T. W., Al-Zaqri, N., & Ali, M. A. (2022). Mechanochemwpical synthesis of ternary heterojunctions TiO2 (A)/TiO_2 (R)/ZnO and TiO_2 (A)/TiO_2 (R)/SnO_2 for effective charge separation in semiconductor photocatalysis: A comparative study. *Environmental Research*, *203*, 111841.
6. Yousefi, R. (2011). Doped ZnO Nanostructures: Growth Properties and Applications, *Encyclopedia of Semiconductor Nanotechnology*, 1–43 ISBN: 1-58883-199-X.
7. Terna, A. D., Elemike, E. E., Mbonu, J. I., Osafile, O. E., & Ezeani, R. O. (2021). The future of semiconductors nanoparticles: Synthesis, properties and applications. *Materials Science and Engineering: B*, *272*, 115363.
8. Ayoub, I., Kumar, V., Abolhassani, R., Sehgal, R., Sharma, V., Sehgal, R.,...& Mishra, Y. K. (2022). Advances in ZnO: Manipulation of defects for enhancing their technological potentials. *Nanotechnology Reviews*, *11*(1), 575–619.
9. Hao, N., Xu, Z., Nie, Y., Jin, C., Closson, A. B., Zhang, M., & Zhang, J. X. (2019). Microfluidics-enabled rational design of ZnO micro-/nanoparticles with enhanced photocatalysis, cytotoxicity, and piezoelectric properties. *Chemical Engineering Journal*, *378*, 122222.

10. Mbrouk, O., Hafez, H., Mozia, S., Othman, A. M., & Abdel Mottaleb, M. S. A. (2022). Stimulated generation of photobiogas by morphologically tuned nanostructured ZnO and ZnO/TiO_2. *BMC Chemistry*, *16*(1), 74.
11. Fan, Z., & Lu, J. G. (2005). Zinc oxide nanostructures: synthesis and properties. *Journal of Nanoscience and Nanotechnology*, *5*(10), 1561–1573.
12. Öztürk, S., Kılınç, N., Taşaltın, N., & Öztürk, Z. Z. (2012). Fabrication of ZnO nanowires and nanorods. *Physica E: Low-Dimensional Systems and Nanostructures*, *44*(6), 1062–1065.
13. Theerthagiri, J., Salla, S., Senthil, R. A., Nithyadharseni, P., Madankumar, A., Arunachalam, P., & Kim, H. S. (2019). A review on ZnO nanostructured materials: energy, environmental and biological applications. *Nanotechnology*, *30*(39), 392001.
14. Mondal, O., & Pal, M. (2011). Strong and unusual violet-blue emission in ring shaped ZnO nanocrystals. *Journal of Materials Chemistry*, *21*(45), 18354–18358.
15. Kumar, S., Bhushan, P., & Bhattacharya, S. (2018). Fabrication of nanostructures with bottom-up approach and their utility in diagnostics, therapeutics, and others. *Environmental, Chemical and Medical Sensors*, 167–198. ISBN: 10. 9811077509.
16. Chan, Y. Y., Pang, Y. L., Lim, S., & Chong, W. C. (2021). Facile green synthesis of ZnO nanoparticles using natural-based materials: properties, mechanism, surface modification and application. *Journal of Environmental Chemical Engineering*, *9*(4), 105417.
17. Singh, S., Thiyagarajan, P., Kant, K. M., Anita, D., Thirupathiah, S., Rama, N., & Rao, M. R. (2007). Structure, microstructure and physical properties of ZnO based materials in various forms: bulk, thin film and nano. *Journal of Physics D: Applied Physics*, *40*(20), 6312.
18. Cruz, D. M., Mostafavi, E., Vernet-Crua, A., Barabadi, H., Shah, V., Cholula-Díaz, J. L., & Webster, T. J. (2020). Green nanotechnology-based zinc oxide (ZnO) nanomaterials for biomedical applications: a review. *Journal of Physics: Materials*, *3*(3), 034005.
19. Gebre, S. H., & Sendeku, M. G. (2019). New frontiers in the biosynthesis of metal oxide nanoparticles and their environmental applications: an overview. *SN Applied Sciences*, *1*, 1–28.
20. Ahmed, S., Chaudhry, S. A., & Ikram, S. (2017). A review on biogenic synthesis of ZnO nanoparticles using plant extracts and microbes: a prospect towards green chemistry. *Journal of Photochemistry and Photobiology B: Biology*, *166*, 272–284.
21. Najib, S., & Erdem, E. (2019). Current progress achieved in novel materials for supercapacitor electrodes: mini review. *Nanoscale Advances*, *1*(8), 2817–2827.
22. Lu, W., Hartman, R., Qu, L., & Dai, L. (2011). Nanocomposite electrodes for high-performance supercapacitors. *The Journal of Physical Chemistry Letters*, *2*(6), 655–660.
23. Selvakumar, M., Bhat, D. K., Aggarwal, A. M., Iyer, S. P., & Sravani, G. (2010). Nano ZnO-activated carbon composite electrodes for supercapacitors. *Physica B: Condensed Matter*, *405*(9), 2286–2289.
24. Kumar, A. (2020). Sol gel synthesis of zinc oxide nanoparticles and their application as nano-composite electrode material for supercapacitor. *Journal of Molecular Structure*, *1220*, 128654.
25. El-Shaarawy, M. G., Khairy, M., & Mousa, M. A. (2020). Structural, electrical and electrochemical properties of ZnO nanoparticles synthesized using dry and wet chemical methods. *Advanced Powder Technology*, *31*(3), 1333–1341.
26. Purushothaman, K. K., Priya, V. S., Nagamuthu, S., Vijayakumar, S., & Muralidharan, G. (2011). Synthesising of ZnO nanopetals for supercapacitor applications. *Micro & Nano Letters*, *6*(8), 668–670.
27. Bishwakarma, H., & Das, A. K. (2020). Synthesis of zinc oxide nanoparticles through hybrid machining process and their application in supercapacitors. *Journal of Electronic Materials*, *49*, 1541–1549.

28. Li, B., Zheng, M., Xue, H., & Pang, H. (2016). High performance electrochemical capacitor materials focusing on nickel based materials. *Inorganic Chemistry Frontiers*, *3*(2), 175–202.
29. Conway, B. E. (2013). *Electrochemical supercapacitors: scientific fundamentals and technological applications*. Springer Science & Business Media.
30. Peng, Z., Freunberger, S. A., Chen, Y., & Bruce, P. G. (2012). A reversible and higher-rate Li-O_2 battery. *Science*, *337*(6094), 563–566.
31. Pan, Q., Qin, L., Liu, J., & Wang, H. (2010). Flower-like ZnO–NiO–C films with high reversible capacity and rate capability for lithium-ion batteries. *Electrochimica Acta*, *55*(20), 5780–5785.
32. Kundu, S., Sain, S., Yoshio, M., Kar, T., Gunawardhana, N., & Pradhan, S. K. (2015). Structural interpretation of chemically synthesized ZnO nanorod and its application in lithium ion battery. *Applied Surface Science*, *329*, 206–211.
33. Xiao, L., Mei, D., Cao, M., Qu, D., & Deng, B. (2015). Effects of structural patterns and degree of crystallinity on the performance of nanostructured ZnO as anode material for lithium-ion batteries. *Journal of Alloys and Compounds*, *627*, 455–462.
34. Laurenti, M., Garino, N., Porro, S., Fontana, M., & Gerbaldi, C. (2015). Zinc oxide nanostructures by chemical vapour deposition as anodes for Li-ion batteries. *Journal of Alloys and Compounds*, *640*, 321–326.
35. Xiao, L., Li, E., Yi, J., Meng, W., Wang, S., Deng, B., & Liu, J. (2018). Enhancing the performance of nanostructured ZnO as an anode material for lithium-ion batteries by polydopamine-derived carbon coating and confined crystallization. *Journal of Alloys and Compounds*, *764*, 545–554.
36. Cauda, V., Pugliese, D., Garino, N., Sacco, A., Bianco, S., Bella, F., & Gerbaldi, C. (2014). Multi-functional energy conversion and storage electrodes using flower-like Zinc oxide nanostructures. *Energy*, *65*, 639–646.
37. Zhang, Q., Dandeneau, C. S., Zhou, X., & Cao, G. (2009). ZnO nanostructures for dye-sensitized solar cells. *Advanced Materials*, *21*(41), 4087–4108.
38. Rathnasamy, R., Thangasamy, P., Thangamuthu, R., Sampath, S., & Alagan, V. (2017). Green synthesis of ZnO nanoparticles using Carica papaya leaf extracts for photocatalytic and photovoltaic applications. *Journal of Materials Science: Materials in Electronics*, *28*, 10374–10381.
39. Sharma, J. K., Srivastava, P., Singh, G., Akhtar, M. S., & Ameen, S. J. M. S. (2015). Green synthesis of Co_3O_4 nanoparticles and their applications in thermal decomposition of ammonium perchlorate and dye-sensitized solar cells. *Materials Science and Engineering: B*, *193*, 181–188.
40. Shashanka, R., Esgin, H., Yilmaz, V. M., & Caglar, Y. (2020). Fabrication and characterization of green synthesized ZnO nanoparticle based dye-sensitized solar cells. *Journal of Science: Advanced Materials and Devices*, *5*(2), 185–191.
41. Alanazi, A. K. (2021). Effect of ZnO nanomaterial and red and green cabbage dyes on the performance of dye-sensitised solar cells. *Coatings*, *11*(9), 1057.
42. Kumar, P. N., Sakthivel, K., Balasubramanian, V., Sengottaiyan, D., & Suresh, J. (2018). Microwave assisted green synthesis of ZnO nanorods for dye sensitized solar cell application. *25*(4), 383–389.
43. Wong, K. K., Ng, A., Chen, X. Y., Ng, Y. H., Leung, Y. H., Ho, K. H., & Phillips, D. L. (2012). Effect of ZnO nanoparticle properties on dye-sensitized solar cell performance. *ACS Applied Materials & Interfaces*, *4*(3), 1254–1261.
44. Wibowo, A., Marsudi, M. A., Amal, M. I., Ananda, M. B., Stephanie, R., Ardy, H., & Diguna, L. J. (2020). ZnO nanostructured materials for emerging solar cell applications. *RSC Advances*, *10*(70), 42838–42859.

45. Gonzalez, V., Lopez, I., Palma, R. M., Pena, Y., & Gómez, I. (2020). Organic-inorganic hybrid solar cells based on 1D ZnO/P3HT active layers and 0D Au as cathode. *Materials Research Express*, *7*(7), 075005.
46. Saboor, A., Shah, S. M., & Hussain, H. (2019). Band gap tuning and applications of ZnO nanorods in hybrid solar cell: Ag-doped verses Nd-doped ZnO nanorods. *Materials Science in Semiconductor Processing*, *93*, 215–225.
47. Beek, W. J., Wienk, M. M., & Janssen, R. A. (2004). Efficient hybrid solar cells from zinc oxide nanoparticles and a conjugated polymer. *Advanced Materials*, *16*(12), 1009–1013.
48. Oosterhout, S. D., Wienk, M. M., Van Bavel, S. S., Thiedmann, R., Jan Anton Koster, L., Gilot, J., & Janssen, R. A. (2009). The effect of three-dimensional morphology on the efficiency of hybrid polymer solar cells. *Nature Materials*, *8*(10), 818–824.
49. Li, S. S., & Chen, C. W. (2013). Polymer–metal-oxide hybrid solar cells. *Journal of Materials Chemistry A*, *1*(36), 10574–10591.
50. W. J. Beek, L. H. Slooff, M. M. Wienk, J. M. Kroon and R. A. Janssen 2005, *Advanced Functional Materials.*, *15*, 1703–1707.
51. W. J. Beek, M. M. Wienk, M. Kemerink, X. Yang and R. A. Janssen. 2005. Hybrid zinc oxide conjugated polymer bulk heterojunction solar cells *The Journal of Physical Chemistry B*, *109*, 9505–9516.
52. W. J. Beek, M. M. Wienk and R. A. Janssen (2006). Hybrid solar cells from regioregular polythiophene and ZnO nanoparticles. *Advanced Functional Materials*, *16*, 1112–1116.
53. D. J. Moet, L. J. A. Koster, B. de Boer and P. W. Blom (2007). Hybrid polymer solar cells from highly reactive diethylzinc: MDMO–PPV versus P3HT. *Chemistry of Materials*, *19*, 5856–5861.
54. Oosterhout, S. D., Wienk, M. M., Van Bavel, S. S., Thiedmann, R., Jan Anton Koster L., Gilot, J., & Janssen, R. A. (2009). The effect of three-dimensional morphology on the efficiency of hybrid polymer solar cells. *Nature Materials*, *8*(10), 818–824.
55. Batmunkh, M., Shearer, C. J., Biggs, M. J., & Shapter, J. G. (2015). Nanocarbons for mesoscopic perovskite solar cells. *Journal of Materials Chemistry A*, *3*(17), 9020–9031.
56. Fakharuddin, A., Vasilopoulou, M., Soultati, A., Haider, M. I., Briscoe, J., Fotopoulos, V., & Nazeeruddin, M. K. (2021). Robust inorganic hole transport materials for organic and perovskite solar cells: insights into materials electronic properties and device performance. *Solar Rrl*, *5*(1), 2000555.
57. Kumar, M. H., Yantara, N., Dharani, S., Graetzel, M., Mhaisalkar, S., Boix, P. P., & Mathews, N. (2013). Flexible, low-temperature, solution processed ZnO-based perovskite solid state solar cells. *Chemical Communications*, *49*(94), 11089–11091.
58. Saparov, B., & Mitzi, D. B. (2016). Organic–inorganic perovskites: structural versatility for functional materials design. *Chemical Reviews*, *116*(7), 4558–4596.
59. Bi, D., Boschloo, G., Schwarzmuller, S., Yang, L., Johansson, E. M., & Hagfeldt, A. (2013). Efficient and stable $CH_3NH_3PbI_3$–sensitized ZnO nanorod array solid-state solar cells. *Nanoscale*, *5*(23), 11686–11691.
60. Gadisa, A., Liu, Y., Samulski, E. T., & Lopez, R. (2012). Minimizing interfacial losses in inverted organic solar cells comprising Al-doped ZnO. *Applied Physics Letters*, *100*(25), 253903.
61. Tian, H., Hu, L., Zhang, C., Chen, S., Sheng, J., Mo, L., & Dai, S. (2011). Enhanced photovoltaic performance of dye-sensitized solar cells using a highly crystallized mesoporous TiO_2 electrode modified by boron doping. *Journal of Materials Chemistry*, *21*(3), 863–868.
62. Mahmood, K., & Park, S. B. (2013). Atmospheric pressure based electrostatic spray deposition of transparent conductive ZnO and Al-doped ZnO (AZO) thin films: effects of Al doping and annealing treatment. *Electronic Materials Letters*, *9*, 161–170.

63. Dong, J., Zhao, Y., Shi, J., Wei, H., Xiao, J., Xu, X., & Meng, Q. (2014). Impressive enhancement in the cell performance of ZnO nanorod-based perovskite solar cells with Al-doped ZnO interfacial modification. *Chemical Communications*, *50*(87), 13381–13384.
64. Zuo, L., Gu, Z., Ye, T., Fu, W., Wu, G., Li, H., & Chen, H. (2015). Enhanced photovoltaic performance of CH3NH3PbI3 perovskite solar cells through interfacial engineering using self-assembling monolayer. *Journal of the American Chemical Society*, *137*(7), 2674–2679.
65. Liu, F., Wong, M. K., Tam, H. W., Djurisic, A. B., Ng, A. M., & Chan, W. K. (2017, February). Effect of ZnO surface defects on efficiency and stability of ZnO-based perovskite solar cells. In *Oxide-based Materials and Devices VIII* (Vol. 10105, pp. 216–224). SPIE.
66. Chen, H., & Yada, R. (2011). Nanotechnologies in agriculture: new tools for sustainable development. *Trends in Food Science & Technology*, *22*(11), 585–594.
67. Lahiani, M. H., Dervishi, E., Chen, J., Nima, Z., Gaume, A., Biris, A. S., & Khodakovskaya, M. V. (2013). Impact of carbon nanotube exposure to seeds of valuable crops. *ACS Applied Materials & Interfaces*, *5*(16), 7965–7973.
68. Singh, N., Khandual, A., Gupta, P. K., & Vaish, S. S. (2016). Preliminary test of functionalized ZnO2 against Bipolaris sorokiniana and other seed associated mycoflora for better wheat germination. *Research Journal of Biotechnology*, *11*(6), 60–73.
69. Azimi, R., Feizi, H., & Hosseini, M. K. (2013). Can bulk and nanosized titanium dioxide particles improve seed germination features of wheatgrass (Agropyron desertorum). *Notulae Scientia Biologicae*, *5*(3), 325–331.
70. Khan, M. A., Khan, T., Riaz, M. S., Ullah, N., Ali, H., & Nadhman, A. (2019). Plant cell nanomaterials interaction: Growth, physiology and secondary metabolism. In *Comprehensive Analytical Chemistry* (Vol. 84, pp. 23–54). Elsevier.
71. Prasad, T. N. V. K. V., Sudhakar, P., Sreenivasulu, Y., Latha, P., Munaswamy, V., Reddy, K. R., & Pradeep, T. (2012). Effect of nanoscale zinc oxide particles on the germination, growth and yield of peanut. *Journal of Plant Nutrition*, *35*(6), 905–927.
72. Selivanov, V. N., & Zorin, E. V. (2001). Sustained Action of ultrafine metal powders on seeds of grain crops. *Perspekt. Materialy*, *4*, 66–69.
73. Kumari, S., & Chauhan, S. (2019). A review on applications of metal oxide nanoparticles in agriculture. *International Journal of Conservation Science*, *7*, 2143–2146.
74. Wan, J., Wang, R., Wang, R., Ju, Q., Wang, Y., & Xu, J. (2019). Comparative physiological and transcriptomic analyses reveal the toxic effects of ZnO nanoparticles on plant growth. *Environmental Science & Technology*, *53*(8), 4235–4244.
75. Awasthi, A., Bansal, S., Jangir, L. K., Awasthi, G., Awasthi, K. K., & Awasthi, K. (2017, December). Effect of ZnO nanoparticles on germination of Triticum aestivum seeds. *In Macromolecular Symposia* (Vol. *376*, No. 1, p. 1700043).
76. Batsmanova, L. M., Gonchar, L. M., Taran, N. Y., & Okanenko, A. A. (2013). *Using a colloidal solution of metal nanoparticles as micronutrient fertiliser for cereals* (Doctoral dissertation, Sumy State University).
77. Mahajan, P., Dhoke, S. K., & Khanna, A. S. (2011). Effect of nano-ZnO particle suspension on growth of mung (Vigna radiata) and gram (Cicer arietinum) seedlings using plant agar method. *Journal of Nanotechnology*, *2011*, 696535.
78. Prasad, T. N. V. K. V., Sudhakar, P., Sreenivasulu, Y., Latha, P., Munaswamy, V., Reddy, K. R., & Pradeep, T. (2012). Effect of nanoscale zinc oxide particles on the germination, growth and yield of peanut. *Journal of Plant Nutrition*, *35*(6), 905–927.
79. Burman, U., Saini, M., & Kumar, P. (2013). Effect of zinc oxide nanoparticles on growth and antioxidant system of chickpea seedlings. *Toxicological & Environmental Chemistry*, *95*(4), 605–612.

80. Raliya, R., & Tarafdar, J. C. (2013). ZnO nanoparticle biosynthesis and its effect on phosphorous-mobilizing enzyme secretion and gum contents in Cluster bean (Cyamopsis tetragonoloba L.). *Agricultural Research*, *2*, 48–57.
81. Zhang, H., Wang, R., Chen, Z., Cui, P., Lu, H., Yang, Y., & Zhang, H. (2021). The effect of zinc oxide nanoparticles for enhancing rice (Oryza sativa L.) yield and quality. *Agriculture*, *11*(12), 1247.
82. Pirzada, T., Chandio, W. A., Talpur, M. M., Ansari, A. M., & Kanhar, F. H. (2022). Synthesis, characterization and role of zinc oxide nanoparticles in wheat (Triticum indicum) seeds germination: role of zinc oxide nanoparticles in wheat, *Biological Sciences-PJSIR*, *65*(2), 167–172.
83. Yan, S., Wu, F., Zhou, S., Yang, J., Tang, X., & Ye, W. (2021). Zinc oxide nanoparticles alleviate the arsenic toxicity and decrease the accumulation of arsenic in rice (Oryza sativa L.). *BMC Plant Biology*, *21*, 1–11.
84. Taheri, M., Qarache, H. A., Qarache, A. A., & Yoosefi, M. (2016). The effects of zinc-oxide nanoparticles on growth parameters of corn (SC704). *STEM Fellowship Journal*, *1*(2), 17–20.
85. Das, A., Wang, D. Y., Leuteritz, A., Subramaniam, K., Greenwell, H. C., Wagenknecht, U., & Heinrich, G. (2011). Preparation of zinc oxide free, transparent rubber nanocomposites using a layered double hydroxide filler. *Journal of Materials Chemistry*, *21*(20), 7194–7200.
86. Sim, L., Ramanan, S. R., Ismail, H., Seetharamu, K. N., & Goh, T. J. (2005). Thermal characterization of Al_2O_3 and ZnO reinforced silicone rubber as thermal pads for heat dissipation purposes. *Thermochimica Acta*, *430*(1–2), 155–165.
87. Yuan, Z., Zhou, W., Hu, T., Chen, Y., Li, F. A. N., Xu, Z., & Wang, X. (2011). Fabrication and properties of silicone rubber/ZnO nanocomposites via in situ surface hydrosilylation. *Surface Review and Letters*, *18*(01n02), 33–38.
88. Heideman, G., Noordermeer, J. W., Datta, R. N., & van Baarle, B. (2004). Zinc loaded clay as activator in sulfur vulcanization: a new route for zinc oxide reduction in rubber compounds. *Rubber Chemistry and Technology*, *77*(2), 336–355.
89. Chatterjee, K., & Naskar, K. (2007). Development of thermoplastic elastomers based on maleated ethylene propylene rubber (m-EPM) and polypropylene (PP) by dynamic vulcanization. *Express Polymer Letters*, *1*, 527–534.
90. Hamed, G. R., & Hua, K. C. (2004). Effect of ZnO particle size on the curing of carboxylated NBR and carboxylated SBR. *Rubber Chemistry and Technology*, *77*(2), 214–226.
91. Przybyszewska, M., & Zaborski, M. (2009). The effect of zinc oxide nanoparticle morphology on activity in crosslinking of carboxylated nitrile elastomer. *Express Polymer Letters*, *3*(9), 542–552.
92. Sabura Begum, P. M., Mohammed Yusuff, K. K., & Joseph, R. (2008). Preparation and use of nano zinc oxide in neoprene rubber. *International Journal of Polymeric Materials*, *57*(12), 1083–1094.
93. Panampilly, B., & Thomas, S. (2013). Nano ZnO as cure activator and reinforcing filler in natural rubber. *Polymer Engineering & Science*, *53*(6), 1337–1346.
94. Lin, Y., Chen, Y., Zeng, Z., Zhu, J., Wei, Y., Li, F., & Liu, L. (2015). Effect of ZnO nanoparticles doped graphene on static and dynamic mechanical properties of natural rubber composites. *Composites Part A: Applied Science and Manufacturing*, *70*, 35–44.
95. Ducháček, V., Kuta, A., & Příbyl, P. (1993). Efficiency of metal activators of accelerated sulfur vulcanization. *Journal of Applied Polymer Science*, *47*(4), 743–746.

96. Qin, X., Xu, H., Zhang, G., Wang, J., Wang, Z., Zhao, Y.,…& Matyjaszewski, K. (2020). Enhancing the performance of rubber with nano ZnO as activators. *ACS Applied Materials & Interfaces*, *12*(42), 48007–48015.
97. Sahoo, S., & Bhowmick, A. K. (2007). Influence of ZnO nanoparticles on the cure characteristics and mechanical properties of carboxylated nitrile rubber. *Journal of Applied Polymer Science*, *106*(5), 3077–3083.
98. Kumar, R., Umar, A., Kumar, G., & Nalwa, H. S. (2017). Antimicrobial properties of ZnO nanomaterials: A review. *Ceramics International*, *43*(5), 3940–3961.
99. Lee, M., Wu, Z., & Li, K. (2015). Advances in ceramic membranes for water treatment. In *Advances in membrane technologies for water treatment* (pp. 43–82). Woodhead Publishing.
100. Dimapilis, E. A. S., Hsu, C. S., Mendoza, R. M. O., & Lu, M. C. (2018). Zinc oxide nanoparticles for water disinfection. *Sustainable Environment Research*, *28*(2), 47–56.
101. Yang, Y., Li, X., Chen, J., Chen, H., & Bao, X. (2003). ZnO nanoparticles prepared by thermal decomposition of β-cyclodextrin coated zinc acetate. *Chemical Physics Letters*, *373*(1–2), 22–27.
102. Spoială, A., Ilie, C. I., Trușcă, R. D., Oprea, O. C., Surdu, V. A., Vasile, B. Ş., & Diţu, L. M. (2021). Zinc oxide nanoparticles for water purification. *Materials*, *14*(16), 4747.
103. Lu, H., Wang, J., Stoller, M., Wang, T., Bao, Y., & Hao, H. (2016). An overview of nanomaterials for water and wastewater treatment. *Advances in Materials Science and Engineering*, *2016*.
104. Ong, C. B., Ng, L. Y., & Mohammad, A. W. (2018). A review of ZnO nanoparticles as solar photocatalysts: Synthesis, mechanisms and applications. *Renewable and Sustainable Energy Reviews*, *81*, 536–551.
105. Ursino, C., Castro-Muñoz, R., Drioli, E., Gzara, L., Albeirutty, M. H., & Figoli, A. (2018). Progress of nanocomposite membranes for water treatment. *Membranes*, *8*(2), 18.
106. Zito, P., & Shipley, H. J. (2015). Inorganic nano-adsorbents for the removal of heavy metals and arsenic: a review. *RSC Advances*, *5*(38), 29885–29907.
107. Angelin, K. B., Siva, S., & Kannan, R. S. (2015). Zinc oxide nanoparticles impregnated polymer hybrids for efficient extraction of heavy metals from polluted aqueous solution. *Asian Journal of Applied Science and Technology*, *6*(12), 2139–2150.
108. Yuvaraja, G., Prasad, C., Vijaya, Y., & Subbaiah, M. V. (2018). Application of ZnO nanorods as an adsorbent material for the removal of As (III) from aqueous solution: kinetics, isotherms and thermodynamic studies. *International Journal of Industrial Chemistry*, *9*, 17–25.
109. Mahdavi, S., Jalali, M., & Afkhami, A. (2014). Removal of heavy metals from aqueous solutions using Fe 3 O 4, ZnO, and CuO nanoparticles. In *Nanotechnology for sustainable development* (pp. 171–188). Springer International Publishing.
110. Ouyang, D., Zhuo, Y., Hu, L., Zeng, Q., Hu, Y., & He, Z. (2019). Research on the adsorption behavior of heavy metal ions by porous material prepared with silicate tailings. *Minerals*, *9*(5), 291.
111. Xie, J., Lin, Y., Li, C., Wu, D., & Kong, H. (2015). Removal and recovery of phosphate from water by activated aluminum oxide and lanthanum oxide. *Powder Technology*, *269*, 351–357.
112. Salmani, M. H., Zarei, S., Ehrampoush, M. H., & Danaie, S. (2013). Evaluations of pH and high ionic strength solution effect in cadmium removal by zinc oxide nanoparticles. *Journal of Applied Sciences and Environmental Management*, *17*(4), 583–593.
113. Zaidi, Z., Siddiqui, S. I., Fatima, B., & Chaudhry, S. A. (2019). Synthesis of ZnO nanospheres for water treatment through adsorption and photocatalytic degradation: modelling and process optimization. *Materials Research Bulletin*, *120*, 110584.

114. Wenjie, O., Ahmed, W., Xiuxian, F., Lu, W., Jiannan, L., Jie, Y., & Mehmood, S. (2022). The Adsorption Potential of Cr from Water by ZnO Nanoparticles Synthesized by Azolla pinnata. *Bioinorganic Chemistry and Applications*, *2022*, 6209013.
115. Alhan, S., Nehra, M., Dilbaghi, N., Singhal, N. K., Kim, K. H., & Kumar, S. (2019). Potential use of ZnO@ activated carbon nanocomposites for the adsorptive removal of Cd2+ ions in aqueous solutions. *Environmental Research*, *173*, 411–418.
116. Dimapilis, E. A. S., Hsu, C. S., Mendoza, R. M. O., & Lu, M. C. (2018). Zinc oxide nanoparticles for water disinfection. *Sustainable Environment Research*, *28*(2), 47–56.
117. Heinlaan, M., Ivask, A., Blinova, I., Dubourguier, H. C., & Kahru, A. (2008). Toxicity of nanosized and bulk ZnO, CuO and TiO2 to bacteria Vibrio fischeri and crustaceans Daphnia magna and Thamnocephalus platyurus. *Chemosphere*, *71*(7), 1308–1316.
118. Brayner, R., Ferrari-Iliou, R., Brivois, N., Djediat, S., Benedetti, M. F., & Fiévet, F. (2006). Toxicological impact studies based on *Escherichia coli* bacteria in ultrafine ZnO nanoparticles colloidal medium. *Nano Letters*, *6*(4), 866–870.
119. Li, Q., Mahendra, S., Lyon, D. Y., Brunet, L., Liga, M. V., Li, D., & Alvarez, P. J. (2008). Antimicrobial nanomaterials for water disinfection and microbial control: potential applications and implications. *Water Research*, *42*(18), 4591–4602.
120. Shinde, S. S. (2015). Antimicrobial activity of ZnO nanoparticles against pathogenic bacteria and fungi. *Sci Med Central*, *3*, 1033.
121. Zhang, L., Li, Y., Liu, X., Zhao, L., Ding, Y., Povey, M., & Cang, D. (2013). The properties of ZnO nanofluids and the role of H_2O_2 in the disinfection activity against Escherichia coli. *Water research*, *47*(12), 4013–4021.
122. Mehrabi, Z., Taheri-Kafrani, A., Asadnia, M., & Razmjou, A. (2020). Bienzymatic modification of polymeric membranes to mitigate biofouling. *Separation and Purification Technology*, *237*, 116464.
123. Li, D., Yan, Y., & Wang, H. (2016). Recent advances in polymer and polymer composite membranes for reverse and forward osmosis processes. *Progress in Polymer Science*, *61*, 104–155.
124. Shi, W., He, B., Ding, J., Li, J., Yan, F., & Liang, X. (2010) .Preparation and characterization of the organic–inorganic hybrid membrane for biodiesel production. *Bioresource Technology*, *101*(5), 1501–1505.
125. Wen, Y., Yuan, J., Ma, X., Wang, S., & Liu, Y. (2019). Polymeric nanocomposite membranes for water treatment: A review. *Environmental Chemistry Letters*, *17*, 1539–1551.
126. Zahid, M., Rashid, A., Akram, S., Rehan, Z. A., & Razzaq, W. J. J. M. S. T. (2018). A comprehensive review on polymeric nano-composite membranes for water treatment. *Journal of Membrane Science and Technology*, *8*(1), 1–20.
127. Awasthi, A., Sharma, P., Jangir, L., Awasthi, G., Awasthi, K. K., & Awasthi, K. (2020). Dose dependent enhanced antibacterial effects and reduced biofilm activity against Bacillus subtilis in presence of ZnO nanoparticles. *Materials Science and Engineering: C*, *113*, 111021.
128. Alsalhy, Q. F., Al-Ani, F. H., Al-Najar, A. E., & Jabuk, S. I. (2018). A study of the effect of embedding ZnO-NPs on PVC membrane performance use in actual hospital wastewater treatment by membrane bioreactor. *Chemical Engineering and Processing-Process Intensification*, *130*, 262–274.
129. Rajabi, H., Ghaemi, N., Madaeni, S. S., Daraei, P., Astinchap, B., Zinadini, S., & Razavizadeh, S. H. (2015). Nano-ZnO embedded mixed matrix polyethersulfone (PES) membrane: Influence of nanofiller shape on characterization and fouling resistance. *Applied Surface Science*, *349*, 66–77.
130. Malini, M., Thirumavalavan, M., Yang, W. Y., Lee, J. F., & Annadurai, G. (2015). A versatile chitosan/ZnO nanocomposite with enhanced antimicrobial properties. *International Journal of Biological Macromolecules*, *80*, 121–129.

131. Lam, S. M., Sin, J. C., Abdullah, A. Z., & Mohamed, A. R. (2012). Degradation of wastewaters containing organic dyes photocatalysed by zinc oxide: a review. *Desalination and Water Treatment*, *41*(1–3), 131–169.
132. Abdullah, F. H., Bakar, N. A., & Bakar, M. A. (2022). Current advancements on the fabrication, modification, and industrial application of zinc oxide as photocatalyst in the removal of organic and inorganic contaminants in aquatic systems. *Journal of Hazardous Materials*, *424*, 127416.
133. Hernández, S., Hidalgo, D., Sacco, A., Chiodoni, A., Lamberti, A., Cauda, V., & Saracco, G. (2015). Comparison of photocatalytic and transport properties of TiO 2 and ZnO nanostructures for solar-driven water splitting. *Physical Chemistry Chemical Physics*, *17*(12), 7775–7786.
134. Dindar, B., & Içli, S. (2001). Unusual photoreactivity of zinc oxide irradiated by concentrated sunlight. *Journal of Photochemistry and Photobiology A: Chemistry*, *140*(3), 263–268.
135. Yan, X., Zou, C., Gao, X., & Gao, W. (2012). ZnO/TiO_2 core–brush nanostructure: processing, microstructure and enhanced photocatalytic activity. *Journal of Materials Chemistry*, *22*(12), 5629–5640.
136. Mousa, H. M., Alenezi, J. F., Mohamed, I. M., Yasin, A. S., Hashem, A. F. M., & Abdal-Hay, A. (2021). Synthesis of TiO_2@ ZnO heterojunction for dye photodegradation and wastewater treatment. *Journal of Alloys and Compounds*, *886*, 161169.
137. Chimupala, Y., Phromma, C., Yimklan, S., Semakul, N., & Ruankham, P. (2020). Dye wastewater treatment enabled by piezo-enhanced photocatalysis of single-component ZnO nanoparticles. *RSC Advances*, *10*(48), 28567–28575.
138. Hariharalakshmanan, R. K., Martinez, J., Ergul-Yilmaz, B., & Karabacak, T. (2023). Suspension of ZnO Nanostructures Synthesized by Hot Water Treatment for Photocatalytic Wastewater Treatment. *Water, Air, & Soil Pollution*, *234*(4), 209.
139. Duncan, T. V. (2011). Applications of nanotechnology in food packaging and food safety: barrier materials, antimicrobials and sensors. *Journal of Colloid and Interface Science*, *363*(1), 1–24.
140. Sahani, S., & Sharma, Y. C. (2021). Advancements in applications of nanotechnology in global food industry. *Food Chemistry*, *342*, 128318.
141. Nile, S. H., Baskar, V., Selvaraj, D., Nile, A., Xiao, J., & Kai, G. (2020). Nanotechnologies in food science: applications, recent trends, and future perspectives. *Nano-micro Letters*, *12*, 1–34.
142. Martirosyan, A., & Schneider, Y. J. (2014). Engineered nanomaterials in food: implications for food safety and consumer health. *International Journal of Environmental Research and Public Health*, *11*(6), 5720–5750.
143. McClements, D. J., & Xiao, H. (2017). Is nano safe in foods? Establishing the factors impacting the gastrointestinal fate and toxicity of organic and inorganic food-grade nanoparticles. *Npj Science of Food*, *1*(1), 6.
144. Bari, M. L., Grumezescu, A., Ukuku, D. O., Dey, G., & Miyaji, T. (2017). New food processing technologies and food safety. *Journal of Food Quality*, *2017*, 1–3.
145. Galstyan, V., Bhandari, M. P., Sberveglieri, V., Sberveglieri, G., & Comini, E. (2018). Metal oxide nanostructures in food applications: Quality control and packaging. *Chemosensors*, *6*(2), 16.
146. Chaudhary, P., Fatima, F., & Kumar, A. (2020). Relevance of nanomaterials in food packaging and its advanced future prospects. *Journal of Inorganic and Organometallic Polymers and Materials*, *30*, 5180–5192.
147. Meng, F., Zheng, H., Sun, Y., Li, M., & Liu, J. (2017). Trimethylamine sensors based on Au-modified hierarchical porous single-crystalline ZnO nanosheets. *Sensors*, *17*(7), 1478.

148. Kim, I., Viswanathan, K., Kasi, G., Thanakkasaranee, S., Sadeghi, K., & Seo, J. (2022). ZnO nanostructures in active antibacterial food packaging: preparation methods, antimicrobial mechanisms, safety issues, future prospects, and challenges. *Food Reviews International*, *38*(4), 537–565.
149. Kim, I., Viswanathan, K., Kasi, G., Thanakkasaranee, S., Sadeghi, K., & Seo, J. (2022). ZnO nanostructures in active antibacterial food packaging: preparation methods, antimicrobial mechanisms, safety issues, future prospects, and challenges. *Food Reviews International*, *38*(4), 537–565.
150. Zhang, Y. R., Nayak T., Hong, H., & Cai, W. (2013). Biomedical applications of zinc oxide nanomaterials. *Current Molecular Medicine*, *13*(10), 1633–1645.
151. Wang, J. X., Sun, X. W., Wei, A., Lei, Y., Cai, X. P., Li, C. M., & Dong, Z. L. (2006). Zinc oxide nanocomb biosensor for glucose detection. *Applied Physics Letters*, *88*(23), 233106.
152. Aini, B. N., Siddiquee, S., Ampon, K., Rodrigues, K. F., & Suryani, S. (2015). Development of glucose biosensor based on ZnO nanoparticles film and glucose oxidase-immobilized eggshell membrane. *Sensing and Bio-Sensing Research*, *4*, 46–56.
153. Frasconi, M., Favero, G., Di Fusco, M., & Mazzei, F. (2009). Polyazetidine-based immobilization of redox proteins for electron-transfer-based biosensors. *Biosensors and Bioelectronics*, *24*(5), 1424–1430.
154. Welch, E. C., Powell, J. M., Clevinger, T. B., Fairman, A. E., & Shukla, A. (2021). Advances in biosensors and diagnostic technologies using nanostructures and nanomaterials. *Advanced Functional Materials*, *31*(44), 2104126.
155. Fu, Y. Q., Luo, J. K., Du, X. Y., Flewitt, A. J., Li, Y., Markx, G. H., & Milne, W. I. (2010). Recent developments on ZnO films for acoustic wave based bio-sensing and microfluidic applications: a review. *Sensors and Actuators B: Chemical*, *143*(2), 606–619.
156. Ali, S. M. U., Aijazi, T., Axelsson, K., Nur, O., & Willander, M. (2011). Wireless remote monitoring of glucose using a functionalized ZnO nanowire arrays based sensor. *Sensors*, *11*(9), 8485–8496.
157. Bashari, A., Koohestani, A. H. S., & Salamatipour, N. (2020). Eco-friendly dual-functional textiles: green water-repellent & anti-bacterial. *Fibers and Polymers*, *21*, 317–323.
158. Vigneshwaran, N., Kumar, S., Kathe, A. A., Varadarajan, P. V., & Prasad, V. (2006). Functional finishing of cotton fabrics using zinc oxide–soluble starch nanocomposites. *Nanotechnology*, *17*(20), 5087.
159. Lim, Z. H., Chia, Z. X., Kevin, M., Wong, A. S. W., & Ho, G. W. (2010). A facile approach towards ZnO nanorods conductive textile for room temperature multifunctional sensors. *Sensors and Actuators B: Chemical*, *151*(1), 121–126.
160. Zhang, F., & Yang, J. (2009). Preparation of nano-ZnO and its application to the textile on antistatic finishing. *International Journal of Chemistry*, *1*(1), 18.
161. Yadav, A., Prasad, V., Kathe, A. A., Raj, S., Yadav, D., Sundaramoorthy, C., & Vigneshwaran, N. (2006). Functional finishing in cotton fabrics using zinc oxide nanoparticles. *Bulletin of Materials Science*, *29*(6), 641–645.
162. Kathirvelu, S., D'Souza, L., & Dhurai, B. (2009). UV protection finishing of textiles using ZnO nanoparticles. *Indian Journal of Fibre & Textile Research*, *34*, 267–273.
163. Rajendra, R., Balakumar, C., Ahammed, H. A. M., Jayakumar, S., Vaideki, K., & Rajesh, E. (2010). Use of zinc oxide nano particles for production of antimicrobial textiles. *International Journal of Engineering, Science and Technology*, *2*(1), 202–208.
164. Mao, Z., Shi, Q., Zhang, L., & Cao, H. (2009). The formation and UV-blocking property of needle-shaped ZnO nanorod on cotton fabric. *Thin Solid Films*, *517*(8), 2681–2686.
165. Ates, E. S., & Unalan, H. E. (2012). Zinc oxide nanowire enhanced multifunctional coatings for cotton fabrics. *Thin Solid Films*, *520*(14), 4658–4661.

166. Nisansala, H. M. D., Rajapaksha, G. K. M., Dikella, D. G. N. V., Dheerasinghe, M. J., Sirimuthu, N. M. S., & Patabendige, C. N. K. (2021). Zinc Oxide Nanostructures in the Textile Industry. *Indian Journal of Science and Technology*, *14*, 3370–3395.
167. Smijs, T. G., & Pavel, S. (2011). Titanium dioxide and zinc oxide nanoparticles in sunscreens: focus on their safety and effectiveness. *Nanotechnology, Science and Applications*, *2011*(4), 95–112.
168. Hasanin, M., Swielam, E. M., Atwa, N. A., & Agwa, M. M. (2022). Novel design of bandages using cotton pads, doped with chitosan, glycogen and ZnO nanoparticles, having enhanced antimicrobial and wounds healing effects. *International Journal of Biological Macromolecules*, *197*, 121–130.
169. Faunce, T. (2010). Exploring the safety of nanoparticles in Australian sunscreens. *International Journal of Biomedical Nanoscience and Nanotechnology*, *1*(1), 87–94.
170. Reinosa, J. J., Leret, P., Álvarez-Docio, C. M., del Campo, A., & Fernández, J. F. (2016). Enhancement of UV absorption behavior in ZnO–TiO_2 composites. *Boletín de la sociedad española de cerámica y vidrio*, *55*(2), 55–62.

4 Piezoelectric Characteristics of ZnO Nanostructures

Anirudh Vashisht, Indu Gupta, and Shivani Singla

4.1 INTRODUCTION

Piezoelectricity, a phenomenon whereby certain materials respond to applied mechanical stress through the generation of an electric charge, has opened up many technological avenues. In the late 19th century, Pierre and Jacques Curie discovered piezoelectricity, and since then, its exploitation has been seen in diverse applications from the simplest lighter ignitions to state-of-the-art medical ultrasound transducers [1–3]. The materials displaying this wonderful feature possess a non-centrosymmetric crystal structure that allows the piezoelectric effects.

Among the piezoelectric materials, there is significant interest in zinc oxide (ZnO), particularly due to its unique combination of semiconductor and piezoelectric properties [4–6]. ZnO possesses a wurtzite structure characterized by a non-centrosymmetric hexagonal lattice that is inherently polar along the c-axis. The significance of this polarity lies in its contribution to the piezoelectric behavior of ZnO, making it a strong applicant for various applications [7].

The success of nanotechnology has enhanced the prospects of ZnO by making it possible to produce ZnO nanostructures (NS) with superior piezoelectric properties [8, 9]. The inherent properties of ZnO at the nanoscale are mechanical flexibility, a higher surface area, and better piezoelectric coefficients than those of its bulk counterpart. For that reason, ZnO NS render themselves favorable for advanced piezoelectric devices such as nanogenerators, sensors, actuators, and energy harvesters [10–13].

ZnO NS, including nanowires, nanorods, and nanotubes, have been found to convert mechanical energy into electrical energy quite widely. This aspect of energy conversion is the backbone of PENGs, have been found to be very useful in powering small electronic devices and other devices that can utilize ambient mechanical energy. Additionally, the high sensitivity of ZnO NS to mechanical and environmental stimuli has opened doors for their application in sophisticated sensor technologies capable of identifying pressure, gases, and biological entities with high precision [14–16].

In this chapter, the fascinating world of piezoelectric ZnO NS has been discussed, delivering a comprehensive overview of their theoretical underpinnings and diverse

DOI: 10.1201/9781003464907-4

applications. Important principles of piezoelectricity are considered in turn, providing sufficient background to allow an appreciation of the behavior of piezoelectric materials. The source of piezoelectricity in ZnO is then discussed, with a focus on its wurtzite crystal structure and intrinsic polarity. Moreover, the applications of ZnO NS in piezoelectric devices, particularly nanogenerators, have recently attracted much attention. A number of design and optimization strategies for ZnO-based nanogenerators are discussed, focusing on hybrid materials, multilayer structures, and novel electrode configurations to optimize device performance. Lastly, challenges and future directions in the field of piezoelectric ZnO NS has been discussed with the aim of finding the need for further research to optimize fabrication techniques, combine ZnO with other materials, and explore novel applications to fully harness its potential.

4.2 FUNDAMENTALS OF PIEZOELECTRICITY

The term "piezoelectricity" is coined from the Greek word "piezein," which means to press or squeeze, suitably describing the process of conversion of mechanical energy into electrical energy [17]. It is only by knowing these basic principles that it is even possible to be able to appreciate the full potential of piezoelectric materials, especially ZnO NS.

4.2.1 The Piezoelectric Effect

The piezoelectric effect may be categorized into two main types, including the direct piezoelectric effect and the converse piezoelectric effect. A pictorial representation of both types is given in Figure 4.1.

Direct piezoelectric effect takes place when a material produces an electric charge as a result of applied mechanical stress. [18]. When stress is applied, the material's internal structure becomes polarized, resulting in a voltage across the material. This effect is utilized in various sensing applications, where mechanical changes need to be detected and converted into electrical signals [19, 20].

In contrast, the piezoelectric effect causes mechanical strain in the material when an electric field is applied [21]. A piezoelectric material undergoes physical distortion when exposed to an electric field; the expansion or contraction is contingent upon the polarity of the applied field. This property is harnessed in actuator applications, where electrical signals are used to induce precise mechanical movements [22].

4.2.2 Piezoelectric Tensor and Coefficients

To describe the piezoelectric properties quantitatively, the piezoelectric tensor and piezoelectric coefficients are used. The **piezoelectric tensor** is a mathematical representation that relates mechanical stress and electric field in a material. For a material with piezoelectric properties, the relationship between the applied stress and the generated electric displacement is given by

$$D_i = d_{ijk}\sigma_{jk} \tag{4.1}$$

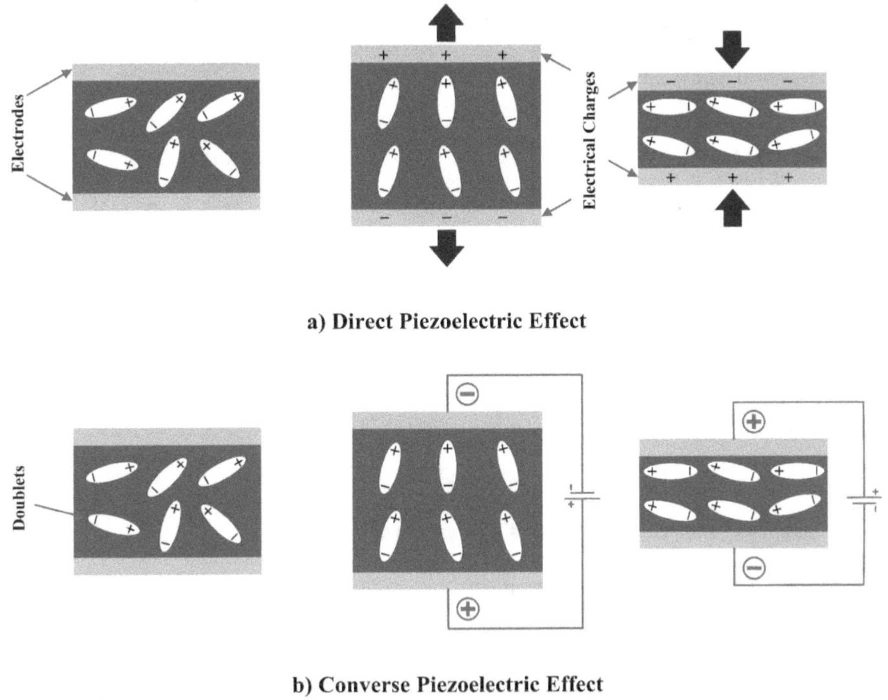

FIGURE 4.1 Schematic representation of (a) direct and (b) converse piezoelectric effect [23].

where D_i is the electric displacement, d_{ijk} is the piezoelectric coefficient tensor, and σ_{jk} is the applied stress tensor. The subscripts i, j, and k denote the spatial coordinates. Similarly, for the converse piezoelectric effect, the relation between generated configurational change and applied electric field can be represented by equation (4.2), represented as follows:

$$\varepsilon_{jk} = d_{ijk} E_i \tag{4.2}$$

where ε_{jk} is the induced strain, d_{ijk} is the piezoelectric coefficient tensor, and E_i is the applied electric field. The subscripts of the piezoelectric constant are generally denoted using the reduced Voigt matrix notation d_{im} [24]. In this system, the index i refers to the component of the electric displacement D or electric field E within the Cartesian coordinate system (x_1, x_2, x_3). Meanwhile, the index m (ranging from 1 to 6) specifies the particular mechanical stress or strain components. The tensor notation captures the anisotropic nature of piezoelectric materials, where the response can vary based on the direction of applied forces or fields.

Another parameter, the **piezoelectric coefficients** (denoted as d-coefficients) are critical parameters that quantify the material's piezoelectric response. These coefficients characterize the amount of electric charge produced per unit of mechanical stress and are parts of the piezoelectric tensor (equation (4.1)). The coefficients can be categorized based on the crystal symmetry of the material and the specific directions of applied stress or electric fields [24].

4.2.3 Crystallographic Requirements for Piezoelectricity

The piezoelectric effect arises from the non-centrosymmetric nature of specific crystal structures. When charge centers are coincident, there is no net polarization in centrosymmetric crystals under mechanical stress. However, in non-centrosymmetric crystals, such as those found in ZnO, mechanical stress causes a relative displacement of charge centers, leading to polarization and the generation of an electric field. For a material to exhibit piezoelectricity, it must dearth a center of symmetry in its crystal structure. Due to this lack of centrosymmetry, an electric field can be produced when the crystal lattice polarizes under mechanical stress.

4.3 ORIGIN OF PIEZOELECTRICITY IN ZnO

ZnO is a prominent piezoelectric material primarily due to its wurtzite crystal structure. The inherent polarity of the wurtzite structure is a key factor in ZnO's piezoelectric behavior. Several factors responsible for this behavior of ZnO are discussed as follows.

4.3.1 Crystallographic Structure of ZnO

ZnO crystallizes in a hexagonal wurtzite structure, as well as in cubic rock salt and zinc blende arrangement [25, 26]. Among these structures, the wurtzite structure is characterized by a hexagonal close-packed arrangement with a space group of $P_{63}m_c$, which makes it non-centrosymmetric in nature and thus conducive to piezoelectricity [27]. The atoms of zinc (Zn) and oxygen (O) are arranged in alternating planes along the c-axis in this structure. The tetrahedral organization of Zn and O atoms creates a lack of inversion symmetry by creating two distinct faces: the Zn-terminated (0001) face and the O-terminated (000$\underline{1}$) face, which is essential for the piezoelectric effect [25].

The wurtzite crystal structure can be designated by the lattice parameters a and c, where a is the distance between adjacent atoms in the basal plane, and c is the height of the unit cell. The typical lattice parameters for ZnO are $a \approx 3.25\ A^{\circ}$ and $c \approx 5.2\ A^{\circ}$ [28]. The ratio of c/a in ZnO is approximately 1.60, slightly deviating from the ideal hexagonal close-packed structure [29].

4.3.2 Polarization and Spontaneous Polarization

In ZnO, the lack of centrosymmetry leads to an impulsive polarization along the c-axis. This intrinsic polarization arises due to the variance in electronegativity between Zn and O atoms, as well as their tetrahedral coordination, which results in

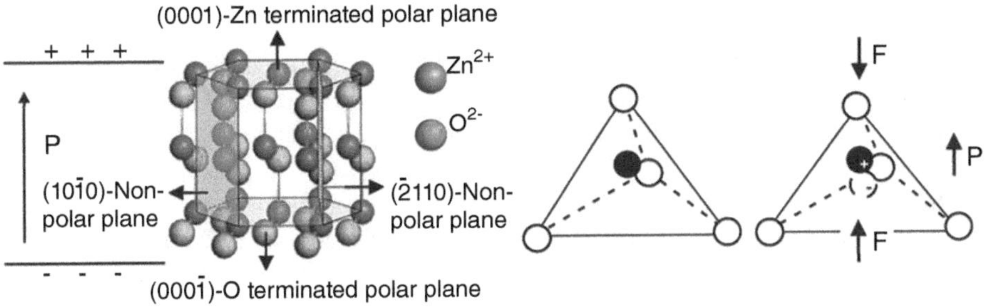

FIGURE 4.2 Wurtzite crystal structure of ZnO and its tetrahedral coordination [25].

a displacement of charge centers [30, 31]. The Zn atoms possess a partial positive charge, while the O atoms carry a partial negative charge. This separation of charges generates a dipole moment, contributing to the spontaneous polarization of the crystal, as shown in Figure 4.2.

When ZnO is subjected to mechanical stress, it further displaces the Zn and O atoms relative to each other, enhancing the polarization [28]. This induced polarization is the basis of the direct piezoelectric effect in ZnO [18]. Conversely, applying an electric field to ZnO results in mechanical deformation, manifesting as the converse piezoelectric effect [21].

4.3.3 Influence of Crystallographic Orientation

The piezoelectric properties of ZnO are profoundly influenced by its crystallographic orientation due to the fact that different crystal faces exhibit varying degrees of piezoelectric response due to the directional dependence of the piezoelectric coefficients. The key crystallographic planes in ZnO are (0001), (10$\underline{1}$0), and (11$\underline{2}$0) [32]. The (0001) is perpendicular to the c-axis and is often referred to as the **basal plane**. It is polar, with one side terminated by Zn atoms and the other by O atoms. Due to alignment with the c-axis, it exhibits the strongest piezoelectric response due to the polarity of the Zn-O bonds along this direction. When mechanical stress is applied along the c-axis, the displacement of Zn and O atoms results in a substantial electric dipole moment [28]. This makes the (0001) plane particularly suitable for applications requiring high piezoelectric sensitivity, such as nanogenerators and sensors [33].

Whereas, (10$\underline{1}$0), and (11$\underline{2}$0) are planes normal to the a-axis (**m-plane**) and b-axis (**a-plane**), respectively. Both planes are non-polar and are often used in studies to eliminate the effects of polarization. In these orientations, the applied mechanical stress does not align with the polar axis, resulting in a weaker piezoelectric response. However, these planes are useful for applications where the influence of polarization needs to be minimized, such as in certain types of sensors and actuators, where intrinsic material properties are desired without the effects of piezoelectric anisotropy [34, 35].

4.3.4 Piezoelectric Tensor and Coefficients in ZnO

For materials with hexagonal symmetry like ZnO, the piezoelectric tensor simplifies, with the primary non-zero components being d_{33}, d_{31}, and d_{15}. These coefficients describe the material's response along different crystallographic axes.

d_{33}: This coefficient represents the piezoelectric response along the c-axis (polar axis). It is typically the largest piezoelectric coefficient in ZnO, reflecting the strong coupling between mechanical and electrical effects along this direction.

d_{31}: This coefficient describes the piezoelectric response in the plane normal to the c-axis. It is generally smaller than d_{33} but still significant.

d_{15}: This coefficient is associated with shear stress and electric fields oriented at an angle to the c-axis.

The typical values for these coefficients in ZnO are $d_{33} \approx 12.4$ pC/N, $d_{31} \approx -5.2$ pC/N, and $d_{15} \approx -9.9$ pC/N [36]. These values highlight the anisotropic nature of ZnO's piezoelectric properties, with the strongest response observed along the c-axis.

4.4 METHODS USED TO HARNESS THE PROPERTY OF PIEZOELECTRICITY IN ZnO

ZnO is being extensively studied for use in piezoelectric energy harvesters, which can be considerably improved at the nanoscale than those at the bulk scale. The piezoelectric strain constant, for bulk ZnO d_{33} (12 pm/V), is noticeably lower than conventional lead zirconate titanate (PZT) ceramics, whose d_{33} coefficient ranges from approximately 300 to 600 pm/V, depending on the nature of PZT [26]. The d_{33} can be improved by simply tuning the morphology of ZnO along with the use of some effective dopants, which is explained in the following section.

4.4.1 Impact of Morphology

In Table 4.1, a comparison of the d_{33} component among different NS of pure ZnO has been listed. It has been noticed that the value of d_{33} can be modified from 12.4 pm/V to 80 pm/V by simply changing the dimensions of the material. More specifically, d_{33} for nanobelts is lowest followed by nanorods and then nanosheets. The observed increase in the d_{33} piezoelectric coefficient with decreasing dimensions of ZnO NS can be attributed to several key factors. As the size of these NS decreases, particularly in terms of diameter, the surface-to-volume ratio increases significantly. As a result, the surface charge density is increased and, hence, stronger the piezoelectric response is observed [37–40]. In addition to this, when the dimensions are smaller, the polarization per unit volume increases because the decrease in size leads to an accumulation of electric dipoles, and hence, the piezoelectric effect increases [41, 42]. As an example, Li et al. [43] discovered that the d_{33} value rose from 18.2 to 46.9 pm/V when the diameter of micropillars was reduced from 7 to 1.5 μm. This increase is due to an increased polarization per unit volume, which is stronger at smaller scales [43, 44].

TABLE 4.1
Comparison for d_{33} Parameter for Pure ZnO NS where D and L Represent the Diameter and Length, Respectively

Morphology	Method of Preparation	d_{33} (pm/V)	Dimensions	Method of Measurement	References
ZnO Bulk	Hydrothermal method	12.4	—	Rotated cut method	[36]
Rods	Hydrothermal method	49.7	D = 250 μm, L = 150 nm	PFM	[50]
	Template-free electrochemical deposition	11.8	D = 150 nm, L = 1.50 μm	PFM	[51]
	template-free low-temperature aqueous chemical growth method	9.2	D = 150 ± 55 nm L = 2300 ± 120 nm	Nanoindentation	[52]
Pillars	Template-directed approach	7.5 ± 0.6	D = 300 nm L = 2 μm	PFM	[53]
Belts	thermal evaporation	14.3 – 26.7	Thickness = 65 nm, Width = 360 nm, Length = tens of nm	PFM	[47]
Sheets	Ionic layer epitaxy	6.48–18.9	Thickness = 1.5–4.5nm length ~ 8 m	AFM	[54]
	Self-limiting approach	80 ± 0.8	Thickness = 1.1 nm, Length several mm	AFM	[55]
Thin Film	Radio-frequency magnetron sputtering	5–5.3	Thickness = 285–710 nm	PFM	[56]

In addition, smaller sizes can result in fewer structural defects and imperfections like dislocations and vacancies, especially if the NS are of high crystallinity [45, 46]. These defects typically become obstacles to the movement of polarized charges, so their decrease can contribute effectively to the enhancement of piezoelectric properties. For instance, a high d_{33} value of 26.7 pm/V was reported in ZnO nanobelts by Zhao et al. while ascribing improvement in piezoelectric behavior to perfect single crystallinity and absence of dislocations [47]. Lastly, quantum confinement effects at these reduced scales can alter the electronic properties of ZnO, potentially leading to increased piezoelectric responses [48, 49].

4.4.2 Role of Doping

Doping plays a vital role in enhancing the piezoelectric properties of ZnO NS (**Table 4.2**), as it addresses several intrinsic limitations and introduces beneficial modifications. One significant reason for doping is to mitigate the internal and external screening effects that reduce the efficiency of piezoelectric energy harvesting [41]. ZnO, being an n-type semiconductor due to inherent defects, has an excess of free electrons. These electrons can neutralize the piezoelectric potential (piezopotential) generated when mechanical stress is applied, a phenomenon known as the internal screening effect [57]. In addition, some electrons may leak from ZnO into the external circuit, known as the external screening effect [25]. Both effects minimize the total piezopotential generated, thereby reducing the efficiency of the device [25, 41, 57]. P-type doping of ZnO tends to reduce these excess free electrons, thereby reducing the screening effects and preserving the generated piezopotential. [58].

Doping also has significant effects on piezoelectric properties, mainly by modifying the crystal structure and introducing strain that can lead to higher polarity and increased piezoelectric coefficients [59–61]. For instance, the incorporation of dopants with a higher ionic charge to ionic radius ratio than Zn^{2+}, such as Y^{3+} or Nd^{3+}, leads to increased polarity in dopant-O bonds [61–63]. Increased polarity allows bonds to bend more easily along the polar c-axis under an electric field, with a greater piezoelectric response. Sinha et al. [63] reported that larger ions compared with Zn^{2+}, with a charge-to-radius ratio over 2.71 e/Å, such as Y^{3+} (2.88 e/Å) and Nd^{3+} (3.05 e/Å), improve piezoelectric performance due to enhanced polarization and bond flexibility, significantly increasing the piezoelectric coefficient. As an example, the d_{33} of Y-doped ZnO nanosheets was measured to be 420 pm/V, and Nd-doped ZnO nanorods showed an impressive d_{33} value at 512 pm/V, similar to that of bulk PZT materials [63].

In addition, doping can sometimes even introduce ferroelectricity in ZnO, thereby further augmenting its piezoelectric properties [64–66]. The specific effects like ferromagnetism, enhanced optical and electrical properties associated with doping depend on dopant type and concentration, and must be carefully controlled to avoid issues like chemical inhomogeneity [67–70]. Properly managed doping can result in significant improvements in the piezoelectric properties of ZnO, making it a potential material for piezoelectric energy harvesters and other related applications. However, it has been noticed that the existence of humidity may reduce the piezoelectric behavior of ZnO which may be controlled while manufacturing [71–73].

TABLE 4.2
Comparison for d_{33} Parameter for Doped ZnO NS

Morphology	Method of Preparation	Dopant	d_{33} (pm/V)	Dimensions	Method of Measurement	Reference
Rods	Hydrothermal	Ag	8.75 - 4	Diameter = 100–150 nm, Length = 1 μm	Nanoindentation	[74]
	Wet chemical co-precipitation	Ce	30	Diameter = 16–20 nm, Length = 80–120 nm	PFM	[75]
	Wet chemical co-precipitation	Eu	43	Diameter = 78.92 nm, Length = 359.07 nm	D-V curve	[76]
	Wet chemical co-precipitation	La	101	Diameter = 123.49 nm, Length = 704.38 nm	D-V curve	[77]
	Wet chemical co-precipitation	Gd	45	Diameter = 59.85 nm, Length = 245.13nm	D-V curve	[77]
	Wet chemical co-precipitation	Nd	512	Diameter = 110 nm, Length = 520 nm	D-V curve	[63]
Films	Radio-frequency magnetron sputtering	Mg	54.1	Thickness = 380 nm	PFM	[78]
	Magnetron sputtering	Cu	6	Thickness = 200 nm	PFM	[79]
	Magnetron sputtering	Ni	10	Thickness = 200 nm	PFM	[79]
	Magnetron sputtering	Co	11	Thickness = 200 nm	PFM	[79]
	Magnetron sputtering	Fe	14	Thickness = 200 nm	PFM	[79]
	Magnetronsputtering	V	110	Thickness = 210 nm	SPM	[80]
	Magnetron sputtering	Cr	120	Thickness = 150 nm	SPM	[81]
	Magnetron sputtering	Fe	127	Thickness = 280 nm	SPM	[82]
	Magnetron sputtering		86	Thickness = 80 nm	AFM	[83]
Micro-flowers	Wet chemical co-precipitation	Eu^{3+}/Gd^{3+} codoped	55	diameter ~ 184 nm	D-V curve	[84]

4.5 PIEZOELECTRIC NANOGENERATORS: DESIGN AND OPTIMIZATION

PENGs harness the piezoelectric properties of materials like ZnO to convert mechanical energy into electrical energy on a nanoscale. ZnO is an ideal material for PENGs due to its high piezoelectric coefficients, mechanical robustness, and ease of fabrication into various NS [41, 42]. The design and optimization of PENGs are critical for enhancing their efficiency and performance in practical applications, such as powering small electronic devices, sensors, and wearable technology [85–88]. The schematic preparation and representation of ZnO PENGs are displayed in Figure 4.3. The core components of a PENG typically include: piezoelectric material (ZnO NS like nanowires, nanorods, nanotubes); electrodes which are conductive materials that collect the generated charges; and substrate which is generally a base material that supports the NS and provides mechanical stability. When the piezoelectric material is exposed to mechanical stress, the displacement of ions within the crystal lattice creates an electric potential difference across the material, which can be harvested as electrical energy. The design of ZnO-based PENGs involves several critical considerations to maximize their efficiency and performance:

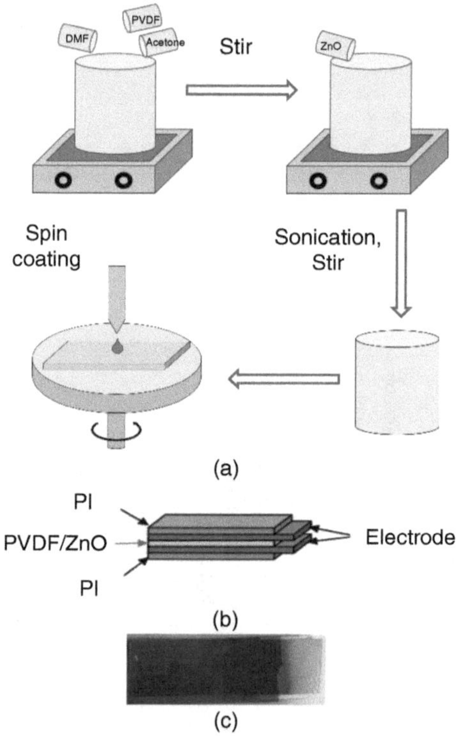

FIGURE 4.3 (a) Preparation method, (b) design, and (c) obtained film [89].

4.5.1 Selection of ZnO Nanostructures

Different forms of ZnO NS (nanowires, nanorods, nanotubes) offer unique advantages in PENG applications. Nanowires and nanorods are commonly used due to their high aspect ratio, which enhances the piezoelectric effect [90–92]. For instance, ZnO nanowires provide high surface area and flexibility, making them potent for applications, where mechanical deformation is significant [93]. Whereas, ZnO nanorods offer structural stability and can be densely packed, leading to higher output power [94]. Le et al. [25] have reviewed the parameters monitoring the performance of ZnO nanorods-based PENGs. However, ZnO nanotubes combine the benefits of nanowires and nanorods with additional surface area, which can enhance the piezoelectric response [95, 96]. Xi et al. [97] reported the improvement in piezoelectric nature with the growth of ZnO tubes on GaN substrate.

4.5.2 Substrate Material and Configuration

The choice of substrate material influences the mechanical and electrical properties of the PENG. For instance, nanorods of diameter 50 nm grown on conductive textile fabric, carbon fiber, and ITO/PTE supply an output of 250, 333 × $10^{3,}$ and 150 nA, respectively [98–100]. Flexible substrates, such as polymers (Polydimethylsiloxane (PDMS), polyethylene terephthalate (PET)), allow for greater mechanical deformation, while rigid substrates, like silicon, provide structural stability [101, 102]. The substrate configuration, together with its thickness and surface roughness, also influences the performance of the PENG. Improving the mechanical coupling between ZnO NS and the substrate can enhance the piezoelectric response achieved through surface modification and patterned substrates. Thinner substrates enable better mechanical coupling with ZnO NS, while surface roughness can enhance adhesion and mechanical interaction [55]. Furthermore, Hsu et al. [103] reported that a slight bend in the substrate may improve the efficacy by four times more than the non-bend substrate. Moreover, creating micro- or nano-patterns on the substrate to improve mechanical interlocking and stress transfer. Zhang et al. [104] examined the impact of the presence of both (micro and nano) patterns on the silicon substrate used to prepare the triboelectric nanogenerator. Choi et al. [105] reported the improvement of the effective contact area of the triboelectric materials by surface functionalization.

4.5.3 Electrode Materials and Layout

The selection and layout of electrodes are crucial for efficient charge collection and minimization of electrical losses. Common electrode materials include metals (gold (Au), silver (Ag), aluminum (Al)) and transparent conductive oxides like indium tin oxide (ITO). Briscoe et al. [57] listed the effect of electrode material on device performance, which is challenging to discuss here. Apart from electrode materials, its layout designs, such as interdigitated and parallel plate configurations, affect the electric field distribution and charge collection efficiency. Recently, interdigitated patterned electrodes are attracting research interests as they allow for a more efficient distribution of the electric field across the piezoelectric material. This configuration maximizes the piezoelectric coupling, particularly when in-plane actuation is required, which is often the case in thin-walled structures. Recently, Motora et al. [106] studied the effect of electrode pattern on device output. It was concluded that presence of electrode with interdigitated configuration improves the harvested output current, voltage, and power density of 1.15 μA, 50 V, and 20.10 mW/m^2, respectively, at 12 N. Moreover, the fabrication of IDE patterns can be achieved through methods, such as screen printing, which is not only cost-effective and straightforward but also scalable for larger production runs. This accessibility in fabrication means that high-performance PENG devices can be produced more easily and at a lower cost, making them more viable for practical applications [107, 108]. Ko et al. [109] reported the increase in output current density by 1.54 times by simply modifying the surface roughness of the Au electrode. Optimizing the electrode pattern can significantly improve the PENG's output performance.

4.5.4 Optimization through Fabrication Techniques

Primarily, the fabrication technique tunes the morphology of NS. Several fabrication techniques are employed to create ZnO NS for PENGs, each with its own advantages in terms of control over structure, orientation, and scalability.

I. **Hydrothermal synthesis**: Hydrothermal synthesis is a widely used method for growing ZnO NS due to its simplicity and ability to produce high-quality, oriented nanowires and nanorods. This technique involves the preparation of precursors where zinc nitrate and hexamethylenetetramine (HMTA) are commonly used, followed by a growth process where the precursors are mixed and heated in an autoclave, and the controlled temperature and pressure promote the growth of ZnO NS on a substrate. Lastly, annealing the grown NS can improve crystallinity and piezoelectric properties [110, 111]. Specific crystallographic orientations may be promoted by adjusting the concentrations of precursors [111], the pH of the solution [112], or even growth time. This is because Mahmud et al. [113] prepared nanoballs and nanorods by simply altering the temperature of the solution. For instance, specific surfactants or additives may enhance the growth of ZnO NS along the c-axis [114, 115].
II. **Vapor–liquid–solid (VLS) method** is used to grow high-quality, dimensionally controlled ZnO nanowires with controlled orientation [116]. This method comprises vapor phase deposition of zinc vapor reacted with oxygen and forms ZnO nanowires. Then comes catalyst-assisted growth wherein a catalyst metal, e.g., gold is used to grow nanowires directly. This process allows for precise monitoring of the growth of the nanowire structures, leading to well-aligned structures with good piezoelectric properties [116]. The preferred crystallographic orientation in the resulting ZnO NS may be influenced by modifying the growth parameters, such as temperature, pressure, and gas flow rates [117, 118].
III. **Epitaxial growth** is the deposition of ZnO onto a substrate in a well-defined crystallographic orientation. The substrate acts as a template for dictating

the orientation of the ZnO NS. Techniques such as Molecular Beam Epitaxy (MBE), and Metal-Organic Chemical Vapor Deposition (MOCVD) are used for epitaxial growth [119, 120]. The material of the substrate and the lattice matching are critical factors to achieving high-quality oriented ZnO NS. Bhadwal et al. [26] compared several composite materials, which can be useful for optimization, concerning their piezoelectric properties.

4.6 CHALLENGES AND FUTURE PERSPECTIVES

There are several challenges in the development and application of ZnO-based piezoelectric NS which need to be overcome to fully tap their potential. Key material challenges include crystallographic defects that affect piezoelectric coefficients and mechanical strength. Achieving uniformity and reproducibility is also difficult when trying to fabricate NSs. Advanced synthesis techniques combined with precise growth methods can minimize defects while ensuring consistency.

It would be very important at the device level to use it to maximize energy conversion efficiency, which can be attained through optimized mechanical coupling, charge collection, electrode configurations, stability under mechanical stress and environmental conditions, develop robust materials and protective coatings, and improve performance over the long term. For integrating ZnO-based PENGs into already existing electronic systems, efficient interface engineering and seamless system integration would be of more importance, especially in cases that apply to flexible electronics and wearable devices.

Future research directions include the development of advanced materials, including composite materials and doped ZnO, which can provide superior piezoelectric properties. Theoretical modeling and simulation can provide much deeper insight and guide experimental efforts. Multifunctional devices – the integration of energy harvesting with sensing and actuation – expands applications to include a self-powered sensor and wearable technology. Environmental and biological applications also have promise. Advances in fabrication techniques to high-precision and scalable manufacturing processes are seen as an avenue for field advancement. Addressing these challenges through interdisciplinary research and innovation will unlock the full potential of ZnO-based piezoelectric NS, leading to technological advancements and sustainable solutions across various applications.

4.7 CONCLUSION

This chapter gives a comprehensive overview of ZnO NS-based piezoelectric properties, including basic mechanisms, fundamental crystallographic orientations, and their extensive applications in energy harvesting and sensing technologies. ZnO is a promising candidate for development into efficient devices like PENGs and sensors, due to its high piezoelectric coefficients, mechanical robustness, and ease of fabrication. Despite these developments, good defect management, uniform fabrication, and integration with electronic devices remain as the significant challenges associated with this field. Therefore, advanced materials research, improved fabrication techniques, and collaboration among interdisciplinary groups of researchers will be an essential step in optimizing the performance and expanding applications of ZnO-based piezoelectric devices. Future research based on multifunctional devices, theoretical modeling with enhanced accuracy, and the ability to scale up in their processing will lead further into innovation, with a potential breakthrough solution emerging toward energy, environmental monitoring, and biomedical fields. Exploration and development of ZnO NS are likely to be a significant contributor toward nanotechnology advancement and sustainable energy applications.

REFERENCES

1. Taylor, P., & De Gennes, P. G. (2011). Pierre Curie and the role of symmetry in physical laws. *Ferroelectrics*, *40*(1), 37–41.
2. Zhu, X. (2017). *Piezoelectric ceramics materials: Processing, properties, characterization, and applications*. Nova Science Publishers.
3. Zhou, Q., Lam, K. H., Zheng, H., Qiu, W., & Shung, K. K. (2015). Piezoelectric single crystals for ultrasonic transducers in biomedical applications. *Progress in Materials Science*, *66*, 87–111.
4. Ramírez-Ortega, D., González, I., & Arroyo, R. (2014). Semiconducting properties of ZnO/TiO_2 composites by electrochemical measurements and their relationship with photocatalytic activity. *Electrochimica Acta*, *140*, 2–11.
5. Momeni, K., Odegard, G. M., & Yassar, R. S. (2012). Finite size effect on the piezoelectric properties of ZnO nanobelts: A molecular dynamics approach. *Acta Materialia*, *60*(15), 5117–5124.
6. Izyumskaya, N., Avrutin, V., Özgür, Ü., Alivov, Y. I., & Morkoç, H. (2007). Preparation and properties of ZnO and devices. *Physica Status Solidi (b)*, *244*(5), 1439–1450.
7. Lee, J., Lee, W., Lee, S., & Kim, M. (2015). Atomic-scale origin of piezoelectricity in wurtzite. *Physical Chemistry Chemical Physics*, *17*(11), 7857–7863.
8. Kumar, S. G., & Rao, K. S. R. K. (2015). Zinc oxide-based photocatalysis: Tailoring surface-bulk structure and related interfacial charge carrier dynamics for better environmental applications. *RSC Advances*, *5*(4), 3306–3351.
9. Viñes, F., & Bromley, S. T. (2017). Size-dependent structural and polymorphic transitions in ZnO: From nanocluster to bulk. *Nanoscale*, *9*(29), 10067–10074.
10. Abubakar, S., Tan, S. T., Ying, J., Liew, C., Talib, Z. A., Sivasubramanian, R.,... & Paiman, S. (2023). Controlled growth of semiconducting ZnO nanorods for piezoelectric energy harvesting-based nanogenerators. *Nanomaterials and Nanotechnology*, *13*, 1025.
11. Sha, R., Basak, A., Maity, P. C., & Badhulika, S. (2022). ZnO nanostructured-based devices for chemical and optical sensing applications. *Sensors and Actuators Reports*, *4*, 100098.
12. Santos, A., Sabino, F., Rovisco, A., Barquinha, P., Martins, R., Igreja, R.,... & Fortunato, E. (2021). Optimization of ZnO nanorods concentration in a micro-structured polymeric composite for nanogenerators. *Chemosensors*, *9*(1), 1–13.
13. Calestani, D., Zha, M., Mosca, R., Zappettini, A., Carotta, M. C., Di Natale, V., & Zanotti, L. (2010). Growth of ZnO tetrapods for nanostructure-based gas sensors. *Sensors and Actuators B: Chemical*, *144*(2), 472–478.
14. Khan, S., Ali, S., & Bermak, A. (2019). Recent developments in printing flexible and wearable sensing electronics for healthcare applications. *Sensors*, *19*(5), 1230.
15. Kumar, R., Kumar, G., & Umar, A. (2015). Zinc oxide nanostructures for NO_2 gas-sensor applications. *Nano-Micro Letters*, *7*(1), 97–120.

16. Ortiz-Casas, B., Galdámez-Martínez, A., Gutiérrez-Flores, J., Ibañez, A. B., Panda, P. K., Santana, G.,… & Mishra, Y. K. (2021). Bio-acceptable 0D and 1D ZnO nanostructures for cancer diagnostics and treatment. *Materials Today*, *50*, 533–569.
17. Banerjee, S., Bairagi, S., & Wazed Ali, S. (2021). A critical review on lead-free hybrid materials for next generation piezoelectric energy harvesting and conversion. *Ceramics International*, *47*(11), 16402–16421.
18. Katsouras, B. I., Asadi, K., Li, M., & Van Driel, T. B. (2015). The negative piezoelectric effect of the ferroelectric polymer poly(vinylidene fluoride). *Nature Materials*, *15*(1), 78–84.
19. Zhang, Z., Kan, J., Cheng, G., Jia, Y., & Wang, H. (2013). Influence of multiple piezoelectric effects on sensors and actuators. *Mechanical Systems and Signal Processing*, *35*(1), 95–107.
20. Moubarak, P. M., Ben-Tzvi, P., & Zaghloul, M. E. (2012). A self-calibrating mathematical model for the direct piezoelectric effect of a new MEMS tilt sensor. *Sensors and Actuators A: Physical*, *12*(5), 1033–1042.
21. Arnau, A., & Soares, D. (2008). *Fundamentals of piezoelectricity*. Springer.
22. Gao, X., Yang, J., Wu, J., Xin, X., Li, Z., & Yuan, X. (2019). Piezoelectric actuators and motors: Materials, designs, and applications. *Advanced Materials Technologies*, *1900716*, 1–26.
23. Kargar, S. M., & Hao, G. (2022). An atlas of piezoelectric energy harvesters. *Sensors*, *22*, 1949.
24. Curie, P. (2021). Fundamentals of piezoelectricity. In *Lead-free piezoelectric materials* (pp. 1–18). WILEY-VCH GmbH.
25. Le, A. T., Ahmadipour, M., & Pung, S. (2020). A review on ZnO-based piezoelectric nanogenerators: Synthesis, characterization techniques, performance enhancement and applications. *Journal of Alloys and Compounds*, *5*, 156172.
26. Bhadwal, N., Ben Mrad, R., & Behdinan, K. (2023). Review of zinc oxide piezoelectric nanogenerators: Piezoelectric properties, composite structures, and power output. *Sensors*, *23*, 3859.
27. Anh, D. T. N., Phuong, H. T. L., Thao, H. T. H., Ha, N. T. C., & Hoan, N. X. (2011). Crystal structures and properties of ZnO nanopowders prepared by ultrasonic method. *Journal of Surface Science and Nanotechnology*, *9*, 482–485.
28. Wang, Z. L., Kong, X. Y., Ding, Y., Gao, P., & Hughes, W. L. (2004). Semiconducting and piezoelectric oxide nanostructures induced by polar surfaces. *Advanced Functional Materials*, *28* (10), 943–956.
29. Version, D., & Materials, F. (2018). Local structure and chemistry of C-doped ZnO @ C core–shell nanostructures with room-temperature ferromagnetism. *Advanced Functional Materials*, *28*, 1704567.
30. Troy, W., Dutta, M., & Stroscio, M. A. (2021). Spontaneous polarization in nanoparticles. *IEEE Electron Device Letters*, *42*, 1838–1840.
31. Troy, W., Dutta, M., & Stroscio, M. (2021). Spontaneous polarization calculations in wurtzite II-oxides, III-nitrides, and SiC polytypes through net dipole moments and the effects of nanoscale layering. *Nanomaterials*, *11*, 1956.
32. Kaneti, Y. V., Zhang, Z., Yue, J., Zakaria, Q. M. D., Chen, C., Jiang, X., & Yu, A. (2014). Crystal plane-dependent gas-sensing properties of zinc oxide nanostructures: Experimental and theoretical studies. *Physical Chemistry Chemical Physics*, *16*, 11471–11480.
33. Leonardi, S. G. (2017). Two-dimensional zinc oxide nanostructures for gas sensor applications. *Chemosensors*, *5*, 17.
34. Lin, B. H., Lin, C. Y., Hsu, S. T., Yang, S., Kuo, C. C., Hsieh, W. F., & Chang, C. S. (2012). Single domain m-plane ZnO grown on m-plane sapphire by radio frequency magnetron sputtering. *Applied Materials and Interfaces*, *4*, 5333–5337.

35. Heinhold, R., Cooli, S. P., Evans, D. A., & Allen, M. W. (2014). Stability of the surface electron accumulation layers on the nonpolar (1010) and (1120) faces of ZnO. *The Journal of Physical Chemistry C*, *118*, 24575–24582.
36. Oitaviani, G. (1968). Dielectric, piezoelectric, and electromechanical coupling constants of zinc oxide crystals. *Proceedings of the IEEE*, *56*, 1967–1968.
37. Krishna, M. S., Singh, S., Batool, M., & Fahmy, M. (2023). A review on 2D-ZnO nanostructure-based biosensors: From materials to devices. *Materials Advances*, *4*, 320–354.
38. Shankar, P., & Rayappan, J. B. B. (2015). Electrospun tailored ZnO nanostructures – Role of chloride ions. *RSC Advances*, *5*, 85363–85372.
39. Yan, Z., & Jiang, L. (2017). Modified continuum mechanics modeling on size-dependent properties of piezoelectric nanomaterials: A review. *Nano-Micro Letters*, *7*, 1–18.
40. Kim, M., & Fan, J. (2021). Piezoelectric properties of three types of PVDF and ZnO nanofibrous composites. *Advanced Fiber Materials*, *3*, 160–171.
41. Pandey, R. K., Dutta, J., Brahma, S., Rao, B., & Liu, C. (2021). Review on ZnO-based piezotronics and piezoelectric nanogenerators: Aspects of piezopotential and screening effect. *Journal of Physics: Materials*, *4*, 044011.
42. Wang, Z. L. (2012). From nanogenerators to piezotronics—A decade-long study of ZnO nanostructures. *MRS Bulletin*, *37*, 814–827.
43. Li, T., Li, Y. T., Qin, W. W., Zhang, P. P., Chen, X. Q., Hu, X. F., & Zhang, W. (2015). Piezoelectric size effects in a zinc oxide micropillar. *Nanoscale Research Letters*, *10*, 394.
44. Agrawal, R., & Espinosa, H. D. (2011). Giant piezoelectric size effects in zinc oxide and gallium nitride nanowires: A first principles investigation. *Nano Letters*, *11*, 786–790.
45. Ayoub, I., Kumar, V., Abolhassani, R., Sehgal, R., & Sharma, V. (2022). Advances in ZnO: Manipulation of defects for enhancing their technological potentials. *Nanotechnology Reviews*, *11*, 575–619.
46. Ischenko, B. V., Polarz, S., Grote, D., Stavarache, V., Fink, K., & Driess, M. (2008). Zinc oxide nanoparticles with defects. *Advanced Functional Materials*, *15*, 1945–1954.
47. Zhao, M., Wang, Z., & Mao, S. X. (2004). Piezoelectric characterization of individual zinc oxide nanobelt probed by piezoresponse force microscope. *American Chemical Society*, *4*, 587–590.
48. Yadav, D., Tyagi, N., Yadav, H., James, A., Sareen, N., Kapoor, M., Kumar, K., Kataria, Y., & Singhal, K. (2023). Effect of various morphologies and dopants on piezoelectric and detection properties of ZnO at the nanoscale: A review. *Journal of Materials Science*, *58*, 10576–10599.
49. Tran, K., Tawfik, S. A., & Spencer, M. J. S. (2023). Restoring piezoelectric properties in 2D zinc oxide nanosheets by surface modifications: Implications for piezoelectric nanogenerators. *ACS Applied Nano Materials*, *6*, 14767–14776.
50. Ghosh, M., Ghosh, S., Seibt, M., Rao, K. Y., Peretzki, P., & Rao, G. M. (2015). Ferroelectric origin in one-dimensional undoped ZnO towards high electromechanical response. *CrystEngComm*, *18*, 622–630.
51. Tamvakos, D., Lepadatu, S., Antohe, V., Tamvakos, A., Weaver, P. M., Piraux, L., Cain, M. G., & Pullini, D. (2015). Piezoelectric properties of template-free electrochemically grown ZnO nanorod arrays. *Applied Surface Science*, *356*, 1214–1220.
52. Broitman, E., Soomro, M. Y., Lu, J., & Willander, M. (2013). Nanoscale piezoelectric response of ZnO nanowires measured using a nanoindentation technique. *Royal Society of Chemistry*, *15*, 11113–11118.
53. Fan, H. J., Lee, W., Hauschild, R., Alexe, M., Rhun, L., Scholz, R., Dadgar, A., Nielsch, K., Kalt, H., Krost, A., Zacharias, M., & Gösele, U. (2006). Template-assisted large-scale ordered arrays of ZnO pillars for optical and piezoelectric applications. *Small*, *4*, 561–568.

54. Carlos, C., Wang, Y., Wang, J., Li, J., & Wang, X. (2021). Thickness-dependent piezoelectric property from quasi-two-dimensional zinc oxide nanosheets with unit cell resolution. *Research*, *2021*, 1–7.
55. Mahmood, N., Khan, H., Tran, K., Kuppe, P., Zavabeti, A., Atkin, P., Ghasemian, M. B., Yang, J., Xu, C., Abdulkader, S., Spencer, M. J. S., Zhen, J., Khoshmanesh, K., Mcconville, C. F., Li, Y., & Kalantar-Zadeh, K. (2021). Maximum piezoelectricity in a few unit-cell thick planar ZnO: A liquid metal-based synthesis approach. *Materials Today*, *44*, 69–77.
56. Laurenti, M., Stassi, S., Lorenzoni, M., Fontana, M., & Canavese, G. (2016). Evaluation of the piezoelectric properties and voltage generation of flexible zinc oxide thin films. *Nanotechnology*, *26*, 1–9.
57. Briscoe, J., & Dunn, S. (2015). Piezoelectric nanogenerators: A review of nanostructured piezoelectric energy harvesters. *Nano Energy*, *14*, 15–29.
58. Goel, S., & Kumar, B. (2019). A review on piezo-/ferro-electric properties of morphologically diverse ZnO nanostructures. *Journal of Alloys and Compounds*, *816*, 152491.
59. Li, Y., Rao, R., Wang, Y., Du, H., Shi, J., & Liu, X. (2023). Effect of chemical inhomogeneity on the dielectric and impedance behaviors of bismuth sodium titanate based relaxors. *ECS Journal of Solid State Science and Technology*, *12*, 13005.
60. Shi, J., Dong, R., He, J., Wu, D., Tian, W., & Liu, X. (2023). Regulating ferroelectric polarization and dielectric properties of BT-based lead-free ceramics. *Journal of Alloys and Compounds*, *933*, 167746.
61. Laurenti, M., Castellino, M., Perrone, D., Asvarov, A., Canavese, G., & Chiolerio, A. (2017). Lead-free piezoelectrics: V3+ to V5+ ion conversion promoting the performances of V-doped zinc oxide. *Scientific Reports*, *7*, 41957.
62. Sinha, N., Goel, S., Joseph, A. J., Yadav, H., Batra, K., & Gupta, M. K. (2018). Y-doped ZnO nanosheets: Gigantic piezoelectric response for an ultra-sensitive flexible piezoelectric nanogenerator. *Ceramics International*, *44*, 8582–8590.
63. Batra, K., Sinha, N., Goel, S., Yadav, H., Joseph, A. J., & Kumar, B. (2018). Enhanced dielectric, ferroelectric, and piezoelectric performance of Nd-ZnO nanorods and their application in flexible piezoelectric nanogenerator. *Journal of Alloys and Compounds*, *767*, 1003–1011.
64. Huang, J., Hu, Y., & Liu, S. (2022). Origin of ferroelectricity in magnesium-doped zinc oxide. *Physical Review B*, *106*, 144106.
65. Wang, X. S., Wu, Z. C., Webb, J. F., & Liu, Z. G. (2003). Ferroelectric and dielectric properties of Li-doped ZnO thin films prepared by pulsed laser deposition. *Applied Physics A*, *77*, 561–565.
66. Awan, S. U., Hasanain, S. K., Anjum, D. H., Awan, M. S., & Shah, S. A. (2024). Room temperature p-type conductivity and coexistence of ferroelectric order in ferromagnetic Li-doped ZnO nanoparticles. *Journal of Applied Physics*, *116*, 164109.
67. Sanyal, B., Knut, R., Grånäs, O., Iusan, D., Karis, O., & Eriksson, O. (2008). Inhomogeneity in Co-doped ZnO diluted magnetic semiconductor. *Journal of Applied Physics*, *103*, 07D131.
68. Kvit, A. V., Yankovich, A. B., Avrutin, V., Liu, H., Izyumskaya, N., Ozgur, U., Morkoc, H., & Voyles, P. M. (2012). Impurity distribution and microstructure of Ga-doped ZnO films grown by molecular beam epitaxy. *Journal of Applied Physics*, *123527*, 1–7.
69. Guo, B., Xing, Z., Yi, J. B., Tao, J. G., Liu, T., Wong, L. M., Zhang, Z., Li, G. P., Wang, S. J., Ding, J., Sum, T. C., Hon, C., Huan, A., & Wu, T. (2008). Comparative study of room-temperature ferromagnetism in Cu-doped ZnO nanowires enhanced by structural inhomogeneity. *Advanced Materials*, *20*, 3521–3527.
70. Yao, Z., Tang, K., Xu, Z., Ye, J., Zhu, S., & Gu, S. (2016). The luminescent inhomogeneity and the distribution of zinc vacancy-related acceptor-like defects in N-doped ZnO microrods. *Nanoscale Research Letters*, *11*, 511.

71. Zhao, T., Fu, Y., Zhao, Y., Xing, L., & Xue, X. (2015). Ga-doped ZnO nanowire nanogenerator as self-powered / active humidity sensor with high sensitivity and fast response. *Journal of Alloys and Compounds*, *648*, 571–576.
72. Zhang, J., & Zhou, J. (2018). Humidity-dependent piezopotential properties of zinc oxide nanowires: Insights from atomic-scale modelling. *Nano Energy*, *50*, 298–307.
73. Zang, W., Wang, W., Zhu, D., Xing, L., & Xue, X. (2014). Humidity-dependent piezoelectric output of Al – as a self-powered active humidity sensor. *RSC Advances*, *4*, 56211–56215.
74. Nour, E. S., Echresh, A., Liu, X., Broitman, E., Willander, M., & Nur, O. (2015). Piezoelectric and opto-electrical properties of silver-doped ZnO nanorods synthesized by low temperature aqueous chemical method. *AIP Advances*, *5*, 077163.
75. Sinha, N., Ray, G., Bhandari, S., Godara, S., & Kumar, B. (2014). Synthesis and enhanced properties of cerium-doped ZnO nanorods. *Ceramics International*, *40*, 12337–12342.
76. Yadav, H., Sinha, N., Goel, S., & Kumar, B. (2016). Eu-doped ZnO nanoparticles for dielectric, ferroelectric, and piezoelectric applications. *Journal of Alloys and Compounds*, *689*, 333–341.
77. Goel, S., Sinha, N., Yadav, H., Abhilash, J., & Kumar, B. (2017). Ferroelectric Gd-doped ZnO nanostructures: Enhanced dielectric, ferroelectric, and piezoelectric properties. *Materials Chemistry and Physics*, *202*, 56–64.
78. Chen, Y. J., Brahma, S., Liu, C. P., & Huang, J. (2017). Enhancement of the piezoelectric coefficient in hexagonal MgxZn1-xO films at lower Mg compositions. *Journal of Alloys and Compounds*, *728*, 1248–1253.
79. Wang, X. B., Song, C., Li, D. M., Geng, K. W., Zeng, F., & Pan, F. (2006). The influence of different doping elements on microstructure, piezoelectric coefficient, and resistivity of sputtered ZnO film. *Applied Surface Science*, *253*, 1639–1643.
80. Yang, Y. C., Song, C., Wang, X. H., Zeng, F., & Pan, F. (2008). Giant piezoelectric d33 coefficient in ferroelectric vanadium-doped ZnO films. *Applied Physics Letters*, *92*, 15–17.
81. Yang, Y. C., Song, C., Wang, X. H., Zeng, F., & Pan, F. (2008). Cr-substitution-induced ferroelectric and improved piezoelectric properties of Zn1−xCrxO films. *Journal of Applied Physics*, *103*, 074107.
82. Luo, J. T., Yang, Y. C., Zhu, X. Y., Chen, G., Zeng, F., & Pan, F. (2010). Enhanced electromechanical response of Fe-doped ZnO films by modulating the chemical state and ionic size of the Fe dopant. *Physical Review B*, *82*, 014116.
83. Feng, P. A. N., Luo, J., Yang, Y., Wang, X., & Zeng, F. (2012). Giant piezoresponse and promising application of environmentally friendly small-ion-doped ZnO. *Science China Technological Sciences*, *55*, 1–16.
84. Goel, S., Sinha, N., Hussain, A., Joseph, A. J., & Kumar, B. (2018). Di-/piezo-/ferroelectric characterizations of 3D hierarchical sisal-like Eu3+/Gd3+ co-doped ZnO microflowers assembled with 1D nanopencils. *Ionics*, *25*, 1373–1386.
85. Li, X., Sun, M., Wei, X., Shan, C., & Chen, Q. (2018). 1D piezoelectric material-based nanogenerators: Methods, materials, and property optimization. *Nanomaterials*, *8*, 188.
86. Kasi, A. K., Najma, B., Kasi, J. K., & Bokhari, M. (2019). Fabrication of piezoelectric nanogenerator using 3D-ZnO nanosheets and optimization of charge storage system. *Materials Research Bulletin*, *123*, 110711.
87. Nourafkan, M., Mohammadi, E., & Manavizadeh, N. (2019). Influence of the ZnO nanostructures shape on piezoelectric energy harvesters performance. *IEEE Transactions on Electron Devices*, *66*, 4989–4996.
88. Peng, F., Xie, Z., Li, H., Kai, X., Wang, W., & Wu, C. (2024). Optimization of piezoelectric CO_2 reduction on ZnO via α-Fe2O3 decoration for enhanced activity and selectivity. *Catalysis Letters*, *154*(10), 5271–5279.

89. Islam, J., Lee, H., Lee, K., Cho, C., & Kim, B. (2023). Piezoelectric nanogenerators fabricated using spin coating of poly(vinylidene fluoride) and ZnO composite. *Nanomaterials*, *13*, 1289.
90. Al-Ruqeishi, M. S., Mohiuddin, T., Al-Habsi, B., Al-Ruqeishi, F., Al-Fahdi, A., & Al-Khusaibi, A. (2022). Piezoelectric nanogenerator based on ZnO nanorods. *Arabian Journal of Chemistry*, *12*, 5173–5179.
91. Kumar, B., Lee, K. Y., Park, H., Chae, S. J., Lee, Y. H., & Kim, S. (2011). Controlled growth of semiconducting nanowire, nanowall, and hybrid nanostructures on graphene for piezoelectric nanogenerators. *ACS Nano*, *5*, 4197–4204.
92. Kammel, R. S., & Sabry, R. S. (2019). Effects of the aspect ratio of ZnO nanorods on the performance of piezoelectric nanogenerators. *Journal of Science: Advanced Materials and Devices*, *4*, 420–424.
93. Desai, A. V., & Haque, M. A. (2014). Mechanical properties of ZnO nanowires. *Sensors and Actuators A: Physical*, *134*, 169–176.
94. Dash, P., Manna, A., Mishra, N. C., & Varma, S. (2018). Synthesis and characterization of aligned ZnO nanorods for visible light photocatalysis. *Physica E: Low-dimensional Systems and Nanostructures*, *107*, 38–46.
95. Xie, Y., Li, Z., Xu, Z., & Zhang, H. (2011). Preparation of coaxial TiO2/ZnO nanotube arrays for high-efficiency photo-energy conversion applications. *Electrochemistry Communications*, *13*, 788–791.
96. Sun, B. Y., Fuge, G. M., Fox, N. A., Riley, D. J., & Ashfold, M. N. R. (2005). Synthesis of aligned arrays of ultrathin ZnO nanotubes on a Si wafer coated with a thin ZnO film. *Advanced Materials*, *17*, 2477–2481.
97. Xi, Y., Song, J., Xu, S., Yang, R., Gao, Z., & Lin, Z. (2009). Growth of ZnO nanotube arrays and nanotube-based piezoelectric nanogenerators. *Journal of Materials Chemistry*, *19*, 9260–9264.
98. Khan, A., Hussain, M., Nur, O., & Willander, M. (2014). Mechanical and piezoelectric properties of zinc oxide nanorods grown on conductive textile fabric as an alternative substrate. *Journal of Physics D: Applied Physics*, *47*, 345102.
99. Liu, C., Zhang, W., Sun, J., Wen, J., Yang, Q., Cuo, H., Ma, X., & Zhang, M. (2014). Piezoelectric nanogenerator based on a flexible carbon-fiber/ZnO–ZnSe bilayer structure wire. *Applied Surface Science*, *322*, 95–100.
100. Yang, D., Qiu, Y., Jiang, Q., Guo, Z., Song, W., Xu, J., Zong, Y., & Feng, Q. (2017). Patterned growth of ZnO nanowires on flexible substrates for enhanced performance of flexible piezoelectric nanogenerators. *Applied Physics Letters*, *110*, 063901.
101. Aleksandrova, M., Kolev, G., Vucheva, Y., & Pathan, H. (2017). Characterization of piezoelectric microgenerator with nanobranched ZnO grown on a polymer coated flexible substrate. *Applied Sciences*, *7*(9), 890.
102. Voiculescu, I., Li, F., Kowach, G., Lee, K., Mistou, N., & Kastberg, R. (1982). Stretchable piezoelectric power generators based on ZnO thin films on elastic substrates. *Micromachines*, *10*, 1–7.
103. Hsu, C.-L., & Chen, K.-C. (2012). Improving piezoelectric nanogenerator comprises ZnO nanowires by bending the flexible PET substrate at low vibration frequency. *Journal of Physical Chemistry C*, *116*, 9351–9355.
104. Zhang, X., Han, M., Wang, R., Zhu, F., Li, Z., Wang, W., & Zhang, H. (2013). Frequency-multiplication high-output triboelectric nanogenerator for sustainably powering biomedical microsystems. *Nano Letters*, *13*, 1168–1172.
105. Choi, M., Murillo, G., Hwang, S., Kim, J. W., Jung, J. H., Chen, C. Y., & Lee, M. (2017). Mechanical and electrical characterization of PVDF-ZnO hybrid structure for application to nanogenerator. *Nano Energy*, *33*, 462–468.

106. Motora, K. G., Wu, C.-M., Rani, G. M., Yen, W.-T., & Lin, K.-S. (2023). Effect of electrode patterns on piezoelectric energy harvesting property of zinc oxide polyvinylidene fluoride-based piezoelectric nanogenerator. *Renewable Energy*, *217*, 119208.
107. Yévenes, C. F. G., Wongkaew, N., Ngamchana, S., & Surareungchai, W. (2022). Exploring interdigitated electrode arrays screen-printed on paper substrates for steady-state electrochemical measurements. *Journal of the Electrochemical Society*, *169*, 103502.
108. Crowley, K., Morrin, A., Shepherd, R. L., Wallace, G. G., Smyth, M. R., & Killard, A. J. (2010). Fabrication of polyaniline-based gas sensors using piezoelectric inkjet and screen printing for the detection of hydrogen sulfide. *IEEE Sensors Journal*, *10*, 1419–1426.
109. Ko, Y. H., Lee, S. H., Yu, J. S., & Su, J. (2014). Performance enhanced piezoelectric ZnO nanogenerators with highly rough Au electrode surfaces on ZnO submicrorod arrays. *Applied Physics Letters*, *103*, 022911.
110. Rovisco, A. I. B., Branquinho, R., Pinto, J. V., Martins, R., Fortunato, E., & Barquinha, P. (2020). Hydrothermal synthesis of zinc tin oxide nanostructures for photocatalysis, energy harvesting, and electronics. In K. Krishnamoorthy (Ed.), *Novel Nanomaterials*. IntechOpen.
111. Pathirane, M. K., & Wong, W. S. (2016). Optical and Electrical Characteristics of Hybrid ZnO Nanowire/a-Si: H Solar Cells on Flexible Substrates under Mechanical Bending. *Small*, *12(19)*, 2554–2558.
112. Lockman, Z., Pet, Y., Wai, T., Ibrahim, K., & Abdul, K. (2010). Formation of self-aligned ZnO nanorods in aqueous solution. *Journal of Alloys and Compounds*, *493*, 699–706.
113. Mahmud, A., Khan, A. A., Voss, P., Das, T., Abdel-Rahman, E., & Ban, D. (2018). A high performance and consolidated piezoelectric energy harvester based on 1D / 2D hybrid zinc oxide nanostructures. *Advanced Materials Interfaces*, *5*, 1801167.
114. Thilagavathi, T., & Geetha, D. (2014). Nano ZnO structures synthesized in presence of anionic and cationic surfactant under hydrothermal process. *Applied Nanoscience*, *4*, 127–132.
115. Zare, M., Namratha, K., Byrappa, K., Surendra, D. M., Yallappa, S., & Hungund, B. (2018). Their antimicrobial and antioxidant properties. *Journal of Materials Science & Technology*, *34*, 1035–1043.
116. Suh, D.-I., Byeon, C. C., & Lee, C.-L. (2010). Synthesis and optical characterization of vertically grown ZnO nanowires in high crystallinity through vapor–liquid–solid growth mechanism. *Applied Surface Science*, *257*, 1454–1456.
117. Poduri, S., & Dutta, M. (2015). Growth and characterization of indium oxide, zinc oxide and cadmium sulfide nanowires by vapor-liquid-solid growth technique. *Applied Physics Research*, *6*, 55–63.
118. Zhao, Q. X., Klason, P., & Willander, M. (2007). Growth of ZnO nanostructures by vapor–liquid–solid method. *Applied Physics A: Materials Science & Processing*, *30*, 27–30.
119. Rogers, D. J., Teherani, F. H., McClintock, R., Bayram, C., Razeghi, M., Drouhin, H. J., Clochard, M. C., Sallet, V., Garry, G., & Falyouni, F. (2009). Comparison of ZnO nanostructures grown using pulsed laser deposition, metal organic chemical vapor deposition, and physical vapor. *American Vacuum Society*, *27*, 1678–1683.
120. Opel, M., Geprägs, S., Althammer, M., Brenninger, T., & Gross, R. (2014). Laser molecular beam epitaxy of ZnO thin films and heterostructures. *Journal of Physics D: Applied Physics*, *47*(3), 034002.

5 Antimicrobial Characteristics of ZnO Nanostructures

Shubham Kumar Patial, Anuj Sharma, Neelam Vishwakarma, and Suman Singh

5.1 INTRODUCTION

Overusing antimicrobial agents has not only led to the alarming rise in drug-resistant microbial strains but also immensely escalated the vulnerability to infectious diseases, such as typhoid fever, dysentery, cholera, and diarrhea, thereby causing mass population deaths worldwide [1]. This rise in microbial resistance has also further intensified the horizontal gene transfer mechanisms [2]. The resulting condition presents an urgent need for developing new treatment strategies and inventions to mitigate the adverse effects of multidrug-resistant microbial strains. Antimicrobial agents also play vital roles in various industries, including food, packaging, cosmetics, medical, and healthcare [3]. The inorganic antimicrobial agents, especially metal oxides, offer advantages over organic agents due to their higher stability and improved safety. There is still a dire need for the evolution of antimicrobial agents that could neutralize multidrug-resistant microbes for a broad spectrum of strains.

In the ever-evolving landscape of nanotechnology, the exploration of antimicrobial properties has emerged as a pivotal area of research. Nanomaterials have revolutionized the fight against microbial pathogens due to their enhanced surface area-to-volume ratio and tailored physicochemical properties [4]. These attributes also endow them with exceptional antimicrobial capabilities, enabling targeted and efficient eradication of a wide range of pathogens, including bacteria, viruses, and fungi, while amplifying their efficacy in comparison to their bulk counterparts. The utilization of nanomaterials has offered promising avenues for combating microbial threats, owing to their unique physicochemical characteristics at the nanoscale. These materials include ZnO, TiO_2 [5], Fe_2O_3 [6], CaO [7], CuO [6], MgO [8], etc. Among these, zinc oxide (ZnO) nanostructures have garnered significant attention for their remarkable antimicrobial efficacy [9].

Zinc oxide nanostructures stand out prominently in the search for antimicrobial agents while offering multifaceted benefits. Their inherent properties, such as high surface area, photocatalytic activity, and biocompatibility, make them ideal candidates for diverse antimicrobial applications [10, 11]. ZnO nanostructures exhibit potent antimicrobial activity against both Gram-positive and Gram-negative bacteria like

DOI: 10.1201/9781003464907-5

Staphylococcus aureus, *Aeromonas hydrophila*, *Salmonella enteritidis*, *Escherichia coli*, etc., as well as viruses and fungi [12]. Moreover, their compatibility with various substrates enables their integration into medical devices, textiles, packaging materials, and environmental remediation systems, thereby presenting versatile solutions for combating microbial proliferation and safeguarding public health [13]. It serves as a filler in rubber composites, an additive in rubber vulcanization, and a component in sunscreens, coatings, optoelectronic devices, sensors, photocatalysts, and antimicrobial wound dressings [14]. In the rubber industry, ZnO enhances sulfur vulcanization efficiency. Its properties as an n-type semiconductor, with zinc interstitials and oxygen vacancies, contribute to a large exciton binding energy (60 meV) and wide bandgap energy (3.37 eV), making it suitable for semiconductor devices. ZnO's high dielectric permittivity also makes it attractive for gate dielectric applications [15].

ZnO's wide bandgap allows it to effectively absorb UV light, providing antibacterial and UV-protection properties beneficial in photodetectors, laser diodes, sensors, and solar cells [16]. ZnO nanoparticles can be integrated into polymers, textiles, and packaging films for UV protection in food and medical applications. Incorporating ZnO into polymers, which are easy to process and have good mechanical properties, results in functional nanocomposites with UV-shielding capabilities [17]. ZnO also exhibits selective toxicity to bacteria, making it practical for antibacterial applications in water disinfection, food preservation, and medical wound dressings. Its multifunctional properties and adaptability make ZnO a crucial material in advancing technology and healthcare.

Several mechanisms explain how ZnO nanomaterials disrupt bacterial cell walls, including metal-ion release, generation of reactive oxygen species (ROS), membrane dysfunction, and nanoparticle penetration [18, 19]. Nanostructured ZnO, with its 3.28 eV bandgap and high exciton binding energy, induces the photogeneration of ROS such as superoxide free radicals, hydroxyl free radicals, and H_2O_2 from adsorbed oxygen and water molecules. These ROS damage cellular components, causing oxidative stress. The antibacterial properties of ZnO nanomaterials are influenced by their size, morphology, and specific surface area, with increased surface area enhancing antibacterial effectiveness [20]. Controlling the morphology with surfactants prevents agglomeration and stabilizes nanoparticles through steric or electrostatic repulsion. Functionalizing the ZnO surface with polar groups further improves interactions with bacterial cell walls [21].

This chapter aims to highlight the antimicrobial characteristics of zinc oxide while providing a comprehensive review of its properties, which offer advantages over existing materials. The chapter focuses on the various methods used to synthesize ZnO nanostructures. By depicting their mode of action, including ROS-mediated damage, membrane disruption, and intracellular effects, the chapter aims to provide insights into the underlying principles driving their antimicrobial efficacy (Figure 5.1). It includes a detailed description of the mechanisms involved in the antimicrobial actions of ZnO and an extensive overview of its wide range of applications. Additionally, this chapter addresses the ethical and safety concerns related to the in vitro and in vivo use of zinc oxide, the challenges faced in practical and real-world applications, and the untapped opportunities for researchers to explore.

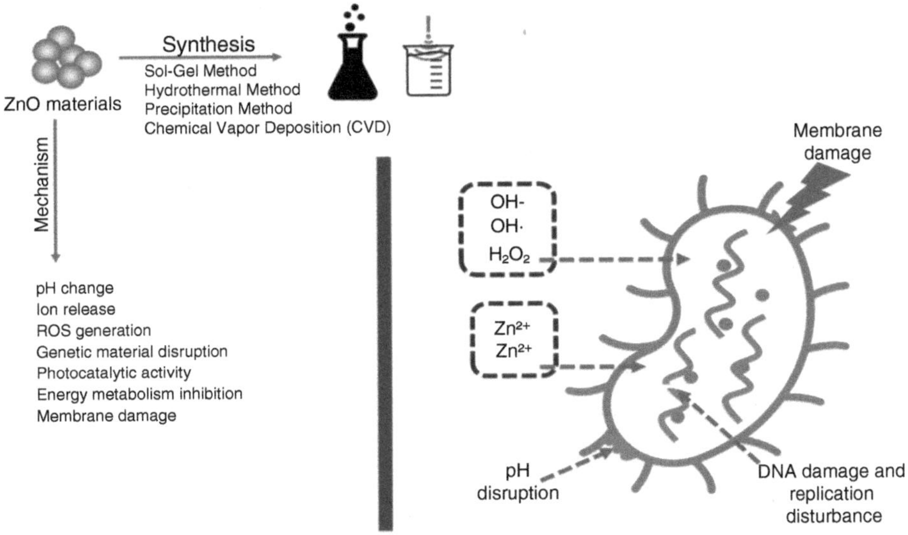

FIGURE 5.1 Schematic of the antimicrobial action of nanoparticles on bacteria. Nanoparticles disrupt the cell wall, generate ROS, and interfere with cellular components, leading to bacterial cell death.

5.2 SYNTHESIS OF ZnO NANOSTRUCTURES

This approach details the methods for fabricating ZnO nanostructures (ZONSs) aimed at antimicrobial applications. Figure 5.2 illustrates a general flow chart depicting various synthesis routes that have been effectively employed to create different types of ZONSs. Synthesis of ZnO nanostructures employs various techniques, each offering distinct advantages for tailoring properties like size, shape, and crystallinity. Some of the techniques are discussed below.

5.2.1 Sol–Gel Method

The sol–gel process is a versatile technique involving the transition of a solution system from a liquid "sol" into a solid "gel" phase. Zinc salts (such as zinc acetate) are initially dissolved in a solvent and reacted with a stabilizing agent, often alcohol. Hydrolysis and polycondensation reactions lead to the formation of a colloidal suspension, which gradually evolves into a gel. This gel is then dried and calcined to produce ZnO nanoparticles. The sol–gel method allows for excellent particle size and morphology control and is suitable for large-scale production [22–24].

5.2.2 Hydrothermal Method

The hydrothermal technique involves crystallizing ZnO nanostructures from aqueous solutions at elevated temperatures and pressures in an autoclave. Zinc salts and a base (like NaOH) are mixed in water and heated above the boiling point,

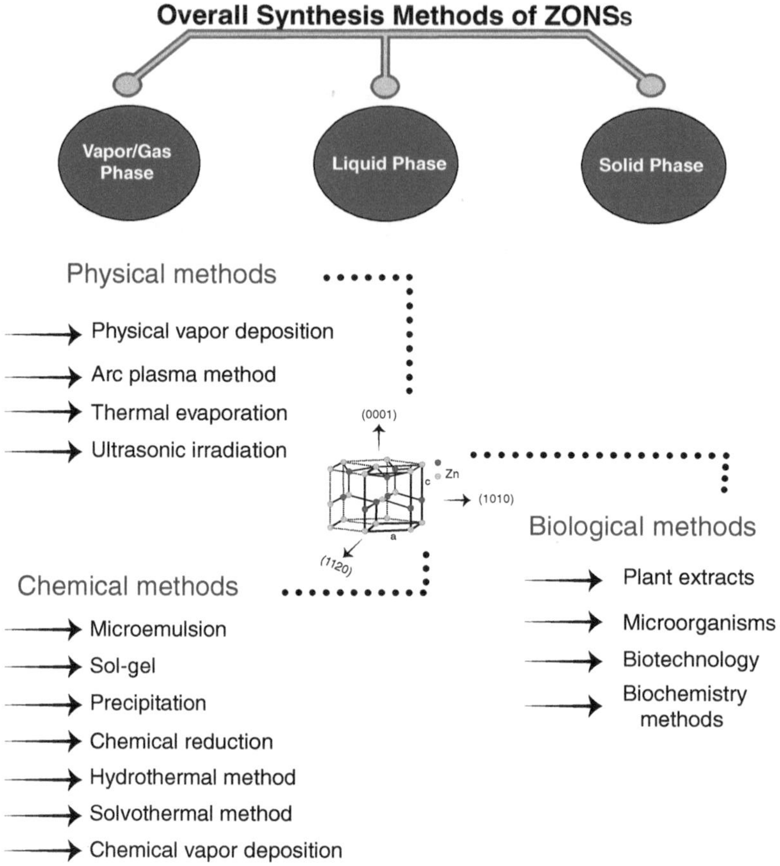

FIGURE 5.2 Overview of synthesis methods for producing ZnO nanostructures (ZONSs) [34].

promoting nucleation and growth of ZnO crystals. This method is particularly effective for producing well-defined nanostructures, such as nanorods, nanowires, and nanoflowers. The hydrothermal method's advantages include low temperature, environmentally friendly conditions, and high crystallinity of the produced nanostructures [25, 26].

5.2.3 Precipitation Method

ZnO nanostructures are formed in the precipitation technique by mixing zinc salt solutions with a precipitating agent (such as NaOH or NH_4OH). The reaction leads to the formation of a zinc hydroxide intermediate, which is then converted into ZnO upon heating. This method is straightforward, cost-effective, and suitable for producing large quantities of ZnO nanoparticles. Control over particle size and morphology can be achieved by adjusting parameters such as concentration, temperature, and pH[27, 28].

5.2.4 Chemical Vapor Deposition (CVD)

The CVD method involves the chemical reaction of gaseous precursors, typically zinc-containing compounds (like diethylzinc) and oxygen, in a heated reactor to form ZnO nanostructures. The reaction occurs on a substrate, allowing the formation of high-purity, crystalline ZnO films, or nanostructures like nanowires. CVD offers precise control over the growth process and is widely used in the semiconductor industry for fabricating high-quality nanostructures [29, 30].

5.3 ANTIMICROBIAL MECHANISMS OF ZnO NANOSTRUCTURES

The antimicrobial activity of ZnO NSs increases with a decrease in size, which is a consequence of light irradiation [31]. Direct contact with the cell wall disrupts the cell wall, releasing antimicrobial Zn^{2+} ions that, in turn, generate ROS. Sawai et al. reported the superior antibacterial activity of ZnO as compared to MgO and CaO [32]. Jin et al. systematically assessed the change in antimicrobial activity based on the morphology of ZnO used. They reported that ZnO NSs have better antimicrobial activity than different forms of ZnO like powder, film, capped, and coated [33]. The antimicrobial activity of ZnO nanostructures, or any other metal oxide, works on physical and chemical mechanisms. The physical mechanism does not require the dissolution of ZnO NSs into Zn^{2+} ions and works on structure attachment on membranes. Chemical mechanisms, on the other hand, require oxidative stress that disrupts cell functioning. The oxidative stress is provided by ZnO dissolution into Zn2+ ions and ROS [35–37]. Figure 5.3 briefly depicts the mechanisms involved in the antimicrobial activity of ZnO nanostructures.

5.3.1 Physical Mechanisms (e.g., Membrane Damage and Cell Penetration)

In physical mechanisms of antimicrobial activity, direct contact of ZnO NSs with the cell wall/membrane by electrostatic interactions destabilizes the membrane structure [38]. The ability to attach and introduce Zn^{2+} ions inside the cell wall, thus disrupting cell wall integrity and penetrating it, constitutes the physical mechanisms by which ZnO demonstrates its antimicrobial activity [39]. Attaching to and fixating on the cell wall leads to a change in the surface tension of the cell wall, which initiates cell wall disruption [40]. ZnO NSs do irreparable damage to cell membranes and, upon penetration, attach themselves to cell organelles, thus disrupting their membranes and contributing to their microbicidal activity. Once the membranes are disrupted, these nanostructures release Zn^{2+} ions, which undergo chemical mechanisms of antimicrobial activity [35, 36, 41]. The attachment of ZnO NSs relies on the cell signaling that responds to zinc availability. As discussed earlier, zinc is an essential metal that supports life up to a certain concentration, above which it shows cytotoxic effects. The attachment is accompanied by penetration of NSs inside the cell and initiation of chemical mechanisms of antimicrobial activity [42]. Apart from the chemical mechanisms, the nanostructures attach to membranes of cell organelles and perform mechanisms similar to those of cell walls. These structures can also attach to the DNA of the microbe, which leads to inhibited replication and DNA damage.

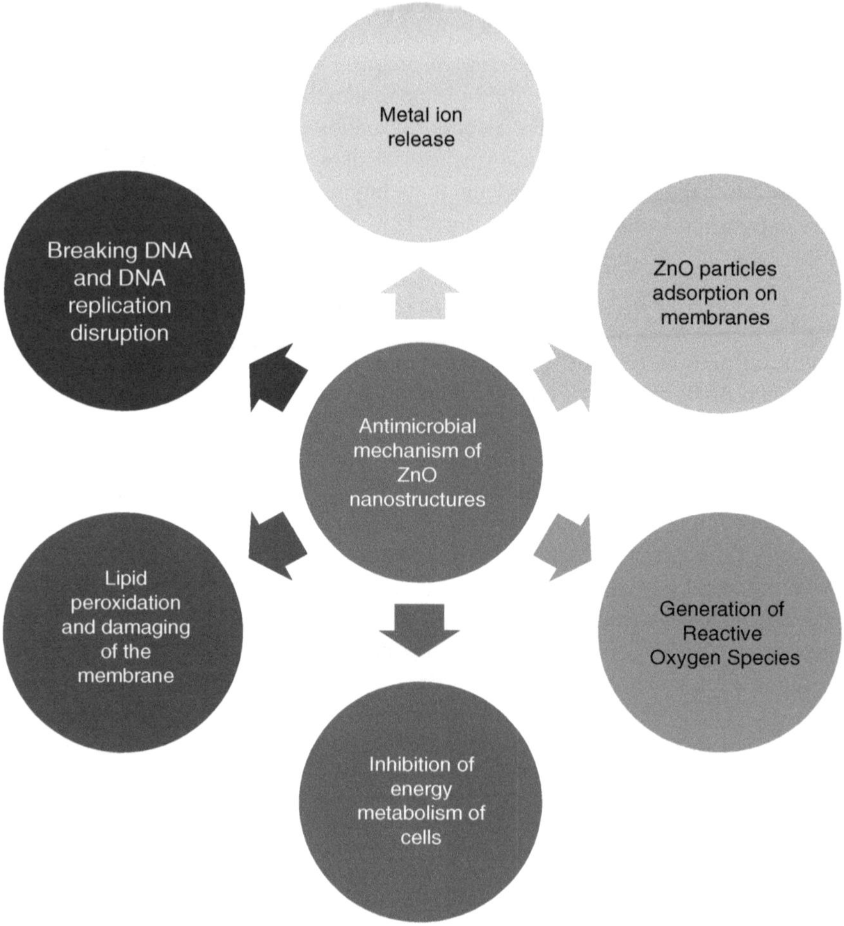

FIGURE 5.3 Mechanism of antimicrobial activity of ZnO nanostructures.

5.3.2 Chemical Mechanisms (e.g., ROS Generation)

Even a little acidic shift in pH values leads to rapid dissolution of ZnO NSs into Zn^{2+} ions. These ions interact and produce ROS, altering the metabolic and enzymatic processes [43]. The oxidative stress generated by ZnO NSs significantly affects their antimicrobial activity. The absorption of Zn^{2+} ions above the threshold limit does irreversible damage to cell organelles and DNA [44]. It generates ROS, which increases oxidative stress and inhibits the respiratory enzymes that lead to cell death. The physical and chemical mechanisms of antimicrobial activity go hand in hand and depend on each other for successful antimicrobial activity [45]. The chemical mechanism of ZnO nanostructures is biocidal and works in the conventional way of generating ROS for cell damage and death [46]. Figure 5.4 shows the collective antimicrobial mechanism of ZnO nanostructures, from attachment to cell death by ROS generation and DNA damage.

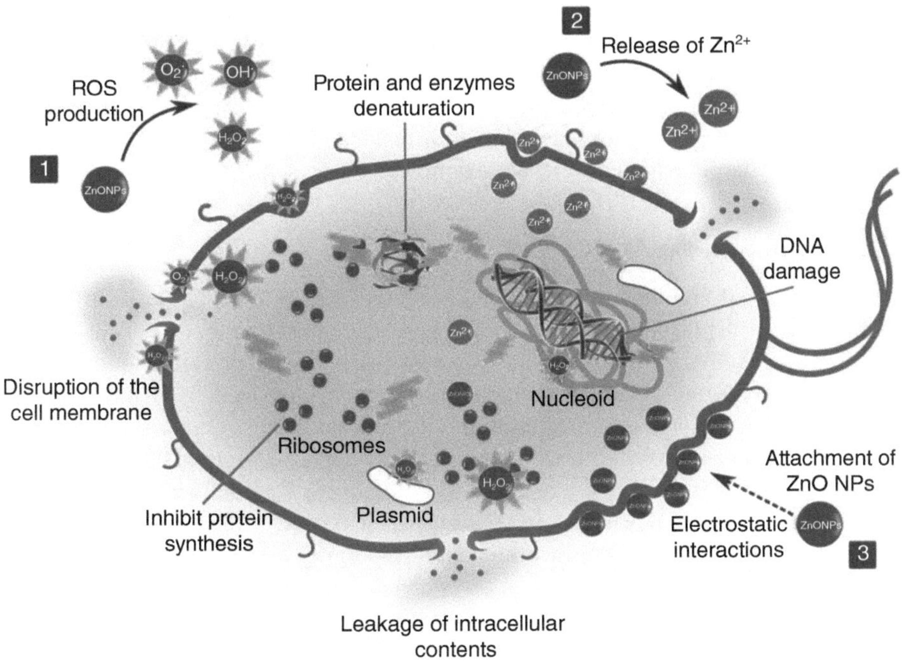

FIGURE 5.4 Illustration of antimicrobial activity of ZnO NSs against a microbial cell. The mechanism involves (i) ROS generation, (ii) dissolution of ZnO NSs into Zn^{2+} ions, and (iii) interaction between NSs and cell membrane [47].

5.4 ANTIMICROBIAL APPLICATIONS OF ZnO NANOSTRUCTURES

Zinc is vital in human physiological pathways like bone metabolism, antioxidative processes, blood clotting, and enzyme activation for protein and nucleic acid synthesis and digestion [48]. At higher concentrations above the limit, Zn^{2+} can show toxicity effects as homeostasis processes the uptake of Zn, but they do not control membrane adsorption of zinc in cells [49]. Thus, increasing Zn concentration above optimal levels disturbs the Zn^{2+} homeostasis, and Zn^{2+} ions infiltrate cells, leading to cytotoxicity and cell death. These reasons constitute the premise of using Zn^{2+} ions as an antimicrobial agent [50]. Furthermore, ZnO NSs are biocompatible and demonstrate high stability under physiological conditions. The antimicrobial activity of ZnO nanostructures is size-dependent, and the size of the structure influences both the kinetics of ion release and the production of ROS in the solution [40]. A study by Raghupati et al. demonstrated the correlation between antimicrobial activity and the size of the ZnO nanostructure [51]. The study systematically assessed seven different sizes of ZnO NSs against Gram-negative and positive bacteria. It concluded that the small size of ZnO NSs corresponds to higher bacterial inhibitions. Another study by Rajat et al. assessed the role of shape in the antimicrobial activity of ZnO nanostructures [52]. They experimented on different shapes of the NSs (flower-like, hexagon, spherical, rod-like, etc.). They found that the antimicrobial activity decreases for Flower-like > Spherical > rod-like > hexagon ZnO NSs. This holds true when UV irradiation is used; the "point effect" phenomenon strengthens the ROS catalytic activity. The antimicrobial application of ZnO nanostructures can be classified under three microbial categories: antibacterial, antifungal, and antiviral [53]. These categories are briefly explained with relevant examples.

5.4.1 Antibacterial Applications

ZnO NSs present significant bactericidal potential against both Gram-positive and negative bacteria [48]. In the presence of green synthesis methods of ZnO nanostructures, structures synthesized via biogenic methods are preferred over chemically synthesized ones [50]. In a study, ZnO NSs synthesized from the biogenic method were used for its antibacterial application, and the structure demonstrated higher inhibition zones for both Gram-positive and negative bacteria than previously reported [54]. Researchers have recently tried to couple ZnO NSs with other materials to develop hybrid systems. In this regard, Rokbani et al. coupled ZnO NSs with polyethylene films to produce bioactive packaging material with antibacterial activity [55]. Similarly, ZnO NSs can be coupled with various materials to provide antimicrobial activity to those materials. Another study led by Hari et al. proposed the incorporation of ZnO NSs with cross-linked pectin-based films [56]. These studies aimed to replace traditional packaging materials with more biocompatible and sustainable materials. The conjugation of ZnO NSs with different films has been demonstrated to be environmentally friendly and sustainable [57]. For instance, ZnO NSs incorporated with polytetrafluoroethylene (PTFE) have been proposed for application in multifunctional smartphone coatings with antireflective, hydrophobic, and antimicrobial properties [58]. These hybrid systems have also seen growth in the biomedical field. Wang et al. proposed the use of carbon-fiber-reinforced poly(ether ether ketone) (CFRPEEK) for orthopedic and dental implants [59]. The inhibition of implant-associated infections was done by distributing ZnO NSs evenly throughout the matrix. Even though ZnO NSs demonstrate good antibacterial activity, resistance toward these structures has been reported. The resistance can be attributed to electrostatic repulsion, ion efflux pumps, expression of extracellular matrices, biofilm adaptation, and gene mutations [60–62].

5.4.2 Antifungal Applications

The ZnO NSs have shown antifungal properties against many harmful fungi and yeasts, but the sensitivity pattern varies due to their distinct biological structures. A study done by Sun et al. demonstrated ROS generation and poisoning effect due to Zn^{2+} release [63]. They concluded that these two mechanisms were the two main contributions to the antifungal activity of ZnO NSs. The fungicidal mechanism of ZnO NSs is attributed to the disruption of cellular structure, prohibition of biological macromolecular activity, prevention of DNA replication, and disruption of the antioxidative system that arises in ROS-Zn^{2+} mediated poisoning [64]. These methods are similar to the antibacterial mechanisms. Table 5.1 depicts the varied use of ZnO

TABLE 5.1
Synthesis of ZnO Nanostructures and Their Antimicrobial Activity for Different Bacterial Strains

S. No	Size (nm)	Synthesis Method	Composition	Concentration	Microbial Species	Exposure Time	Medium	Ref.
1	3	Sol–gel method	Metal oxide core surrounded by a shell of surface groups	0.5–1.5 M	*E. coli, P. aeruginosa, S. aureus*	24 hrs	TCS	[66]
2	<50	Chemical synthesis followed by heat treatment	ZnO	0.78–50 mM	*V. cholerae, E. coli, C. jejuni, S. aureus (MRSA)*	48 hrs	LB medium, MHA	[67]
3	50–70	ZnO, chitosan	Microwave heating using chitosan as a stabilizing agent	30 µL/ml	*S. aureus, E. coli*	24 hrs	TSA	[68]
4	18	ZnO	The sol–gel method followed by heat treatment	10–100 µg/ml	*E. coli, S. aureus, P. aeruginosa, B. subtilis,*	24 hrs	-	[69]
5	10	ZnO	Chemical deposition method	1–64 µg/ml	*M. tuberculosis*	24 hrs	LJ medium	[70]
6	50; 1000	ZnO; spindle-shaped GO covered with ZnO crystals	Modified hammers method	100 µM	*E. coli, S. typhimurium, B. subtilis, E. faecalis*	12 hrs	LB media	[71]
7	18	ZnO	Chemical method using *A. indica* (water extract) followed by heat treatment	50, 100, 200 µg/ml	*E. coli, S. typhimurium, B. subtilis, P. mirabilis, E. coli, C. albicans, C. tropicalis*	24 hrs	MHA for bacteria, sabouraud dextrose agar media for fungi	[72]
8	10	ZnO	Chemical synthesis	10–5187 µM	*S. aureus, P. aeruginosa, E. coli, C. albicans, A. niger, T. rubrum*	5 days-3 weeks for fungi, 18 hrs for bacteria	SDA, BHI, MH broth, SD broth	[73]

(*Continued*)

TABLE 5.1 (CONTINUED)
Synthesis of ZnO Nanostructures and Their Antimicrobial Activity for Different Bacterial Strains

S. No	Size (nm)	Synthesis Method	Composition	Concentration	Microbial Species	Exposure Time	Medium	Ref.
9	33	ZnO	Chemical followed by mechanical and heat treatment	10 mg/ml	*E. coli, S. choleraesuis, B. subtilis*	48 hrs	MHA	[74]
10	20	ZnO; ZnO + gelatin	Chemical followed by heat treatment	25; 50 μg/ml	*P. aeruginosa, E. faecalis, C. albicans*	24 hrs	NA	[75]
11	23–26	ZnO	Chemical synthesis	62.5–100 μg/ml	*K. pneumonia, S. aureus, C. albicans, P. notatum*	-	MHA	[76]
12	60–70	ZnO	Chemical method followed by heat treatment	125–1028 μg/ml	*S. aureus, P. aeruginosa, E. coli*	18 hrs	MHA	[77]
13	22	ZnO	Sol–gel method	0.5 mg	*B. subtilis, S. mutans, S. aureus, E. coli, P. aeruginosa, K. oxytoca*	24 hrs	MHA	[78]
14	20–25	ZnO	Chemical deposition	20–100 μg/ml	*S. aureus, Styphimurium, A. flavus, A. fumigates*	24 hrs	MHB, LB broth	[79]
15	42–64	ZnO	Biosynthesis of ZnO nanoparticles using *A. hydrophila* as a reducing agent	1.2–25 μg/ml	*A. hydrophila, E. coli, S. aureus, P. aeruginosa, E. faecalis, S. pyogenes, A. flavus, A. niger, C. albicans*	24	MHA	[80]
16	5	ZnO-NP bound to polystyrene film	Sol–gel method followed by heat treatment	0.1–0.5 mg/ml	*L. monocytogenes, S. enteritidis, E. coli*	48 hrs	LEW, BHI, TSB	[81]
17	5	ZnO	Chemical precipitation	0.09–1.5 mg/ml	*S. aureus, E. coli*	24 hrs	TSB	[82]
18	20	ZnO, Ag	Chemical synthesis using *T. vulgaris* leaf extract	20–70 μg/ml	*E. coli, S. aureus*	24 hrs	NA, LB broth	[83]
19	125	ZnO	Chemical synthesis using *T. grandis* extract	20–100 μg/ml	*S. aureus, B. subtilis, E. coli, S. paratyphi*	24 hrs	MHA	[84]
20	10–60	ZnO	Sol–gel method using aqueous extract of *C. procera* leaves	1–5 mM	*A. baumannii*	12 hrs	LB broth	[85]

nanostructures and their antimicrobial activity against different species of bacteria [65]. A study by Alhazmi et al. demonstrated the antifungal activity, which proved an impactful antifungal effect against isolates of azole-resistant *A. flavus*. In varied applications, ZnO-based materials have been used as an antifungal coating for stone artworks in the cultural heritage field.

5.4.3 Antiviral Applications

Viruses cannot spread outside the body, and antiviral supplication of material is solely studied for medicinal/biomedical applications [51]. ZnO nanostructures exhibit antiviral activity against a variety of viruses. The mode of action is a simulated so-called adaptive immune response that governs the toll-like receptor signaling pathway and protein downstream [51]—the mechanism of antiviral activity results in proinflammatory cytokine secretion, which inhibits viral proliferation. The antiviral activity of ZnO NSs majorly depends on ZnO penetration for Zn^{2+} ion generation. Zn^{2+} ions actively participate in ROS generation and DNA damage that constitutes the antiviral activity of ZnO NSs as a whole [86]. ZnO nanostructures can be collaborated with other materials for efficient surface disinfection. Surface disinfection decreases surface disinfection and provides early containment of viruses by inactivating virus entry in host cells, ROS generation, hindering viral replication, and inactivating viral proteins [87]. A study done on Rotavirus by Inge et al. reported the inhibition of two essential proteins when ZnO nanostructure was introduced to the viral complex [88]. The RNA-dependent RNA polymerase (RdRp) and the 3C-like proteinase (3CL), which are involved in viral replication, were inhibited by Zn^{2+} ions. The advantage of using ZnO NSs instead of Zn^{2+} ions lies in the controlled and long-lasting Zn^{2+} release by ZnO NSs. The efficacy of antimicrobial activity depends on various factors, which are discussed in the upcoming section.

5.5 FACTORS INFLUENCING ANTIMICROBIAL EFFICACY

The antimicrobial efficacy of ZnO nanostructures is influenced by several factors, including their size, shape, surface area, and crystallinity, as well as the presence of defects and doping elements.

5.5.1 Size and Concentration

The antimicrobial efficacy of ZnO nanostructures is significantly influenced by particle size and concentration. Numerous studies have shown that the antibacterial activity of ZnO nanoparticles (ZnO-NPs) correlates directly with their concentration and inversely with their size [89]. Smaller ZnO-NPs have a larger surface area, which enhances their ability to penetrate bacterial membranes and interact with microbial cells, thus increasing their antibacterial efficiency [90].

Research by Padmavathy and Vijayaraghavan demonstrated that smaller ZnO-NPs generate more hydrogen peroxide (H_2O_2), a key factor in their antibacterial activity, due to their higher surface area. Conversely, larger particles tend to be less effective [91]. For example, Yamamoto found that ZnO-NPs with sizes ranging from

100 to 800 nm showed increased antibacterial activity against *S. aureus* and *E. coli* as their size decreased [92].

ZnO-NP concentration is crucial; higher concentrations generally lead to increased antibacterial activity. This is attributed to the greater availability of ROS and enhanced disruption of bacterial cell membranes at higher nanoparticle concentrations [93]. Studies have shown that ZnO-NPs at 50–100 mg/L concentrations can significantly disrupt bacterial growth by interfering with mitochondrial function and causing cell membrane damage [94].

ZnO-NPs with higher concentrations exhibit the most potent antimicrobial effects, making them effective candidates for various antibacterial applications, including in food packaging and biomedical devices.

5.5.2 Surface Morphology

Recent research has highlighted the significant impact of ZnO nanostructure shapes on their toxicity and antimicrobial efficacy. Synthesis conditions, including solvents, precursor types, temperature, pH, and shape-directing agents, determine the morphology of ZnO-NPs. Different morphologies exhibit varying antibacterial activities due to the presence of specific active facets [95]. For instance, rod-shaped ZnO-NPs possess (111) and (100) facets, while spherical ones mainly have (100) facets. High-atom-density (111) facets are associated with enhanced antibacterial properties [96]. Studies have shown that nanostructures with various shapes, like rods and wires, can penetrate bacterial cell walls more effectively than spherical ZnO-NPs [97]. Additionally, flower-shaped ZnO-NPs demonstrate higher biocidal activity against *S. aureus* and *E. coli* than spherical and rod-shaped forms [95]. The enhanced antibacterial activity is also attributed to polar facets with more oxygen vacancies, which increase ROS generation and improve photocatalytic performance. Specifically, ZnO morphologies with highly exposed (0001)-Zn terminated polar facets show superior antibacterial results [98]. Nanostructured ZnO materials with higher surface areas provide more sites for interaction with microbial cells, thereby enhancing antimicrobial efficacy [99]. Surface defects, such as oxygen vacancies and zinc interstitials, can promote ROS generation and contribute to antimicrobial activity [100]. Furthermore, surface functionalization or modification can be employed to tailor the interaction between ZnO nanostructures and microbial cells, enhancing their antimicrobial properties [101].

5.5.3 Minimum Inhibitory Concentration (MIC)

To discuss the minimum inhibitory concentration (MIC) and minimum bactericidal concentration (MBC) measurements, comparisons were drawn between studies by Emami-Karvani and Chehrazi [102], and Reddy et al. [103], an agar diffusion test with ZnO nanoparticles (ZnO-NPs) on *E. coli* and *S. aureus* was performed. Colony-forming units (CFU) were counted post-24-hour incubation to assess bacterial growth rates. MIC, the lowest nanoparticle concentration preventing bacterial growth, was found to be 1.5 mg/mL for *S. aureus* and 3.1 mg/mL for *E. coli*, aligning closely with Reddy et al.'s results of 1 mg/mL and 3.4 mg/mL, respectively. This highlights a higher susceptibility of Gram-positive bacteria than Gram-negative bacteria due to cell wall structure and metabolism differences. Xie et al. [104] reported MIC values of 0.05–0.25 mg/mL for *C. jejuni*, significantly lower than for *E. coli* O157:H7 and *S. enterica* (0.4 mg/mL). This demonstrates the concentration-dependent efficacy of ZnO-NPs, confirmed through varied methodologies like growth curve tests and INT reduction assays used by Salem et al. [105], revealing higher antimicrobial activity of ZnO-NPs compared to Ag-NPs.

5.5.4 Doping and Environmental Factors

Doping ZnO nanostructures with foreign ions, such as transition metals or non-metals, can modify their electronic structure and enhance antimicrobial properties [106]. For example, doping with elements like silver or copper can increase ROS generation and improve antimicrobial efficacy [107]. Environmental conditions, such as pH, temperature, and humidity, can influence the antimicrobial activity of ZnO nanostructures. Optimal conditions for ROS generation and microbial cell interaction should be considered to maximize antimicrobial efficacy [108].

5.5.5 Surface Defects

Various factors, including surface defects, surface charges, and particle orientation, influence the antimicrobial efficacy of ZnO-NPs [109]. Surface defects, such as edges and corners, enhance the reactivity of ZnO-NPs, contributing to their toxicity by causing membrane damage in bacteria. Controlling these defects and associated charge carriers can optimize ZnO-NPs' antimicrobial properties [92]. Orientation of ZnO-NPs also affects their biocidal activity, with random configurations showing higher antibacterial action than regular structures [110]. Additionally, doping ZnO with other metal nanoparticles like Ag can significantly enhance its antibacterial activity. Silver-loaded ZnO has been shown to increase ZnO antimicrobial effectiveness, suggesting that such composites could serve as potent antibacterial agents [111].

5.6 SAFETY AND TOXICITY CONSIDERATIONS

ZnO nanostructures exhibit significant antimicrobial properties but pose safety risks due to potential migration and toxicity. Comprehensive evaluation and careful control of ZnO-NP applications are necessary to mitigate health risks while harnessing their beneficial properties. Further research is needed to fully understand their interaction with biological systems and ensure safe usage in various applications.

5.6.1 Safety Issues

ZnO nanoparticles (ZnO-NPs) enhance materials' mechanical, barrier, and antimicrobial properties, but they also pose safety concerns due to their potential migration into human tissues. Migration depends on factors such as nanoparticle characteristics (concentration, size, shape), environmental conditions (mechanical stress, temperature), material properties (viscosity, structure), and duration of contact. There is no standard protocol to assess migration comprehensively, so each application must be evaluated individually.

5.6.2 *Migration*

Migration is the unintended transfer of substances from materials into the human body. It is categorized into three types.

Overall Migration Limit (OML): Measures the total amount of non-volatile substances that can migrate.
Specific Migration Limit (SML): Measures the concentration of a specific substance in a simulant.
Maximum Permitted Quantity (QM): Measures the residual substance level in contact materials.

Regulations limit zinc migration from materials, with different thresholds set by the European Commission and the National Institute of Health imposing 40 mg/day on zinc daily consumption for the human body [41]. Studies show that proper control can keep ZnO-NP migration within safe limits.

5.6.3 Toxicity of ZnO Nanoparticles

ZnO-NPs can enter cells directly or via endocytosis, generating ROS and oxidative stress, which damage biological systems. Factors influencing toxicity include nanoparticle size, shape, aggregation, and surface chemistry. ZnO-NPs' large surface area increases reactivity, leading to lipid peroxidation, protein damage, inflammation, and DNA damage. Excess ROS overwhelms cellular antioxidant defenses, causing significant oxidative stress and inflammation. Neurotoxicity arises from disrupted metal-ion homeostasis and increased intracellular zinc ion concentration [112, 113].

5.6.4 Human Health Impact

ZnO-NPs can enter the body through inhalation, skin penetration, and ingestion. Their small size allows them to cross cellular barriers, causing inflammation and oxidative damage. Once in the bloodstream, they distribute to various organs, including the brain and reproductive system. Understanding how nanomaterials interact with organs, their metabolism, and elimination is crucial for safety. Some nanomaterials, like ZnO-NPs, do not penetrate human or animal skin easily, reducing some risks.

5.7 FUTURE DIRECTIONS AND CHALLENGES

5.7.1 Emerging Trends in ZnO Nanostructure Research

ZnO nanostructures are rapidly evolving, with several promising trends emerging. One major trend is the development of ZnO nanostructures with tailored properties for specific applications. Researchers are focusing on manipulating the size, shape, and surface chemistry of ZnO nanoparticles to enhance their antimicrobial efficacy while minimizing toxicity. Another significant trend is the integration of ZnO nanostructures with other nanomaterials to create composite materials with synergistic properties. For instance, combining ZnO with silver or graphene has been shown to improve antimicrobial activity and stability [114, 115].

The use of ZnO nanostructures in innovative materials is also gaining traction. These materials can respond to environmental stimuli, such as pH, temperature, or light, to release antimicrobial agents on demand. This approach could lead to more effective and sustainable antimicrobial treatments.

5.7.2 Challenges to Overcome for Practical Applications

Despite the promising advances, several challenges need to be addressed to realize the full potential of ZnO nanostructures in practical applications. One of the primary challenges is ensuring the safety and biocompatibility of ZnO nanoparticles. While ZnO is generally considered safe, its nanoscale form can exhibit different properties, including potential toxicity. Comprehensive toxicity studies and the development of safe dosage guidelines are essential.

Another challenge is the scalability and reproducibility of ZnO nanostructure synthesis. Producing ZnO nanoparticles with consistent quality and properties on a large scale remains a significant hurdle. This issue is crucial for their reliable application in the healthcare and food packaging industries.

The stability of ZnO nanostructures in various environments also poses a challenge. Factors such as humidity, temperature, and exposure to light can affect their antimicrobial properties. Developing strategies to enhance the stability and longevity of ZnO nanostructures is necessary for their practical use.

5.7.3 Opportunities for Further Exploration and Innovation

Despite the challenges, there are numerous opportunities for further exploration and innovation in ZnO nanostructures. One promising area is the development of multifunctional ZnO-based nanomaterials that combine antimicrobial activity with other beneficial properties, such as UV protection, self-cleaning, and photocatalytic activity. These multifunctional materials could find applications in various fields, from healthcare to environmental protection.

Another exciting avenue is exploring new synthesis methods to produce ZnO nanostructures with novel morphologies and enhanced properties. Techniques such as green synthesis, which uses biological molecules for nanoparticle production, offer a sustainable and environmentally friendly alternative to traditional chemical methods.

Applying ZnO nanostructures in emerging fields, such as nanomedicine and wearable technology, holds significant potential. For instance, ZnO nanoparticles could be used in drug delivery systems to target specific pathogens or in wearable sensors to detect and respond to infections in real time.

5.8 CONCLUSION

The synthesis and characterization of ZnO nanostructures give out pristine structures of known sizes. Like any other metal/metal oxide, ZnO nanostructures have been subjected to tremendous research for their antimicrobial activity. Various methods, such as biogenic and chemical, can synthesize ZnO nanostructures. The morphology and size of these nanostructures govern their antimicrobial activity. The mechanism of microbial growth inhibition and microbicidal activity is divided into physical and chemical mechanisms. The cytotoxic effects of Zn^{2+} ions are the major reason behind chemical mechanisms, and the size of nanostructures is the major reason behind physical mechanisms. The smaller the size, the greater the penetrating strength and the greater the damage caused to the cell. Above that, Zn^{2+} ions facilitate ROS generation that causes oxidative stress, leading to cell damage and death. The advantage of using ZnO NSs over the direct application of Zn^{2+} ions lies in the stability of these structures and the controlled release of Zn^{2+} ions over long periods of time. The idea for accurately assessing the cytotoxic and detrimental effects of ZnO NSs lies in the fabrication of nanostructure formulations with known and regulated size, surface charge, and shell composition. Nonetheless, ultrasmall nanoparticles should be applied directly for antimicrobial purposes. ZnO NSs can be permanently immobilized in protective or embedded coatings to avoid direct NS release into biofluids or the environment.

The conjugation of ZnO NSs with other materials gives birth to antimicrobial films with varied applications. Whether it be packaging or application in filter technology, these nanostructures provide antimicrobial activity wherever they are applied. The primary feature of ZnO-NPs' antimicrobial packaging systems is their biocidal efficiency against both Gram-positive and Gram-negative bacteria, as well as fungi. In creating ZnO NSs-coated antimicrobial packaging, the surface of packaging materials is coated, or the NSs are added to them using solution casting and extrusion techniques. Based on the production of ROS, the disruption of cellular metabolism, and cell rupture, these systems can give ZnO-NPs antibacterial characteristics. The use of ZnO NSs has been established to be advantageous, but the overuse can turn out to be a peril after some time. As discussed earlier, these NSs demonstrate significant antimicrobial activity, and the biogenic synthesis methods promote using these materials with little to no repercussions. This short-time solution leads to cytotoxic effects in plants and humans, too.

Furthermore, the growing AMR issue outruns the discovery or development of novel antimicrobial molecular medications/systems. It has been established that NSs work better against bacteria than fungi and viruses. They even work for bacteria that are resistant to antibiotics. However, if real-world items containing NSs are utilized without rules or protocols, the nano-resistance problem may become a burden. Proper regulation and monitoring of the use of NSs will diminish or at least limit the resistance problem. As a result, in order to provide accurate and repeatable results, further mechanistic investigations into the cytotoxicity and antibacterial activity of ZnO NSs should be carried out utilizing established procedures. By optimizing the synthesized nanomaterial, it will be feasible to create safer and more effective ZnO nanostructures that are less vulnerable to antimicrobial resistance.

REFERENCES

1. Llor, C. and L. Bjerrum, Antimicrobial resistance: risk associated with antibiotic overuse and initiatives to reduce the problem. *Therapeutic Advances in Drug Safety*, 2014. **5**(6): p. 229–241.
2. Christaki, E., M. Marcou, and A. Tofarides, Antimicrobial resistance in bacteria: mechanisms, evolution, and persistence. *Journal of Molecular Evolution*, 2020. **88**(1): p. 26–40.
3. Ogunsona, E.O., et al., Engineered nanomaterials for antimicrobial applications: A review. *Applied Materials Today*, 2020. **18**: p. 100473.
4. Sánchez-López, E., et al., Metal-based nanoparticles as antimicrobial agents: an overview. *Nanomaterials*, 2020. **10**(2): p. 292.
5. Gupta, K., et al., Photocatalytic antibacterial performance of TiO2 and Ag-doped TiO2 against S. aureus. P. aeruginosa and E. coli. *Beilstein Journal of Nanotechnology*, 2013. **4**(1): p. 345–351.
6. Chang, Y.-N., et al., The toxic effects and mechanisms of CuO and ZnO nanoparticles. *Materials*, 2012. **5**(12): p. 2850–2871.
7. Allahverdiyev, A.M., et al., Antimicrobial effects of TiO2 and Ag2O nanoparticles against drug-resistant bacteria and leishmania parasites. *Future Microbiology*, 2011. **6**(8): p. 933–940.
8. Jin, T. and Y. He, Antibacterial activities of magnesium oxide (MgO) nanoparticles against foodborne pathogens. *Journal of Nanoparticle Research*, 2011. **13**: p. 6877–6885.
9. Hackenberg, S., et al., Functional responses of human adipose tissue-derived mesenchymal stem cells to metal oxide nanoparticles in vitro. *Journal of Biomedical Nanotechnology*, 2013. **9**(1): p. 86–95.
10. Rodrigo Salinas, R., et al., Molecular Basis of Cardioprotective Effect of Antioxidant Vitamins in Myocardial Infarction. 2013, **2013**(1): p. 437613(1-15).
11. Neto, V., et al., The effect of megavoltage radiation on polymeric materials to be used in biomedical devices. *Journal of Nanoscience and Nanotechnology*, 2012. **12**(8): p. 6779–6784.
12. Luo, Z., et al., Selectively enhanced antibacterial effects and ultraviolet activation of antibiotics with ZnO nanorods against Escherichia coli. *Journal of Biomedical Nanotechnology*, 2013. **9**(1): p. 69–76.
13. Malhotra, S.P.K. and T. Mandal, Zinc oxide nanostructure and its application as agricultural and industrial material. *Contamin. Agric. Environ. Health Risks Remed*, 2019. **1**: p. 216.
14. Shafiee, P., et al., Sol-gel zinc oxide nanoparticles: advances in synthesis and applications. *Synthesis and Sintering*, 2021. **1**(4): p. 242–254.
15. Puzyn, T., et al., Using nano-QSAR to predict the cytotoxicity of metal oxide nanoparticles. *Nature Nanotechnology*, 2011. **6**(3): p. 175–178.
16. Raha, S. and M. Ahmaruzzaman, ZnO nanostructured materials and their potential applications: progress, challenges and perspectives. *Nanoscale Advances*, 2022. **4**(8): p. 1868–1925.
17. Patterson, B.A. and H.A. Sodano, Enhanced interfacial strength and UV shielding of aramid fiber composites through ZnO nanoparticle sizing. *ACS Applied Materials & Interfaces*, 2016. **8**(49): p. 33963–33971.
18. Zhang, L., et al., The properties of ZnO nanofluids and the role of H_2O_2 in the disinfection activity against Escherichia coli. *Water Research*, 2013. **47**(12): p. 4013–4021.
19. Upadhyay, P., et al., Synthesis and applications of ZnO nanoparticles in biomedicine. *Research Journal of Pharmacy and Technology*, 2020. **13**(4): p. 1636–1644.

20. Forman, H.J., M. Maiorino, and F. Ursini, Signaling functions of reactive oxygen species. *Biochemistry*, 2010. **49**(5): p. 835–842.
21. Gangadoo, S., et al., Probing nanoscale interactions of antimicrobial zinc oxide quantum dots on bacterial and fungal cell surfaces. *Advanced Materials Interfaces*, 2022. **9**(3): p. 2101484.
22. Caratto, V., et al., Different sol–gel preparations of iron-doped TiO2 nanoparticles: characterization, photocatalytic activity and cytotoxicity. *Journal of Sol-Gel Science and Technology*, 2016. **80**(1): p. 152–159.
23. Chibac, A.L., et al., Preparation of photocrosslinked sol-gel composites based on urethane-acrylic matrix, silsesquioxane sequences, TiO2, and Ag/Au Nanoparticles for use in photocatalytic applications. *Journal of Polymer Science Part A: Polymer Chemistry*, 2015. **53**(10): p. 1189–1204.
24. Cai, H., et al., Sol–gel synthesis highly porous titanium dioxide microspheres with cellulose nanofibrils-based aerogel templates. *Inorganic Chemistry Communications*, 2015. **51**: p. 71–74.
25. Lu, P., et al., Abnormal room temperature ferromagnetism in CuO/ZnO nanocomposites via hydrothermal method. *Applied Surface Science*, 2017. **399**: p. 396–402.
26. Marlinda, A.R., et al., Highly efficient preparation of ZnO nanorods decorated reduced graphene oxide nanocomposites. *Materials Letters*, 2012. **80**: p. 9–12.
27. Mahmood, N.B., et al., Synthesis and characterization of zinc oxide nanoparticles via oxalate co-precipitation method. *Materials Letters: X*, 2022. **13**: p. 100126.
28. Kotresh, M.G., M.K. Patil, and S.R. Inamdar, Reaction temperature based synthesis of ZnO nanoparticles using co-precipitation method: Detailed structural and optical characterization. *Optik*, 2021. **243**: p. 167506.
29. Laurenti, M., et al., Zinc oxide nanostructures by chemical vapour deposition as anodes for Li-ion batteries. *Journal of Alloys and Compounds*, 2015. **640**: p. 321–326.
30. Ding, J., et al., Aerosol assisted chemical vapour deposition of nanostructured ZnO thin films for NO_2 and ethanol monitoring. *Ceramics International*, 2020. **46**(10, Part A): p. 15152–15158.
31. Qi, K., et al., Review on the improvement of the photocatalytic and antibacterial activities of ZnO. *Journal of Alloys and Compounds*, 2017. **727**: p. 792–820.
32. Sawai, J., et al., Hydrogen peroxide as an antibacterial factor in zinc oxide powder slurry. *Journal of Fermentation and Bioengineering*, 1998. **86**(5): p. 521–522.
33. Raybaudi-Massilia, R.M., J. Mosqueda-Melgar, and O. Martín-Belloso, Antimicrobial activity of malic acid against Listeria monocytogenes, Salmonella Enteritidis and Escherichia coli O157: H7 in apple, pear and melon juices. *Food Control*, 2009. **20**(2): p. 105–112.
34. Król, A., et al., Zinc oxide nanoparticles: Synthesis, antiseptic activity and toxicity mechanism. *Advances in Colloid and Interface Science*, 2017. **249**: p. 37–52.
35. Babayevska, N., et al., ZnO size and shape effect on antibacterial activity and cytotoxicity profile. *Scientific Reports*, 2022. **12**(1): p. 8148.
36. Hassan, H.S., et al., Assessment of antimicrobial, cytotoxicity, and antiviral impact of a green zinc oxide/activated carbon nanocomposite. *Scientific Reports*, 2022. **12**(1): p. 8774.
37. Izzi, M., et al., Synthesis and antimicrobial applications of ZnO nanostructures: a review. *ACS Applied Nano Materials*, 2023. **6**(13): p. 10881–10902.
38. da Silva, B.L., et al., Increased antibacterial activity of ZnO nanoparticles: Influence of size and surface modification. *Colloids and Surfaces B: Biointerfaces*, 2019. **177**: p. 440–447.

39. Zare, M., et al., Surfactant assisted solvothermal synthesis of ZnO nanoparticles and study of their antimicrobial and antioxidant properties. *Journal of Materials Science & Technology*, 2018. **34**(6): p. 1035–1043.
40. Iqbal, T., et al., Facile synthesis and antimicrobial activity of CdS-Ag2S nanocomposites. *Bioorganic Chemistry*, 2019. **90**: p. 103064.
41. Kim, I., et al., ZnO nanostructures in active antibacterial food packaging: preparation methods, antimicrobial mechanisms, safety issues, future prospects, and challenges. *Food Reviews International*, 2022. **38**(4): p. 537–565.
42. Palanikumar, L., S.N. Ramasamy, and C. Balachandran, Size-dependent antimicrobial response of zinc oxide nanoparticles. *IET Nanobiotechnology*, 2014. **8**(2): p. 111–117.
43. Jones, N., et al., Antibacterial activity of ZnO nanoparticle suspensions on a broad spectrum of microorganisms. *FEMS Microbiology Letters*, 2008. **279**(1): p. 71–76.
44. Berube, D.M., Rhetorical gamesmanship in the nano debates over sunscreens and nanoparticles. *Journal of Nanoparticle Research*, 2008. **10**: p. 23–37.
45. Bintsis, T., E. Litopoulou-Tzanetaki, and R.K. Robinson, Existing and potential applications of ultraviolet light in the food industry–a critical review. *Journal of the Science of Food and Agriculture*, 2000. **80**(6): p. 637–645.
46. Brayner, R., et al., Toxicological impact studies based on Escherichia coli bacteria in ultrafine ZnO nanoparticles colloidal medium. *Nano Letters*, 2006. **6**(4): p. 866–870.
47. Mohd Yusof, H., et al., Microbial synthesis of zinc oxide nanoparticles and their potential application as an antimicrobial agent and a feed supplement in animal industry: a review. *Journal of Animal Science and Biotechnology*, 2019. **10**: p. 1–22.
48. Emami-Karvani, Z. and P. Chehrazi, Antibacterial activity of ZnO nanoparticle on gram-positive and gram-negative bacteria. *African Journal of Microbiology Research*, 2011. **5**(12): p. 1368–1373.
49. Fortuny, A., et al., Bimetallic catalysts for continuous catalytic wet air oxidation of phenol. *Journal of Hazardous Materials*, 1999. **64**(2): p. 181–193.
50. Gajjar, P., et al., Antimicrobial activities of commercial nanoparticles against an environmental soil microbe, Pseudomonas putida KT2440. *Journal of Biological Engineering*, 2009. **3**: p. 1–13.
51. Raghupathi, K.R., R.T. Koodali, and A.C. Manna, Size-dependent bacterial growth inhibition and mechanism of antibacterial activity of zinc oxide nanoparticles. *Langmuir*, 2011. **27**(7): p. 4020–4028.
52. Saha, R.K., et al., Antibacterial and nonlinear dynamical analysis of flower and hexagon-shaped ZnO microstructures. *Scientific Reports*, 2020. **10**(1): p. 2598.
53. Kachynski, A.V., et al., Zinc oxide nanocrystals for nonresonant nonlinear optical microscopy in biology and medicine. *The Journal of Physical Chemistry C*, 2008. **112**(29): p. 10721–10724.
54. Ahmed, T., et al., Bioinspired green synthesis of zinc oxide nanoparticles from a native Bacillus cereus strain RNT6: characterization and antibacterial activity against rice panicle blight pathogens Burkholderia glumae and B. gladioli. *Nanomaterials*, 2021. **11**(4): p. 884.
55. Rokbani, H., F. Daigle, and A. Ajji, Long-and short-term antibacterial properties of low-density polyethylene-based films coated with zinc oxide nanoparticles for potential use in food packaging. *Journal of Plastic Film & Sheeting*, 2019. **35**(2): p. 117–134.
56. Hari, K.D., et al., Improvement of the UV barrier and antibacterial properties of cross-linked pectin/zinc oxide bionanocomposite films. *Polymers*, 2021. **13**(15): p. 2403.
57. Khalid, N., et al., Enhanced photocatalytic activity of Al and Fe co-doped ZnO nanorods for methylene blue degradation. *Ceramics International*, 2019. **45**(17): p. 21430–21435.

58. Kim, T.W., et al., Chemical bonding character and physicochemical properties of mesoporous zinc oxide-layered titanate nanocomposites. *The Journal of Physical Chemistry C*, 2007. **111**(4): p. 1658–1664.
59. Wang, X., et al., Multifunctionalized carbon-fiber-reinforced polyetheretherketone implant for rapid osseointegration under infected environment. *Bioactive Materials*, 2023. **24**: p. 236–250.
60. Kumar, K.M., et al., Synthesis and characterisation of flower shaped zinc oxide nanostructures and its antimicrobial activity. *Spectrochimica Acta Part A: Molecular and Biomolecular Spectroscopy*, 2013. **104**: p. 171–174.
61. Li, Q., et al., Antimicrobial nanomaterials for water disinfection and microbial control: potential applications and implications. *Water Research*, 2008. **42**(18): p. 4591–4602.
62. Padmavathy, N. and R. Vijayaraghavan, Enhanced bioactivity of ZnO NPLs-an antibacterial study. *Science and Technology of Advanced Materials*, 2008. **9**: p. 432–438.
63. Sun, Q., J. Li, and T. Le, Zinc oxide nanoparticle as a novel class of antifungal agents: current advances and future perspectives. *Journal of Agricultural and Food Chemistry*, 2018. **66**(43): p. 11209–11220.
64. Padmavathy, N. and R. Vijayaraghavan, Enhanced bioactivity of ZnO nanoparticles—an antimicrobial study. *Science and Technology of Advanced Materials*, 2008, **9**(3): p. 035004(1-7).
65. Pulit-Prociak, J., et al., Functionalization of textiles with silver and zinc oxide nanoparticles. *Applied Surface Science*, 2016. **385**: p. 543–553.
66. Dadi, R., et al., Antibacterial activity of ZnO and CuO nanoparticles against gram positive and gram negative strains. *Materials Science and Engineering: C*, 2019. **104**: p. 109968.
67. Siddique, S., et al., Preparation, characterization and antibacterial activity of ZnO nanoparticles on broad spectrum of microorganisms. *Acta Chimica Slovenica*, 2013. **60**(3): p. 660–665.
68. Yusof, N.A.A., N.M. Zain, and N. Pauzi, Synthesis of ZnO nanoparticles with chitosan as stabilizing agent and their antibacterial properties against Gram-positive and Gram-negative bacteria. *International Journal of Biological Macromolecules*, 2019. **124**: p. 1132–1136.
69. Azam, A., et al., Antimicrobial activity of metal oxide nanoparticles against Gram-positive and Gram-negative bacteria: a comparative study. *International Journal of Nanomedicine*, 2012, **7**: p. 6003–6009.
70. Heidary, M., et al., The anti-mycobacterial activity of Ag, ZnO, and Ag-ZnO nanoparticles against MDR-and XDR-Mycobacterium tuberculosis. *Infection and Drug Resistance*, 2019, **12**: p. 3425–3435.
71. Zhong, L., et al., Synthesis of ZnO nanoparticles-decorated spindle-shaped graphene oxide for application in synergistic antibacterial activity. *Journal of Photochemistry and Photobiology B: Biology*, 2018. **183**: p. 293–301.
72. Elumalai, K. and S. Velmurugan, Green synthesis, characterization and antimicrobial activities of zinc oxide nanoparticles from the leaf extract of Azadirachta indica (L.). *Applied Surface Science*, 2015. **345**: p. 329–336.
73. Dobrucka, R., J. Dlugaszewska, and M. Kaczmarek, Cytotoxic and antimicrobial effects of biosynthesized ZnO nanoparticles using of Chelidonium majus extract. *Biomedical Microdevices*, 2018. **20**: p. 1–13.
74. Al-Hada, N.M., et al., Down-top nanofabrication of binary (CdO) x (ZnO) 1–x nanoparticles and their antibacterial activity. *International Journal of Nanomedicine*, 2017, **12**: p. 8309–8323.
75. Divya, M., et al., Biopolymer gelatin-coated zinc oxide nanoparticles showed high antibacterial, antibiofilm and anti-angiogenic activity. *Journal of Photochemistry and Photobiology B: Biology*, 2018. **178**: p. 211–218.

76. Janaki, A.C., E. Sailatha, and S. Gunasekaran, Synthesis, characteristics and antimicrobial activity of ZnO nanoparticles. *Spectrochimica Acta Part A: Molecular and Biomolecular Spectroscopy*, 2015. **144**: p. 17–22.
77. Dobrucka, R. and J. Długaszewska, Biosynthesis and antibacterial activity of ZnO nanoparticles using Trifolium pratense flower extract. *Saudi Journal of Biological Sciences*, 2016. **23**(4): p. 517–523.
78. Precious Ayanwale, A. and S.N.Y. Reyes-López, ZrO2–ZnO nanoparticles as antibacterial agents. *ACS Omega*, 2019. **4**(21): p. 19216–19224.
79. Shinde, S.S., Antimicrobial activity of ZnO nanoparticles against pathogenic bacteria and fungi. *Sci Med Central*, 2015. **3**: p. 1033.
80. Jayaseelan, C., et al., Novel microbial route to synthesize ZnO nanoparticles using Aeromonas hydrophila and their activity against pathogenic bacteria and fungi. *Spectrochimica Acta Part A: Molecular and Biomolecular Spectroscopy*, 2012. **90**: p. 78–84.
81. Jin, T., et al., Antimicrobial efficacy of zinc oxide quantum dots against Listeria monocytogenes, Salmonella enteritidis, and Escherichia coli O157: H7. *Journal of Food Science*, 2009. **74**(1): p. M46–M52.
82. Saliani, M., R. Jalal, and E.K. Goharshadi, Effects of pH and temperature on antibacterial activity of zinc oxide nanofluid against Escherichia coli O157: H7 and Staphylococcus aureus. *Jundishapur Journal of Microbiology*, 2015. **8**(2): p. e17115.
83. Zare, M., et al., Novel green biomimetic approach for synthesis of ZnO-Ag nanocomposite; antimicrobial activity against food-borne pathogen, biocompatibility and solar photocatalysis. *Scientific Reports*, 2019. **9**(1): p. 8303.
84. Senthilkumar, N., et al., Synthesis of ZnO nanoparticles using leaf extract of Tectona grandis (L.) and their anti-bacterial, anti-arthritic, anti-oxidant and in vitro cytotoxicity activities. *New Journal of Chemistry*, 2017. **41**(18): p. 10347–10356.
85. Tiwari, V., et al., Mechanism of anti-bacterial activity of zinc oxide nanoparticle against carbapenem-resistant Acinetobacter baumannii. *Frontiers in Microbiology*, 2018. **9**: p. 1218.
86. Rana, S., et al., Antimicrobial function of Nd3+-doped anatase titania-coated nickel ferrite composite nanoparticles: a biomaterial system. *Acta Biomaterialia*, 2006. **2**(4): p. 421–432.
87. Sawai, J., et al., *c, and* M. Shimizu, "Antibacterial Characteristics of Magnesium Oxide Powder". *Journal of Chemical Engineering of Japan*, 1996. **29**: p. 556.
88. Erk, I., et al., A zinc ion controls assembly and stability of the major capsid protein of rotavirus. *Journal of Virology*, 2003. **77**(6): p. 3595–3601.
89. Zhang, L., et al., Investigation into the antibacterial behaviour of suspensions of ZnO nanoparticles (ZnO nanofluids). *Journal of Nanoparticle Research*, 2007. **9**(3): p. 479–489.
90. Sawai, J., et al., Detection of Active Oxygen Generated from Ceramic Powders Having Antibacterial Activity. *Journal of Chemical Engineering of Japan*, 1996. **29**(4): p. 627–633.
91. Yamamoto, O., Influence of particle size on the antibacterial activity of zinc oxide. *International Journal of Inorganic Materials*, 2001. **3**(7): p. 643–646.
92. Padmavathy, N. and R. Vijayaraghavan, Enhanced bioactivity of ZnO nanoparticles—an antimicrobial study. *Science and Technology of Advanced Materials*, 2008. **9**(3): p. 035004.
93. Franklin, N.M., et al., Comparative Toxicity of Nanoparticulate ZnO, Bulk ZnO, and ZnCl2 to a Freshwater Microalga (Pseudokirchneriella subcapitata): The Importance of Particle Solubility. *Environmental Science & Technology*, 2007. **41**(24): p. 8484–8490.

94. Jeng, H.A. and J. Swanson, Toxicity of Metal Oxide Nanoparticles in Mammalian Cells *Journal of Environmental Science and Health, Part A*, 2006. **41**(12): p. 2699–2711.
95. Talebian, N., S.M. Amininezhad, and M. Doudi, Controllable synthesis of ZnO nanoparticles and their morphology-dependent antibacterial and optical properties. *Journal of Photochemistry and Photobiology B: Biology*, 2013. **120**: p. 66–73.
96. Pal, S., K. Tak Yu, and M. Song Joon, Does the Antibacterial Activity of Silver Nanoparticles Depend on the Shape of the Nanoparticle? A Study of the Gram-Negative Bacterium Escherichia coli. *Applied and Environmental Microbiology*, 2007. **73**(6): p. 1712–1720.
97. Yang, H., et al., Comparative study of cytotoxicity, oxidative stress and genotoxicity induced by four typical nanomaterials: the role of particle size, shape and composition. *Journal of Applied Toxicology*, 2009. **29**(1): p. 69–78.
98. Tong, G.-X., et al., Polymorphous ZnO complex architectures: selective synthesis, mechanism, surface area and Zn-polar plane-codetermining antibacterial activity. *Journal of Materials Chemistry B*, 2013. **1**(4): p. 454–463.
99. Yin, H., H.P. Too, and G.M. Chow, The effects of particle size and surface coating on the cytotoxicity of nickel ferrite. *Biomaterials*, 2005. **26**(29): p. 5818–5826.
100. Esthappan, S., A. Nair, and R. D. B. Joseph, Effect of crystallite size of zinc oxide on the mechanical, thermal and flow properties of polypropylene/zinc oxide nanocomposites. *Composites Part B: Engineering*, 2015. **69**: p. 145–153.
101. Ansari, S.A., et al., Designing and surface modification of zinc oxide nanoparticles for biomedical applications. *Food and Chemical Toxicology*, 2011. **49**(9): p. 2107–2115.
102. Emami-Karvani, Z. and P. Chehrazi, Antibacterial activity of ZnO nanoparticle on Gram-positive and Gram-negative bacteria. *African Journal of Microbiology Research*, 2012. **5**.
103. Reddy, K.M., et al., Selective toxicity of zinc oxide nanoparticles to prokaryotic and eukaryotic systems. *Applied Physics Letters*, 2007. **90**(21): p. 213902.
104. Xie, Y., et al., Antibacterial Activity and Mechanism of Action of Zinc Oxide Nanoparticles against Campylobacter jejuni. *Applied and Environmental Microbiology*, 2011. **77**(7): p. 2325–2331.
105. Salem, W., et al., Antibacterial activity of silver and zinc nanoparticles against Vibrio cholerae and enterotoxic Escherichia coli. *International Journal of Medical Microbiology*, 2015. **305**(1): p. 85–95.
106. Lee, S.H. and B.H. Jun, Silver Nanoparticles: Synthesis and Application for Nanomedicine. *International Journal of Molecular Sciences*, 2019. **20**(4): p. 1–24.
107. Huang, Z., et al., Toxicological Effect of ZnO Nanoparticles Based on Bacteria. *Langmuir*, 2008. **24**(8): p. 4140–4144.
108. Stoimenov, P., et al., Metal Oxide Nanoparticles as Bactericidal Agents. *Langmuir*, 2002. **18**: p. 6679–6686.
109. Sirelkhatim, A., et al., Review on Zinc Oxide Nanoparticles: Antibacterial Activity and Toxicity Mechanism. *Nano-Micro Letters*, 2015. **7**(3): p. 219–242.
110. Wang, X., et al., A study on the antibacterial activity of one-dimensional ZnO nanowire arrays: effects of the orientation and plane surface. *Chemical Communications*, 2007(**42**): p. 4419–4421.
111. Kim, J.S., et al., Antimicrobial effects of silver nanoparticles. *Nanomedicine: Nanotechnology, Biology and Medicine*, 2007. **3**(1): p. 95–101.
112. Deng, X., et al., Nanosized zinc oxide particles induce neural stem cell apoptosis. *Nanotechnology*, 2009. **20**(11): p. 115101.

113. Premanathan, M., et al., Selective toxicity of ZnO nanoparticles toward Gram-positive bacteria and cancer cells by apoptosis through lipid peroxidation. *Nanomedicine: Nanotechnology, Biology and Medicine*, 2011. **7**(2): p. 184–192.
114. Wang, Z.L., Zinc oxide nanostructures: growth, properties and applications. *Journal of Physics: Condensed Matter*, 2004. **16**(25): p. R829.
115. Moritz, M. and M. Geszke-Moritz, The newest achievements in synthesis, immobilization and practical applications of antibacterial nanoparticles. *Chemical Engineering Journal*, 2013. **228**: p. 596–613.

6 ZnO Nanostructures for Ultraviolet Photodetectors

Shubham Kumar Patial and Suman Singh

6.1 INTRODUCTION

Global energy consumption and population have surged in the last decade, creating an urgent need for more energy-efficient systems [1, 2]. Optical energy detection systems, particularly ultraviolet (UV) photodetectors (PDs), have become increasingly important due to their wide range of applications in fields such as environmental monitoring, healthcare, and defense [3, 4]. UV PDs operate primarily through the photoconductivity effect. This process begins with the absorption of UV light by a wide band gap semiconductor, creating electron-hole (e^-–h^+) pairs if the photon energy (hv) surpasses the material's bandgap (E_g). The core steps involved include absorption of UV energy, generating and separating electron-hole pairs, and collecting these charges as photocurrent, which can be enhanced either by external bias (conventional methods) or through self-powering from built-in electric fields [5, 6]. Figure 6.1 shows a schematic of a reconfigurable photodetector incorporating a plasmonic structure and a gold diffraction grating.

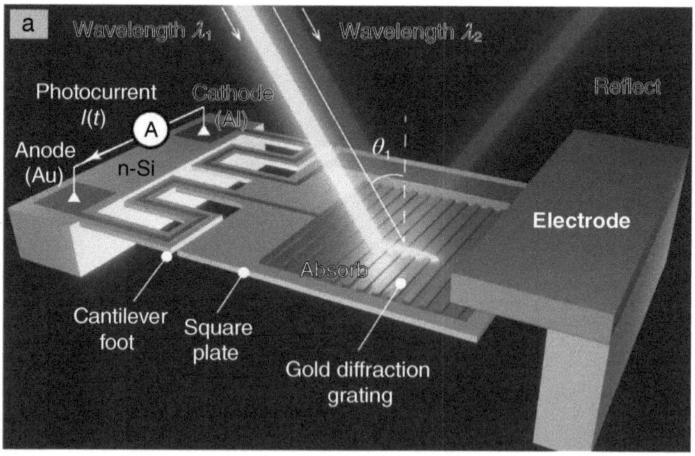

FIGURE 6.1 Schematic of a photodetector featuring a plasmonic structure.

DOI: 10.1201/9781003464907-6

6.1.1 Background on Ultraviolet Photodetectors

Photodetection within the UV spectrum is increasingly significant due to its varied applications across industries, scientific instruments, and everyday life. UV light is categorized into four distinct spectral ranges: UV-A (400-320 nm), UV-B (320-280 nm), UV-C (280-200 nm), and extreme UV (200-10 nm), which approaches the lower energy limit of X-rays [7–9]. The Earth's ozone layer absorbs most of the sun's UV radiation, allowing only light with wavelengths longer than 280 nm to reach the surface. This makes detectors designed to be highly sensitive to UV-C and extreme UV radiation while insensitive to longer wavelengths, which is crucial for monitoring solar radiation accurately. These are termed "solar-blind" detectors. Detectors sensitive to the entire UV spectrum and not the visible range (400–700 nm) are known as "visible-blind." UV detectors are pivotal in many fields, including chemical, environmental, biological analysis, flame and radiation monitoring, astronomy, and optical communications. An optimal UV PD must meet the five critical standards of sensitivity, signal-to-noise ratio, spectral selectivity, speed, and stability—collectively known as the 5S criteria [10]. Detectors must also demonstrate exceptional thermal stability for specific applications like flame detection in high-temperature environments.

6.1.2 Importance and Applications of ZnO Nanostructures

Among various materials used for UV photodetection, zinc oxide (ZnO) stands out due to its excellent properties suitable for UV PD applications. ZnO possesses a direct wide bandgap of approximately 3.37 eV, exhibits n-type semiconductor behavior, and has a high exciton binding energy of 60 meV at room temperature, making it highly effective in UV photodetection [11, 12]. ZnO's most stable crystal form under normal conditions is the wurtzite structure. Its hexagonal pattern and non-centrosymmetric tetrahedral bonding confer piezoelectric and pyroelectric properties, enhancing its photoconductive efficiency [13–15]. Wurtzite ZnO has a hexagonal structure belonging to the $P6_3mc$ space group with the lattice parameters of $a = 0.3296$ and $c = 0.520\ 65$ nm [16].

Nanotechnology has revolutionized UV PDs by introducing nanostructured materials like ZnO, which allow for a high surface-to-volume ratio, increasing the interaction with photons and improving sensitivity. These nanomaterials exhibit unique optical and electrical properties and offer versatility in applications ranging from electronic to biomedical due to their biocompatibility. Novel ZnO nanostructures (ZnO NSs) have been developed in various dimensional forms from 0D to 3D, each providing specific advantages in photodetection.

6.1.3 Overview of Chapter Content

This chapter delves into the significant advancements in UV PDs using ZnO NSs. It starts with exploring the photodetection parameters and the underlying mechanisms of photoconduction using ZnO. The evolution of device fabrication techniques from traditional methods to innovative self-powered systems is reviewed comprehensively.

Special attention is given to novel synergistic effects such as piezoelectric, pyroelectric, and plasmonic enhancements integrated into ZnO UV PDs to boost their performance. A schematic overview of recent developments in ZnO NSs for UV photodetection illustrates current applications and future possibilities.

In this chapter, we discuss in detail the latest progress in ZnO-based UV PDs, highlighting the pivotal role of nanotechnology in advancing these systems. The continuous development of ZnO NSs promises significant improvements in UV PDs' efficiency and application scope, paving the way for next-generation devices offering greater functionality and integration into various industrial, environmental, and health-related areas.

6.2 CHARACTERISTIC PARAMETERS OF UV PHOTODETECTORS

To detect a light signal effectively, the bandgap of a photosensitive material must either be smaller than or equal to the energy of the incoming photons. Analyzing the spectral photoresponse of a UV PD offers crucial insights into its ability to detect UV signals. Separating the photogenerated electrons and holes is essential for efficient photocurrent generation to prevent their recombination upon UV exposure [17]. The response current of a PD refers to the current it generates solely due to UV illumination, providing a measure of its sensitivity to UV light. Sensitivity, in turn, indicates the signal-to-noise ratio at a given UV light intensity, with higher response current and sensitivity being desirable for high-performance UV PDs. Another critical parameter is the photodetection speed, which determines how quickly a PD can detect or recover from a UV signal. Response time and recovery time are the primary parameters defining photodetection speed, with smaller values indicating faster response times essential for rapid UV PDs [17].

The response current generated per unit of incident UV illumination power offers insights into a PD's efficiency in converting UV light into electrical signals. Specific detectivity, on the other hand, quantifies a PD's ability to detect weak UV signals while considering the noise levels inherent to the device [17]. The specific detectivity is closely related to the device's noise characteristics, including shot noise, Johnson noise, and flicker noise. Shot and Johnson's noise stems from thermal fluctuations, with flicker noise also playing a significant role in high-frequency and very low-frequency applications of PDs [17]. Table 6.1 overviews typical UV PD parameters and the mathematical relationships to assess overall photodetection performance.

6.3 FUNDAMENTALS OF ZnO NANOSTRUCTURES

ZnO NSs are renowned for their versatility across various scientific disciplines, standing out among semiconducting oxides such as Al_2O_3, TiO_2, and CuO. These NSs are typically composed of ZnO and can also include combinations with other substances, featuring at least one dimension within the nanometer scale. The development of ZnO NSs has significantly benefited from cost-effective and straightforward synthesis methods, making them accessible for various applications [18, 19].

ZnO NSs exist in several forms, from zero-dimensional (0D) to three-dimensional (3D) structures, each offering unique properties and benefits. The 0D structures

TABLE 6.1
The Characteristic Parameters and Mathematical Relationships of UV PDs [17]

S. No.	Parameter	Mathematical Relation	Unit
1.	Response current	$\Delta I = I_{light} - I_{dark}$, where I_{dark} and I_{light} are the currents in the absence and presence of UV	Ampere (A)
2.	Sensitivity	$S = \frac{I_{Light}}{I_{Dark}}$	Unit less
3.	Response time	$T_r = t_{90\%} - t_{10\%}$, where $t_{10\%}$ and $t_{90\%}$ are the time at 10% and 90% of the saturation value	Second (s)
4.	Recovery time	$T_r = t_{10\%} - t_{90\%}$	Second (s)
5.	Responsivity	$R_s = \frac{\Delta I}{P} A_o = \frac{\Delta I}{P_o}$, where P is the UV signal intensity, A_o is the UV signal exposed area, and P_o is the power of the UV signal	Ampere per watt ($A\ W^{-1}$)
6.	Specific detectivity	$D* = R_s \times \frac{A_o}{2eI_{dark}}$, where e is the absolute value of the elementary charge	Jones

provide increased surface area, enhancing catalytic and surface-related properties. One-dimensional (1D) NSs facilitate rapid electron transport due to their linear configuration, making them ideal for electronic and optoelectronic devices. Two-dimensional (2D) NSs feature large polar surfaces effective in multitasking environments, such as simultaneous analyte detection. The 3D NSs combine these attributes, offering extensive surface areas for various functionalities due to their volumetric configurations [20]. These NSs are predominantly recognized for their hexagonal wurtzite crystal structure, which is crucial for their piezoelectric and pyroelectric properties due to the non-central symmetry and polar (0001) surfaces [21]. Such characteristics, such as sensors and actuators, are instrumental in applications that leverage mechanical and thermal changes into electrical energy. ZnO NSs show superior performance in photodegradation processes, particularly in comparison to TiO_2 [22]. This is attributed to their efficient solar spectrum absorption and non-toxic and economical nature. Moreover, ZnO direct bandgap semiconductor properties and a significant exciton binding energy make it highly suitable for optoelectronic applications. The high ionization energy of oxygen in its lattice contributes to a robust interaction between the Zn (3d) and O (2p) orbitals, further enhancing its electronic properties [23]. ZnO applications are extensive and diverse, ranging from UV blockers in sunscreen lotions to antibacterial agents and from catalysts to additives in industrial products like paint and rubber. Its unique properties also make it a candidate for incorporation into solar cells, liquid crystal displays, and as a biocompatible material in medical applications [24].

6.3.1 Physiochemical Properties of ZnO Nanostructures

ZnO is a material of significant interest, mainly due to its hexagonal wurtzite structure and its most stable form under moderate temperature and pressure conditions. ZnO adopts a cubic structure at higher pressures, transforming into an indirect band gap semiconductor. The hexagonal lattice of ZnO belongs to the space group $P6_3mc$ or $C_6^4\upsilon$, with lattice constants a = b = 0.32539 nm and c = 0.52098 nm [25]. In its crystalline structure, each oxygen anion (O^{2-}) is surrounded by four zinc cations (Zn^{2+}) at the corners of a tetrahedron. Additionally, ZnO exhibits basal plane and polar surfaces, including stable polar planes [0001] and non-polar surfaces [2$\underline{1}$ $\underline{1}$0] and [01$\underline{1}$0] [26].

Regarding optical properties, ZnO behavior under light interaction is of particular interest. Semiconductor materials like ZnO provide valuable insights into their electronic and vibrational states, defects, and impurities through optical studies. Intrinsic properties of semiconductors involve interactions between electrons in the conduction band (CB) and holes in the valence band (VB), along with excitonic effects resulting from Coulomb interactions. On the other hand, extrinsic optical characteristics are influenced by dopants and defects within the semiconductor, which introduce discrete electronic states between the VB and CB [26]. Photoluminescence (PL) spectroscopy is a widely used technique to study the optical properties of ZnO structures. PL spectra of various ZnO NSs typically exhibit UV emissions and one or two bands of visible emissions induced by defects and impurities, such as vacancies and interstitials [27, 28]. ZnO has a band gap of 3.37 eV and a high exciton energy of 60 meV at room temperature, making it a promising material for photonic applications, particularly in the blue-UV region. ZnO nanorods (NRs), for instance, exhibit near-band edge (NBE) UV emissions and broader band emissions associated with deep-level defects [29–31]. The magnetic properties of semiconductor materials, including ZnO, are influenced by the alignment of electron spins within the host lattice. ZnO can be categorized into magnetic semiconductors, diluted magnetic semiconductors (DMS), and nonmagnetic semiconductors, depending on the periodic alignment of magnetic elements or the introduction of transition or rare-earth (RE) ions. Doping ZnO with transition or RE ions, such as Fe, Co, and Mn, can induce local magnetic moments in the crystal structure, leading to intriguing magnetic properties. The magnetic behavior of transition metal-doped ZnO materials depends on doping concentration, carrier density, and crystal quality [32–34].

6.3.2 ZnO Nanostructures in Photodetection

In a ZnO UV PD, the process of photoconduction involves the separation of photogenerated electrons and holes when exposed to UV light, facilitated either by an external driving force (conventional PD) or an internal one (self-powered PD). Figure 6.2 illustrates the photoconduction mechanism in a ZnO nanowire (NW)-based ultraviolet PD (UV PD). Initially, in the absence of light, oxygen molecules adsorb onto the ZnO NW surface, leading to an upward band bending in the valence and conduction bands due to surface trap states. This is depicted schematically with the oxygen molecules capturing electrons from the ZnO NW. When the ZnO NW is exposed to UV light with photon energy equal to or greater than its bandgap,

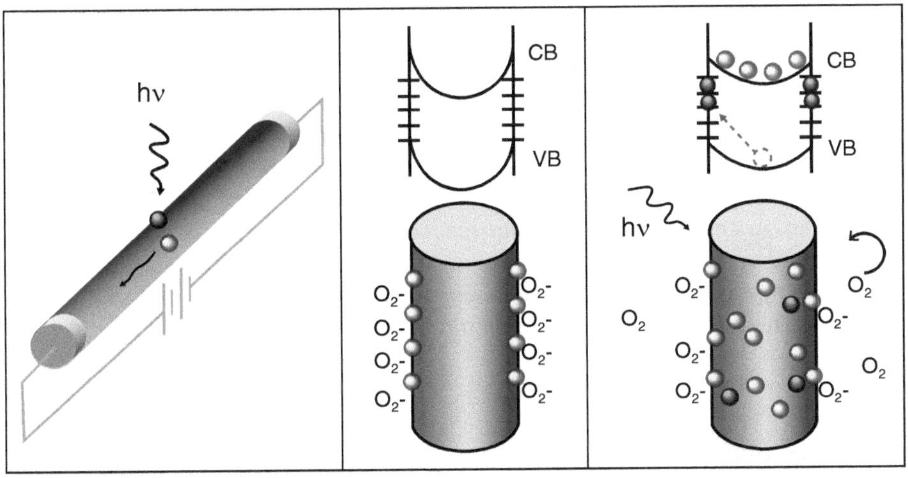

FIGURE 6.2 Photoconduction mechanism in a ZnO nanowire (NW)-based UV PD [57].

electron-hole pairs are generated. Some of the photogenerated holes migrate to the surface, causing the desorption of the adsorbed oxygen molecules, which reduces the depletion width and enhances conductivity. The remaining unpaired electrons, along with the free photogenerated holes, contribute to the increased photocurrent under an applied external voltage. This separation process's efficiency greatly influences the PD's overall photoresponse behavior. Various ZnO nanomaterials, including NWs, NRs, nanotubes (NTs), nanobelts (NBs), nanoflakes (NFs), nanoparticles (NPs), and quantum dots (QDs), have been employed to enhance the performance of UV PDs [35–41]. Due to their large surface-to-volume ratios, these nanomaterials exhibit superior photoconduction compared to their bulk and thin film counterparts [42, 43].

Different dimensional ZnO NSs, including 3D structures like microflowers [44], hierarchical superstructures [45], and superstructures [45], as well as 2D structures such as nanosheets [46], NFs [47], and nanodisks [48], 1D structures like NWs [49], NRs [50], NTs [51], and NBs [52], and 0D structures like QDs [53], NPs [54], core-shell structures [55], and hollow spheres [56] can be synthesized using wet-chemical processes. This versatility in synthesis methods and resulting NSs offers promising avenues for enhancing the performance and functionality of ZnO-based photodetection devices.

6.4 ZnO-BASED UV PHOTODETECTORS

ZnO-based UV PDs come in various types, each with unique characteristics and applications. ZnO-based UV PDs are divided into two categories: first, Conventional UV PDs typically rely on the wide bandgap of ZnO to detect UV light, offering high sensitivity and fast response times, and second, self-powered UV PDs, often using p-n junction or heterojunction design, generate their own electrical power upon UV exposure, eliminating the need for external power sources and enabling energy-efficient operation in remote or portable applications.

6.4.1 Conventional UV Photodetectors

Conventional ZnO UV PDs typically require an external voltage or current source to operate. When exposed to UV light, this external driving force mobilizes photogenerated electrons and holes, generating photocurrent. The applied external force is crucial in influencing the recombination, separation, and transportation of these photogenerated carriers, thus affecting the overall photoresponse of the PD. Depending on their design and operating mechanisms, conventional ZnO UV PDs can be categorized into various types. It is essential to highlight that recent years have seen remarkable advancements in conventional ZnO NS UV PDs. Table 6.2 summarizes some of the high-performance conventional UV PDs developed.

TABLE 6.2
A Summary of the Photoresponse Characteristics Observed in Conventional UV PDs Utilizing ZnO Nanostructure Materials

S. No.	Substrate Use in UV PDs	Bias	UV Light Wavelength	Response Time & Recovery Time	Response Current	Ref.
1.	ZnO NWs	1 V	360 nm	2 s &100 s	34 nA	[83]
2.	ZnO NWs	3 V	310 nm	229 s & 547 s	5.13 mA	[84]
3.	ZnO NW/graphene foam	5 V	365 nm	9.5 s & 38 s	~700 μA	[71]
4.	ZnS/ZnO bilayer Film	5 V	350 nm	50 s & 50 s	~180 μA	[85]
5.	ZnO NP-graphene core-shell	20 V	375 nm	9 ms & 11 ms	—	[86]
6.	Graphene/ZnO NW/ graphene	−5 V	365 nm	3 s & 0.47 s	~300 μA	[77]
7.	Monolayer graphene/ZnO NRs	−1 V	365 nm	0.7 ms & 3.6 ms	—	[87]
8.	Graphene-ZnO QDs-graphene	2 V	365 nm	0.29 s & 0.29 s	~720 μA	[88]
9.	ZnO NR network	4 V	365 nm	2 s & —	—	[89]
10	ZnO spatial tetrapod network	1 V	365 nm	0.4 s & 0.3 s	—	[89]
11.	NiO/ZnO	−5 V	310 nm	—	—	[90]
12.	ZnS-coated ZnO array	3 V	310 nm	229 s & 547 s	162 μA	[84]
13.	ZnS-ZnO bilayer film	5 V	370 nm	50 s & 50 s	~16 μA	[91]
14.	Electrospun ZnO NWs/ PbS QDs	10 V	350 nm	1.02 s & 15.5 s	~25 mA	[92]
15.	ZnO film with Ag NPs	5 V	380 nm	~15 s & ~330 s	—	[93]
16.	Al NPs-ZnO NRs	5 V	252 nm	— & 15 s	674 μA	[94]
17.	ZnO QD/graphene nanocomposite	—	365 nm	1.5 s & 25.6 s	109 nA	[95]
18.	ZnO NPs film	5 V	350 nm	0.467 ms & 0.94 ms	2. 55 μA	[56]
19.	Sn-doped ZnO NRs	5 V	UV lamp	—	17.7 μA	[96]
20.	Al-doped ZnO NWs	1 V	365 nm	0.1 s & 20 s	62 μA	[97]

6.4.1.1 Metal-Semiconductor-Metal (MSM) Photodetectors

In the past decade, pristine ZnO nanomaterials have been extensively used in conventional metal-semiconductor-metal (MSM) PDs for UV light detection due to their straightforward fabrication process. These MSM PDs typically feature symmetric back-to-back ohmic contacts, symmetric back-to-back Schottky contacts, or asymmetric electrodes, depending on the metal electrodes' work function and ZnO electron affinity. When an external voltage is applied across the metal electrodes, photogenerated electrons and holes are mobilized, contributing to the photocurrent. However, pristine ZnO NS-based PDs often exhibit slow photoresponse speeds, ranging from seconds to minutes, due to low-charge carrier mobility, high resistivity, and inherent defects [58–60]. To enhance UV photoresponse, significant research efforts have been directed toward doping ZnO NSs with elements like hydrogen [61], titanium [62], fluorine [62], chlorine [63], aluminum [64], gallium [65], indium [66], and europium [67] to improve carrier mobility. For instance, doping ZnO with copper has enhanced UV and visible light photosensitivity [68].

Despite these improvements, pristine or doped ZnO nanomaterials in conventional MSM PDs still experience high recombination rates of photogenerated carriers, necessitating a high bias voltage for effective UV detection. Different MSM device configurations, ranging from non-rectifying to rectifying electrodes, have been explored to achieve optimal photoconductivity at low bias voltages. Rectifying junction-based PDs generally exhibit lower dark currents, faster-switching behavior, and more controllable photoresponse performance than ohmic junctions. Recently, integrating atomically thin 2D graphene with ZnO has shown promise for efficient charge collection in UV PDs, thanks to graphene's high mobility, transparency, mechanical stability, and flexibility. The built-in electric field at the graphene-ZnO interface facilitates efficient charge separation and transport, enhancing photoconduction [69, 70]. For example, ZnO NWs grown on 3D graphene foam have demonstrated rapid photodetection capabilities with significant current response improvements under UV illumination [71]. Additionally, incorporating carbon nanotubes (CNTs) with ZnO NSs can provide rapid charge carrier transport, further reducing recombination rates and enhancing photodetection performance [72, 73]. These advancements highlight the potential of MSM PDs in achieving high sensitivity and fast response times while addressing the limitations of traditional ZnO-based devices.

6.4.1.2 Schottky Barrier Diodes

Conventional MSM-based devices with non-rectifying electrodes typically exhibit higher dark currents, slower photodetection speeds (in terms of response and recovery times), and increased recombination rates of photogenerated charge carriers under UV illumination. In contrast, Schottky junction-based UV PDs, or Schottky barrier photodiodes, offer significant improvements. These devices demonstrate relatively low dark currents, fast switching behavior, and controllable photoresponse performance under an external driving force [74, 75]. Numerous Schottky barrier-based ZnO UV PDs have been explored, with particular interest in those utilizing graphene and ZnO due to graphene's unique properties. The work function difference between graphene and ZnO creates a rectifying barrier that facilitates efficient charge transport and Schottky barrier formation, enhancing carrier separation.

For instance, Nie and colleagues fabricated a UV PD using a PMMA-supported monolayer graphene film on a ZnO NR array. The device's current-voltage (I-V) characteristics confirmed rectifying behavior with a turn-on voltage of 1.0 V. The work function disparity between graphene (4.6 eV) and ZnO (4.4 eV) leads to an upward energy band bending, promoting efficient electron and hole separation under UV exposure. This results in exceptional photoconduction, with performance three orders of magnitude higher under UV illumination compared to conventional ZnO-based PDs. The device achieved a high responsivity of 113 A W^{-1} at −1 V, with rapid photodetection speeds of 0.7 ms response time and 3.6 ms recovery time [76]. Similarly, Zhang et al. reported a graphene-ZnO Schottky junction PD that demonstrated enhanced photoconduction under reverse bias by reducing dark current and increasing photocurrent with UV illumination [77]. The device exhibited response and recovery times of less than 1 s and 22 s, respectively, with high responsivity and specific detectivity. However, the presence of defect states in ZnO contributed to a slower response under UV light. These Schottky junction-based UV PDs highlight the advantages of using built-in electric fields for efficient charge separation and transport. However, the presence of material defects can still affect performance.

6.4.1.3 Homojunction/Heterojunction-Based UV Photodetectors

In recent years, significant advancements have been made in ZnO nanomaterial-based UV PDs, particularly those utilizing homojunction and heterojunctions. These junction-based devices leverage the built-in electric field generated at the semiconductor interface to enhance photoconductivity by efficiently separating photogenerated electron-hole pairs. In ZnO homojunction PDs, p-type and n-type ZnO NSs form p-n junctions. When exposed to UV light, the internal electric field drives photogenerated electrons toward the n-type region and holes toward the p-type region. For instance, Leung and colleagues developed a visible-blind UV PD using n-type ZnO NWs covered with a p-type Al, N co-doped ZnO film fabricated through hydrothermal and sol-gel synthesis methods [78]. The device exhibited diode-like I-V characteristics, with a notable increase in current under UV illumination, confirming its p-n junction behavior. Another example includes a homojunction UV PD made from Sb-doped p-type ZnO NWs and n-type ZnO film, demonstrating a UV-sensitive response and improved photoconductivity due to the internal electric field [79].

Heterojunction-based ZnO UV PDs involve the integration of ZnO with other semiconductors to form either p-n or n-n junctions, depending on the materials used. These heterojunctions rely on both internally generated and externally applied electric fields to separate photogenerated carriers under UV illumination. The choice of semiconductor material significantly influences the strength of the internal electric field. For example, Tsai and colleagues fabricated a p-NiO/n-ZnO heterojunction UV PD, where the built-in electric field created due to Fermi level alignment enhanced the device's photoconductivity. The current increased in both forward and reverse bias directions upon UV exposure, demonstrating high sensitivity [80].

Another example is the ZnO/SnO_2 heterojunction UV PD, which showed a substantial increase in current under UV light, attributed to the efficient separation of photogenerated carriers facilitated by the internal electric field. The device exhibited a high sensitivity, particularly at a wavelength of 300 nm [81]. Additionally, Fang's group developed a ZnS/ZnO biaxial NB-based UV PD, which demonstrated superior performance compared to pure ZnO or ZnS structures due to the synergistic effects of the heterojunction [82].

Homojunction and heterojunction UV PDs offer significant advantages, including enhanced photoconductivity and improved device performance due to internal electric fields. However, the complexity of fabricating these junctions and the potential for defects affecting performance are notable challenges. Despite these drawbacks, the advancements in these junction-based PDs highlight their potential for high-performance UV detection applications.

6.4.2 Self-Powered UV Photodetectors

Conventional UV PDs need an external driving force to separate and transport photogenerated electrons and holes, leading to photoconduction. Consequently, these PDs can consume significant energy and have a reduced lifespan due to internal heat generation from the external bias voltage. Junction-based PDs, however, create internal built-in electric fields that efficiently separate and transport electron-hole pairs without needing an external power source, making them self-powered [98]. This has spurred considerable research into the development of self-powered UV PDs. These devices offer benefits such as minimal power consumption, addressing energy-related issues, and reducing the overall size compared to conventional PDs. Additionally, self-powered PDs exhibit faster photoconduction because their internal electric fields rapidly separate and transport photogenerated charge carriers under UV illumination. Researchers have made consistent efforts and dedicated significant research toward developing a range of self-powered UV PDs based on ZnO NSs. Table 6.3 summarizes the photoresponse performances of these diverse ZnO nanostructure/heterostructure-based self-powered PDs as reported in the literature.

6.4.2.1 Schottky Junction-Based Self-Powered UV Photodetectors

Schottky junction-based self-powered UV PDs exploit the intrinsic properties of ZnO to operate without external power sources. These devices benefit from an internal built-in electric field that arises at the metal-semiconductor junction, enabling efficient charge carrier separation and transportation upon UV exposure. For instance, a notable study by Zhang's group demonstrated a self-powered PD using an Sb-doped ZnO NB, which spanned between an ohmic (Ag) and a Schottky (Au) contact [38]. This device showcased enhanced photocurrent generation from 1 to 23 nA under UV light without any bias voltage, attributed to the high donor impurity density from Sb doping that strengthens the electric field across the device. Another approach by Duan and colleagues involved graphene and aluminum-doped ZnO NR array film, which formed a Schottky junction with graphene [99]. This configuration was optimized by doping graphene with Au^{3+} and treating the surface of the NR film to

TABLE 6.3
An Overview of the Photoresponse Behaviors Exhibited by ZnO Junction-based Self-powered UV PDs, all Recorded Under Zero Bias Voltage Conditions (V = 0 V)

S. No.	Substrate Use in UV PDs	UV Light Wavelength	Response Time & Recovery Time	Response Current	Ref.
1.	H:VZnO NFs/PEDOT:PSS	365 nm	23 ms & 26 ms	100 nA	[39]
2.	Cl-ZnO NRs/ DMSO-PEDOT:PSS	365 nm	30 ms & 32 ms	300 nA	[105]
3.	rGO-ZnO nanostructure	334 nm	0.2 s & 0.2 s	20 nA	[109]
4.	PEDOT:PSS/ZnO micro/NW	325 nm	1 s & 1 s	1.2 nA	[110]
5.	p-NiO/ZnO NRs	355 nm	0.23 s & 0.21 s	300 nA	[111]
6.	Ag-ZnO NWs	365 nm	0.14 s & 0.52 s	71 μA	[112]
7.	Cl:ZnO NRs/PEDOT:PSS	365 nm	28 ms & 23 ms	564 nA	[113]
8.	n-ZnO/i-MgO/p-GaN	365 nm	3.2 s & 3.84 s	—	[114]
9.	n-ZnO/p-CuSCN	325 nm	0.5 μs & 6.7 μs	—	[115]
10.	Pt/ZnO NWs/Al	355 nm	81 ms & 95 ms	—	[116]
11.	n-ZnO/p-GaN	325 nm	20 μs & 219 μs	~2 μA	[117]
12.	Sb-doped ZnO nanobelt/Au	325 nm	100 ms & 100 ms	22 nA	[39]
13.	ZnO/Spiro-MeOTAD	365 nm	0.16 s & 0.20 s	~0.2 μA	[118]
14.	ZnO/Ga_2O_3	251 nm	100 μs & 900 μs	~6 nA	[119]
15.	ZnO & Pt/FTO PECC	365 nm	0.06 s & 0.06 s	~30 nA	[107]
16.	ZnO & Pt/FTO PECC	365 nm	0.15 s & 0.5 s	21.61 μA cm^{-2}	[112]
17.	ZnO NRs & Pt/FTO PECC	365 nm	0.09 s & 0.31 s	~140 mA	[120]
18.	ZnO/Au	325 nm	0.1 s & 0.1 s	—	[101]
19.	ZnO/perovskite	325 nm	—	~1 mA	[121]
20.	n-ZnO/p-PANI/n-ZnO	365 nm	—	0.5 μA	[122]

adjust the Schottky barrier height, resulting in efficient self-powering capabilities and improved device performance under UV illumination.

The asymmetric design of Schottky junctions also plays a pivotal role in enhancing the photodetection capabilities of these devices. Chen's group reported on devices with varying widths of asymmetric Au electrodes, observing that an increase in the asymmetry ratio significantly boosts the rectification behavior and self-powered response under UV exposure [100]. This is due to the altered electric potential distribution, facilitating better charge separation at zero bias voltage. The most effective configuration demonstrated a responsivity of about 20 mA W^{-1} with ultra-fast response and recovery times of 710 ns and 4 ms, respectively. On another front, Zhang's group also explored the piezo-phototronic effect in a ZnO/Au Schottky junction-based UV PD mounted on a flexible substrate [101]. Applying external strain increased the piezo potential across the device, dramatically enhancing the photocurrent and sensitivity upon UV exposure. These advancements highlight the potential of Schottky junction-based UV PDs in energy-efficient and high-performance applications. However, challenges such as device stability and scalability persist, and further exploration is needed to harness these innovative technologies in practical settings fully.

6.4.2.2 p–n Junction-Based Self-Powered UV Photodetectors

Inorganic homojunction and heterojunction-based self-powered UV PDs capitalize on the spontaneous diffusion of electrons and holes to generate a built-in electric field at the p-n junction interface. This internal field is instrumental in separating and transporting photogenerated carriers under UV illumination without any external driving force, facilitating self-powered operation. These devices are distinguished by their ability to operate independently of external power sources, showcasing a significant reduction in energy consumption and device size. The performance of such devices is evidenced in various studies, including Bie and colleagues' work on an n-ZnO NW/p-GaN junction, which demonstrated enhanced photoresponse in UV light with fast response times [102].

The recent advancements by the Fang group and others in fabricating transparent self-powered UV PDs using materials like Ag-doped ZnO nanofibers highlight the ongoing innovations in this field [103]. These devices maintain clarity and efficiency and exhibit fast photoresponse times due to the effective charge carrier dynamics facilitated by the internal electric fields. However, while the self-powered nature of these devices minimizes energy consumption, the reliance on the intrinsic properties of materials may limit the intensity and efficiency of the photoresponse, as seen in pristine ZnO and NiO, where minimal responses were observed without external bias [104]. This underscores the critical balance between material properties and device architecture in optimizing the performance of UV PDs.

Organic-inorganic heterostructure-based self-powered UV PDs utilize junctions formed between organic and inorganic semiconductors to operate without an external bias. These devices capitalize on the built-in electric fields generated by Fermi energy level alignments, which efficiently separate and propel photogenerated charges. In this design, the inorganic semiconductor, such as ZnO, is light-sensitive, while the organic semiconductor facilitates direct charge conduction. Typically, p-type organic semiconductors like PEDOT: PSS are preferred due to their compatibility with the n-type behavior of ZnO, enabling effective internal electric field generation [105]. Fabrication involves optimizing materials and interfaces, as exemplified by doping ZnO with chlorine and modifying PEDOT: PSS with dimethyl sulfoxide to enhance photoresponse. These self-powered devices offer advantages, including low energy consumption, mitigating external power dependency, and improved charge separation and transport efficiency.

However, challenges persist, such as optimization of material properties, interface engineering, and device scalability. Nonetheless, Basak's group's recent advancements demonstrate promising outcomes. By constructing ZnO@CdS core-shell NSs coupled with PEDOT: PSS, they achieved enhanced self-powered photoresponse through precise control of CdS coating thickness [106]. These developments enhance sensitivity and exhibit faster response times, highlighting the potential of organic-inorganic heterostructures for self-powered UV photodetection, with prospects for broader application in visible light detection.

6.4.2.3 Photoelectrochemical Cell-Based Self-Powered UV Photodetectors

Photoelectrochemical cell (PECC)-based self-powered UV PDs consist of three main components: a ZnO nanostructures/heterostructure-based photoanode, an electrolyte for capturing photogenerated holes, and a counter electrode for receiving photogenerated electrons. When exposed to UV light, photogenerated holes migrate to the ZnO/electrolyte interface and trap electron donors from the electrolyte. At the same time, the counter electrode collects photogenerated electrons through the external circuit. The oxidized redox species diffuse through the electrolyte and are neutralized by photogenerated electrons at the counter electrode. This mechanism, depicted in energy band diagrams, enables continuous photocurrent generation without an external driving force. Notably, Li et al. employed water as the electrolyte in a PECC, utilizing ZnO nanoneedle arrays as the photoanode and platinum-coated fluorine-doped tin oxide (FTO) as the counter electrode [107]. Their device exhibited significant self-powered photoconduction, demonstrated by a short circuit current of 0.8 mA and an open circuit voltage of 0.5 V under UV light exposure. The PECC displayed a maximum responsivity of 0.022 A W^{-1} and fast response and recovery times of around 0.1 s.

Another study by Xie's group utilized TiO_2-coated ZnO nanostrawberry core-shell photoanodes in PECC-type self-powered UV PDs. Photogenerated holes migrated to the TiO_2/electrolyte interface, while electrons were transported to the counter electrode through ZnO and an external circuit. This configuration yielded enhanced self-biased photoresponse, with a sensitivity of approximately 37,900 and a maximum responsivity of 0.38 A W^{-1} at 330 nm [108]. However, despite their advantages, such as low power consumption and compact size, PECC-based PDs may have limited applications due to the involvement of liquid electrolytes, which could restrict their usability in specific environments or applications sensitive to leakage or evaporation. Nonetheless, these devices showcase promising capabilities in self-powered UV photodetection, offering the potential for various practical applications in sensing and monitoring.

6.5 APPLICATIONS OF ZnO UV PHOTODETECTORS

ZnO UV PDs are emerging as a versatile and efficient technology for various applications, from environmental monitoring and industrial safety to biomedical devices and security systems. The ZnO has excellent optoelectronic properties, including high exciton binding energy (60 meV), intrinsic n-type conductivity, and the ability to form diverse NSs further enhancing its appeal for UV photodetection applications [123]. The unique properties of ZnO, combined with its capability to form diverse NSs, provide significant advantages in terms of sensitivity, response time, and durability. Figure 6.3 provides a schematic summary of ZnO NSs, encompassing the evolution of UV PDs from conventional designs to self-powered configurations. It also highlights the potential applications of these advancements.

6.5.1 Environmental Monitoring

ZnO UV PDs are crucial in monitoring UV radiation levels in the environment. This is particularly important due to the harmful effects of excessive UV exposure on

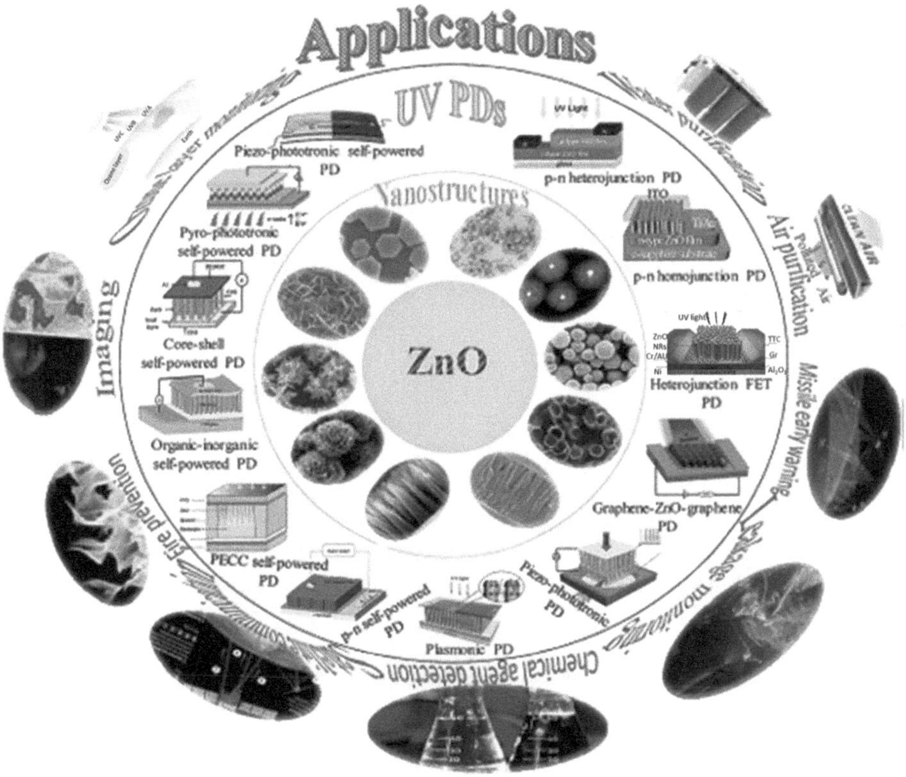

FIGURE 6.3 The potential applications of ZnO UV PDs [17].

human health and ecosystems. Devices based on ZnO can detect UV radiation from the sun, helping to track ozone layer depletion and the consequent increase in UV exposure. Such monitoring is essential for developing protective measures and policies to mitigate UV radiation's impact [123].

6.5.2 Industrial Applications

In industrial settings, ZnO UV PDs are used for flame detection and combustion monitoring. These PDs can quickly and accurately detect UV emissions from flames, which is critical for ensuring safety in industrial processes. The high sensitivity and fast response time of ZnO-based detectors make them ideal for real-time monitoring and control in combustion systems, preventing accidents and improving efficiency [124].

6.5.3 Astronomy and Space Exploration

UV PDs are invaluable in astronomy and space exploration for studying celestial objects that emit UV radiation. With their high sensitivity and stability in harsh environments, ZnO UV PDs are used in telescopes and space probes to detect and analyze UV light from stars, galaxies, and other cosmic phenomena. This data helps astronomers understand the universe's composition, behavior, and evolution [125].

6.5.4 Biomedical Applications

ZnO UV PDs are also finding applications in the biomedical field, particularly in sterilization and disinfection systems. UV-C light (around 254 nm) is known for its germicidal properties, and ZnO PDs can be used to monitor and control UV-C light sources in sterilization equipment. This application is crucial in healthcare settings, food processing, and water purification systems, where maintaining sterile conditions is essential [125].

6.5.5 Security and Sensing

Security systems benefit from ZnO UV PDs through their application of UV-based imaging and detection systems. These systems can detect invisible UV markers or signatures, which are used in various security and anti-counterfeiting measures [126]. Additionally, ZnO PDs are used in chemical and biological sensing to detect specific UV-absorbing substances, enhancing security measures in sensitive areas [127].

6.6 CHALLENGES AND LIMITATIONS

ZnO-based UV detectors face several challenges and limitations that hinder their performance and practical applications. One of the primary issues is the presence of trap states on the ZnO surface, which impede charge transportation [128]. The nano-fabrication processes required to integrate ZnO in UV polarization-sensitive detectors with metallic gratings are highly precise, making it challenging to achieve high polarization ratios [129]. Optimizing ZnO NSs and their assembly with other materials is crucial for enhancing the performance of these detectors [130].

6.6.1 Material Stability and Durability

Material stability and durability are significant concerns, especially considering oxygen's adsorption and desorption behavior on ZnO surfaces. This interaction can reduce dark current, enhance responsiveness, and lead to long rise and recovery times, limiting practical applications. Surface passivation and structural design improvements are essential to address these issues [131].

6.6.2 Sensitivity and Selectivity Issues

Another challenge is the fabrication of high-quality p-type ZnO-based materials. P-type doping in ZnO remains problematic, which is a significant barrier to achieving high-performance UV PDs. Although p-n junction PDs offer high sensitivity, fast response speeds, and low dark current, the lack of stable and reproducible p-type ZnO materials hampers their development. Stability, reproducibility, and carrier concentration, and mobility control are ongoing challenges [131].

6.6.3 Integration with Other Electronic Components

Integrating ZnO with other electronic components also poses difficulties, particularly in fabricating $Zn_{1-x}Mg_xO$ films with varying x values to achieve different detection wavelengths. The ZnO-MgO binary system has a wide miscibility gap due to structural differences and significant lattice mismatches, leading to phase segregation and low crystal quality in MgZnO films [131].

ZnO-based PDs, especially photoconductors, often suffer from slower response speeds due to the adsorption and desorption of oxygen molecules near the semiconductor surface. Improving the response speed is critical to meeting future demands in various fields. Addressing these challenges through material optimization, advanced fabrication techniques, and innovative design strategies is essential for maximizing the efficiency and sensitivity of ZnO-based UV detectors [132].

6.7 CONCLUSION AND FUTURE DIRECTIONS

This chapter has underscored significant advancements in developing ZnO nanostructures and heterostructures for UV PDs, ranging from conventional systems to innovative self-powered designs. These PDs are critical due to their applications in various fields, such as medical diagnostics, environmental monitoring, and optical communications. With its wide bandgap, ZnO is particularly suited for UV detection, offering enhanced sensitivity and stability. The intrinsic properties of ZnO, combined with nanostructuring, have led to enhanced photoresponse behaviors due to increased surface-to-volume ratios and quantum confinement effects. Various NSs, including NPs, NWs, and QDs, have been explored to exploit these effects fully. Integrating these NSs into device architectures has shown promising improvements in PDs' speed, sensitivity, and efficiency.

Despite significant advancements in ZnO UV PDs, several challenges remain that need to be addressed to enhance their performance further. One major challenge is enhancing photodetection speed; the presence of native defects in ZnO often hinders efficient charge carrier dynamics, thus affecting response and recovery times. Advanced material processing techniques, such as reducing these defects, are crucial for achieving faster photodetection speeds. Another challenge lies in minimizing lattice mismatch in heterostructures, where mismatches can introduce defect states that trap charge carriers and slow device response. Precisely engineering the interfaces with better lattice matching could significantly improve performance.

Self-powered detectors offer the advantage of not requiring external power sources; their responsibilities are typically lower than those of their powered counterparts. Enhancing the built-in electric fields within these devices could boost their efficiency. Optimizing surface roughness is also critical; tuning the roughness of ZnO NSs can enhance light absorption and charge carrier generation, necessitating precise control over NS morphology during synthesis.

To overcome these obstacles and advance ZnO UV PD technology, future research should focus on several key areas. Material engineering is paramount, with techniques like atomic layer deposition or molecular beam epitaxy offering the control needed to minimize defects and optimize interfaces. Exploring novel device architectures that leverage ZnO's unique properties more efficiently, including integrating other optoelectronic materials in hybrid structures, is also essential. Moreover, implementing advanced characterization techniques will provide deeper insights into the interactions at the nano-bio interface, charge carrier dynamics, and the role of surface states in device performance. Finally, building and testing prototype devices incorporating the latest advancements in ZnO NSs will help assess their real-world performance under various environmental conditions, moving the field closer to realizing highly efficient, cost-effective UV PDs.

The ultimate goal is to develop ZnO-based UV PDs that are highly efficient, fast, cost-effective, and easy to integrate into existing systems. Achieving this will require a multidisciplinary approach, combining insights from materials science, nanotechnology, physics, and engineering. As research progresses, the next generation of ZnO UV PDs looks promising, with potential applications expanding into new, innovative areas, including wearable technology and integrated optical systems. Through concerted research and development efforts, ZnO NSs will be crucial in advancing PD technology to meet future demands.

REFERENCES

1. Teng, F., et al., Photoelectric Detectors Based on Inorganic p-Type Semiconductor Materials. *Advanced Materials*, 2018. **30**(35): p. 1706262.
2. Bhandari, S., et al., Biomolecule-derived quantum dots for sustainable optoelectronics. *Nanoscale Advances*, 2019. **1**(3): p. 913–936.
3. Chen, H., et al., Highly Desirable Photodetectors Derived from Versatile Plasmonic Nanostructures. *Advanced Functional Materials*, 2017. **27**(45): p. 1704181.
4. Ouyang, W., et al., Enhancing the Photoelectric Performance of Photodetectors Based on Metal Oxide Semiconductors by Charge-Carrier Engineering. *Advanced Functional Materials*, 2019. **29**(9): p. 1807672.
5. Li, X., et al., Carbon and Graphene Quantum Dots for Optoelectronic and Energy Devices: A Review. *Advanced Functional Materials*, 2015. **25**(31): p. 4929–4947.
6. Thorkelsson, K., P. Bai, and T. Xu, Self-assembly and applications of anisotropic nanomaterials: A review. *Nano Today*, 2015. **10**(1): p. 48–66.
7. Lin, C.-H. and C.W. Liu, Metal-insulator-semiconductor photodetectors. *Sensors*, 2010. **10**(10): p. 8797–8826.
8. Monroy, E., F. Omnès, and F. Calle, Wide-bandgap semiconductor ultraviolet photodetectors. *Semiconductor science and technology*, 2003. **18**(4): p. R33.
9. Omnès, F., Wide bandgap UV photodetectors: A short review of devices and applications, Integrated Optoelectronic Devices 2007. *International Society for Optics and Photonics*, 2007, **6473**: p. 111–125.
10. Liao, M., et al., Comprehensive investigation of single crystal diamond deep-ultraviolet detectors. *Japanese Journal of Applied Physics*, 2012. **51**(9R): p. 090115.
11. Özgür, Ü., et al., A comprehensive review of ZnO materials and devices. *Journal of Applied Physics*, 2005. **98**(4), 041301 (1-103).
12. Wang, Z.L., ZnO nanowire and nanobelt platform for nanotechnology. *Materials Science and Engineering: R: Reports*, 2009. **64**(3): p. 33–71.
13. Wang, Z.L., Progress in Piezotronics and Piezo-Phototronics. *Advanced Materials*, 2012. **24**(34): p. 4632–4646.
14. Yang, Y., et al., Pyroelectric Nanogenerators for Harvesting Thermoelectric Energy. *Nano Letters*, 2012. **12**(6): p. 2833–2838.

15. Gao, Z., et al., Effects of piezoelectric potential on the transport characteristics of metal-ZnO nanowire-metal field effect transistor. *Journal of Applied Physics*, 2009. **105**(11): p. 113707.
16. Zhong Lin, W., Zinc oxide nanostructures: growth, properties and applications. *Journal of Physics: Condensed Matter*, 2004. **16**(25): p. R829.
17. Deka Boruah, B., Zinc oxide ultraviolet photodetectors: rapid progress from conventional to self-powered photodetectors. *Nanoscale Advances*, 2019. **1**(6): p. 2059–2085.
18. Look, D. C. (2001). Recent advances in ZnO materials and devices. *Materials science and engineering: B, 80*(1-3), 383–387.
19. Klingshirn, C., et al., ZnO rediscovered — once again!? *Superlattices and Microstructures*, 2005. **38**(4): p. 209–222.
20. Napi, M.L.M., et al., Electrochemical-based biosensors on different zinc oxide nanostructures: A review. *Materials*, 2019. **12**(18): p. 2985.
21. Cauda, V., et al., Nanostructured ZnO materials: Synthesis, properties and applications. *Handbook of Nanomaterials Properties*, 2014: p. 137–177.
22. Ong, C.B., L.Y. Ng, and A.W. Mohammad, A review of ZnO nanoparticles as solar photocatalysts: Synthesis, mechanisms and applications. *Renewable and Sustainable Energy Reviews*, 2018. **81**: p. 536–551.
23. Ellmer, K., A. Klein, and B. Rech, Transparent conductive zinc oxide: basics and applications in thin film solar cells. *Springer Series in Materials Science*, 2007, **104**, 446.
24. Zhang, L., et al., Investigation into the antibacterial behaviour of suspensions of ZnO nanoparticles (ZnO nanofluids). *Journal of Nanoparticle Research*, 2007. **9**(3): p. 479–489.
25. Morkoç, H. and Ü. Özgür, *Zinc oxide: fundamentals, materials and device technology*. 2008: John Wiley & Sons.
26. Gao, P.X., et al., Nanowire Piezoelectric Nanogenerators on Plastic Substrates as Flexible Power Sources for Nanodevices. *Advanced Materials*, 2007. **19**(1): p. 67–72.
27. Janotti, A. and C.G. Van de Walle, Fundamentals of zinc oxide as a semiconductor. *Reports on Progress in Physics*, 2009. **72**(12): p. 126501.
28. Djurišić, A.B. and Y.H. Leung, Optical Properties of ZnO Nanostructures. *Small*, 2006. **2**(8-9): p. 944–961.
29. Cui, J., Zinc oxide nanowires. *Materials Characterization*, 2012. **64**: p. 43–52.
30. Wang, Z.L., Zinc oxide nanostructures: Growth, properties and applications. *Journal of Physics Condensed Matter*, 2004. **16**(25): p. R829–R858.
31. Willander, M., et al., Zinc oxide nanorod based photonic devices: recent progress in growth, light emitting diodes and lasers. *Nanotechnology*, 2009. **20**(33): p. 332001.
32. Han, S.J., et al., Magnetism in Mn-doped ZnO bulk samples prepared by solid state reaction. *Applied Physics Letters*, 2003. **83**(5): p. 920–922.
33. Chen, J., et al., Room-temperature ferromagnetic Co-doped ZnO nanoneedle array prepared by pulsed laser deposition. *Applied Physics Letters*, 2005. **87**(17), 173119(1-3).
34. Dreyhsig, J. and B. Litzenburger, Nature of optical transitions in the charge-transfer region of ZnS:Co and ZnSe:Co. *Physical Review B - Condensed Matter and Materials Physics*, 1996. **54**(15): p. 10516–10524.
35. Liu, X., et al., All-printable band-edge modulated ZnO nanowire photodetectors with ultra-high detectivity. *Nature Communications*, 2014. **5**(1): p. 4007.
36. Chai, G., et al., Crossed zinc oxide nanorods for ultraviolet radiation detection. *Sensors and Actuators A: Physical*, 2009. **150**(2): p. 184–187.
37. Chantarat, N., et al., Enhanced UV photoresponse in nitrogen plasma ZnO nanotubes. *Nanotechnology*, 2009. **20**(39): p. 395201.
38. Yang, Y., et al., Self-powered ultraviolet photodetector based on a single Sb-doped ZnO nanobelt. *Applied Physics Letters*, 2010. **97**(22), 223113, 1–5.

39. Deka Boruah, B. and A. Misra, Energy-efficient hydrogenated zinc oxide nanoflakes for high-performance self-powered ultraviolet photodetector. *ACS Applied Materials & Interfaces*, 2016. **8**(28): p. 18182–18188.
40. Jin, Y., et al., Solution-processed ultraviolet photodetectors based on colloidal ZnO nanoparticles. *Nano Letters*, 2008. **8**(6): p. 1649–1653.
41. Yang, H.Y., et al., Enhancement of the photocurrent in ultraviolet photodetectors fabricated utilizing hybrid polymer-ZnO quantum dot nanocomposites due to an embedded graphene layer. *Organic Electronics*, 2010. **11**(7): p. 1313–1317.
42. Sun, Y., et al., Synthesis and photoluminescence of ultra-thin ZnO nanowire/nanotube arrays formed by hydrothermal growth. *Chemical Physics Letters*, 2006. **431**(4-6): p. 352–357.
43. Liang, J., et al., Hydrothermal growth and optical properties of doughnut-shaped ZnO microparticles. *The Journal of Physical Chemistry B*, 2005. **109**(19): p. 9463–9467.
44. Liang, Y., et al., Preparation of porous 3D Ce-doped ZnO microflowers with enhanced photocatalytic performance. *RSC Advances*, 2015. **5**(74): p. 59887–59894.
45. Zhang, T., et al., Site-specific nucleation and growth kinetics in hierarchical nanosyntheses of branched ZnO crystallites. *Journal of the American Chemical Society*, 2006. **128**(33): p. 10960–10968.
46. Gupta, M.K., et al., Two-dimensional vanadium-doped ZnO nanosheet-based flexible direct current nanogenerator. *ACS Nano*, 2013. **7**(10): p. 8932–8939.
47. Van Ngoc, H. and D.J. Kang, Flexible, transparent and exceptionally high power output nanogenerators based on ultrathin ZnO nanoflakes. *Nanoscale*, 2016. **8**(9): p. 5059–5066.
48. Alenezi, M.R., et al., From 1D and 2D ZnO nanostructures to 3D hierarchical structures with enhanced gas sensing properties. *Nanoscale*, 2014. **6**(1): p. 235–247.
49. Jalali, N., et al. Passivation of zinc oxide nanowires for improved piezoelectric energy harvesting devices. In *Journal of Physics: Conference Series*. 2013. IOP Publishing.
50. Zhang, A.Q., et al., Morphology-controllable synthesis of ZnO nano-/micro-structures by a solvothermal process in ethanol solution. *Crystal Research and Technology*, 2013. **48**(11): p. 947–955.
51. Yu, Q., et al., Gas sensing properties of self-assembled ZnO nanotube bundles. *Rsc Advances*, 2013. **3**(37): p. 16619–16625.
52. Lu, C., et al., Hydrothermal growth of large-scale micropatterned arrays of ultralong ZnO nanowires and nanobelts on zinc substrate. *Chemical Communications*, 200633: p. 3551–3553.
53. Boruah, B.D. and A. Misra, ZnO quantum dots and graphene based heterostructure for excellent photoelastic and highly sensitive ultraviolet photodetector. *RSC Advances*, 2015. **5**(110): p. 90838–90846.
54. Wu, W., et al., Controllable synthesis and optical properties of connected zinc oxide nanoparticles. *Chemistry–An Asian Journal*, 2010. **5**(2): p. 315–321.
55. Majhi, S.M., P. Rai, and Y.-T. Yu, Facile approach to synthesize Au@ ZnO core–shell nanoparticles and their application for highly sensitive and selective gas sensors. *ACS Applied Materials & Interfaces*, 2015. **7**(18): p. 9462–9468.
56. Chen, M., et al., ZnO hollow-sphere nanofilm-based high-performance and low-cost photodetector. *Small (Weinheim an der Bergstrasse, Germany)*, 2011. **7**(17): p. 2449–2453.
57. Soci, C., et al., ZnO Nanowire UV Photodetectors with High Internal Gain. *Nano Letters*, 2007. **7**(4): p. 1003–1009.
58. Manekkathodi, A., et al., Direct growth of aligned zinc oxide nanorods on paper substrates for low-cost flexible electronics. *Advanced Materials*, 2010. **36**(22): p. 4059–4063.
59. Boruah, B.D., et al., Few-layer graphene/ZnO nanowires based high performance UV photodetector. *Nanotechnology*, 2015. **26**(23): p. 235703.

60. Alenezi, M.R., S.J. Henley, and S. Silva, On-chip fabrication of high performance nanostructured ZnO UV detectors. *Scientific Reports*, 2015. **5**(1): p. 8516.
61. Huang, X., et al., Correlating the enhancement of UV luminescence from solution-grown ZnO nanorods with hydrogen doping. *CrystEngComm*, 2012. **14**(16): p. 5163–5165.
62. Chang, L.-W., et al., Enhanced optoelectronic performance from the Ti-doped ZnO nanowires. *Journal of Applied Physics*, 2011. **109**(7).
63. Cui, J., et al., Low-temperature growth and characterization of Cl-doped ZnO nanowire arrays. *The Journal of Physical Chemistry C*, 2008. **112**(12): p. 4475–4479.
64. Lee, J.H., et al., Growth-controlled surface roughness in Al-doped ZnO as transparent conducting oxide. *Nanotechnology*, 2009. **20**(39): p. 395704.
65. Escobedo-Morales, A. and U. Pal, Defect annihilation and morphological improvement of hydrothermally grown ZnO nanorods by Ga doping. *Applied Physics Letters*, 2008. **93**(19), 193120 (1-3).
66. Mahmood, K., S.B. Park, and H.J. Sung, Retracted Article: Enhanced photoluminescence, Raman spectra and field-emission behavior of indium-doped ZnO nanostructures. *Journal of Materials Chemistry C*, 2013. **1**(18): p. 3138–3149.
67. Wang, D., et al., Defects-mediated energy transfer in red-light-emitting Eu-doped ZnO nanowire arrays. *The Journal of Physical Chemistry C*, 2011. **115**(46): p. 22729–22735.
68. Kouklin, N., Cu-doped ZnO nanowires for efficient and multispectral photodetection applications. *Advanced Materials*, 2008. **20**(11): p. 2190–2194.
69. Tang, R., et al., Size-controlled graphene nanodot arrays/ZnO hybrids for high-performance UV photodetectors. *Advanced Science*, 2018. **5**(1): p. 1700334.
70. Dang, V.Q., et al., Ultrahigh responsivity in graphene–ZnO nanorod hybrid UV photodetector. *Small*, 2015. **11**(25): p. 3054–3065.
71. Boruah, B.D., et al., Highly dense ZnO nanowires grown on graphene foam for ultraviolet photodetection. *ACS Applied Materials & Interfaces*, 2015. **7**(19): p. 10606–10611.
72. Ok, J.G., et al., Rapid anisotropic photoconductive response of ZnO-coated aligned carbon nanotube sheets. *ACS Applied Materials & Interfaces*, 2014. **6**(2): p. 874–881.
73. Boruah, B.D. and A. Misra, Conjugated assembly of colloidal zinc oxide quantum dots and multiwalled carbon nanotubes for an excellent photosensitive ultraviolet photodetector. *Nanotechnology*, 2016. **27**(35): p. 355204.
74. Liang, S., et al., ZnO Schottky ultraviolet photodetectors. *Journal of Crystal Growth*, 2001. **225**(2-4): p. 110–113.
75. Guo, D., et al., Oxygen vacancy tuned Ohmic-Schottky conversion for enhanced performance in β-Ga2O3 solar-blind ultraviolet photodetectors. *Applied Physics Letters*, 2014. **105**(2), 023507(1-3).
76. Nie, B., et al., Monolayer graphene film on ZnO nanorod array for high-performance Schottky junction ultraviolet photodetectors. *Small*, 2013. **9**(17): p. 2872–2879.
77. Zhang, T.-F., et al., A sensitive ultraviolet light photodiode based on graphene-on-zinc oxide Schottky junction. *Nanophotonics*, 2017. **6**(5): p. 1073–1081.
78. Leung, Y.H., et al., ZnO nanowires array pn homojunction and its application as a visible-blind ultraviolet photodetector. *Applied Physics Letters*, 2010. **96**(5), 053102(1-3).
79. Wang, G., et al., ZnO homojunction photodiodes based on Sb-doped p-type nanowire array and n-type film for ultraviolet detection. *Applied Physics Letters*, 2011. **98**(4), 041107(1-3).
80. Tsai, S.-Y., M.-H. Hon, and Y.-M. Lu, Fabrication of transparent p-NiO/n-ZnO heterojunction devices for ultraviolet photodetectors. *Solid-State Electronics*, 2011. **63**(1): p. 37–41.
81. Tian, W., et al., Low-cost fully transparent ultraviolet photodetectors based on electrospun ZnO-SnO2 heterojunction nanofibers. *Advanced Materials*, 2013. **25**(33): p. 4625–4630.

82. Hu, L., et al., An optimized ultraviolet-a light photodetector with wide-range photoresponse based on ZnS/ZnO biaxial nanobelt. *Advanced Materials*, 2012. **24**(17): p. 2305–2309.
83. Weng, W., et al., A lateral ZnO nanowire photodetector prepared on glass substrate. *Journal of The Electrochemical Society*, 2009. **157**(2): p. K30.
84. Bera, A. and D. Basak, Photoluminescence and photoconductivity of ZnS-coated ZnO nanowires. *ACS Applied Materials & Interfaces*, 2010. **2**(2): p. 408–412.
85. Shao, D., et al., High responsivity, fast ultraviolet photodetector fabricated from ZnO nanoparticle–graphene core–shell structures. *Nanoscale*, 2013. **5**(9): p. 3664–3667.
86. Boruah, B.D., A. Mukherjee, and A. Misra, Sandwiched assembly of ZnO nanowires between graphene layers for a self-powered and fast responsive ultraviolet photodetector. *Nanotechnology*, 2016. **27**(9): p. 095205.
87. Shin, D.H. and S.-H. Choi, Graphene-based semiconductor heterostructures for photodetectors. *Micromachines*, 2018. **9**(7): p. 350.
88. Yingying, L., et al., Facile fabrication of UV photodetectors based on ZnO nanorod networks across trenched electrodes. *Journal of Semiconductors*, 2009. **30**(6): p. 063004.
89. Hao, Y., et al., Facile fabrication of UV photodetector based on spatial network of tetrapod ZnO nanostructures. *Micro & Nano Letters*, 2012. **7**(3): p. 200–203.
90. Park, N., et al., High efficiency NiO/ZnO heterojunction UV photodiode by sol–gel processing. *Journal of Materials Chemistry C*, 2013. **1**(44): p. 7333–7338.
91. Hu, L., et al., Stacking-order-dependent optoelectronic properties of bilayer nanofilm photodetectors made from hollow ZnS and ZnO microspheres. *Advanced Materials (Deerfield Beach, Fla.)*, 2012. **24**(43): p. 5872–5877.
92. Zheng, Z., et al., An enhanced UV–Vis–NIR and flexible photodetector based on electrospun ZnO nanowire array/PbS quantum dots film heterostructure. *Advanced Science*, 2017. **4**(3): p. 1600316.
93. Wang, X., et al., Highly wavelength-selective enhancement of responsivity in Ag nanoparticle-modified ZnO UV photodetector. *ACS applied materials & interfaces*, 2017. **9**(6): p. 5574–5579.
94. Lu, J., et al., Improved UV photoresponse of ZnO nanorod arrays by resonant coupling with surface plasmons of Al nanoparticles. *Nanoscale*, 2015. **7**(8): p. 3396–3403.
95. Ick Son, D., et al., Photoresponse mechanisms of ultraviolet photodetectors based on colloidal ZnO quantum dot-graphene nanocomposites. *Applied Physics Letters*, 2013. **102**(2), 021105(1-3).
96. Boruah, B.D., Zinc oxide ultraviolet photodetectors: rapid progress from conventional to self-powered photodetectors. *Nanoscale Advances*, 2019. **1**(6): p. 2059–2085.
97. Wang, R.C., et al., Fabrication of a large-area Al-doped ZnO nanowire array photosensor with enhanced photoresponse by straining. *Advanced Functional Materials*, 2012. **22**(18): p. 3875–3881.
98. Peng, L., L. Hu, and X. Fang, Low-dimensional nanostructure ultraviolet photodetectors. *Advanced Materials*, 2013. **25**(37): p. 5321–5328.
99. Chang, S.-J., et al., Low-frequency noise characteristics of in-doped ZnO ultraviolet photodetectors. *IEEE Photonics Technology Letters*, 2013. **25**(21): p. 2043–2046.
100. Chen, H.-Y., et al., Realization of a self-powered ZnO MSM UV photodetector with high responsivity using an asymmetric pair of Au electrodes. *Journal of Materials Chemistry C*, 2014. **2**(45): p. 9689–9694.
101. Lu, S., et al., Piezotronic interface engineering on ZnO/Au-based Schottky junction for enhanced photoresponse of a flexible self-powered UV detector. *ACS Applied Materials & Interfaces*, 2014. **6**(16): p. 14116–14122.

102. Bie, Y.Q., et al., Self-powered, ultrafast, visible-blind UV detection and optical logical operation based on ZnO/GaN nanoscale p-n junctions. *Advanced Materials*, 2011. **23**(5): p. 649–653.
103. Ning, Y., et al., Novel transparent and self-powered UV photodetector based on crossed ZnO nanofiber array homojunction. *Small*, 2018. **14**(13): p. 1703754.
104. Zhang, Z., Y. Ning, and X. Fang, From nanofibers to ordered ZnO/NiO heterojunction arrays for self-powered and transparent UV photodetectors. *Journal of Materials Chemistry C*, 2019. **7**(2): p. 223–229.
105. Boruah, B.D., S.N. Majji, and A. Misra, Surface photo-charge effect in doped-ZnO nanorods for high-performance self-powered ultraviolet photodetectors. *Nanoscale*, 2017. **9**(13): p. 4536–4543.
106. Sarkar, S. and D. Basak, Self powered highly enhanced dual wavelength ZnO@ CdS core–shell nanorod arrays photodetector: an intelligent pair. *ACS Applied Materials & Interfaces*, 2015. **7**(30): p. 16322–16329.
107. Li, Q., et al., ZnO nanoneedle/H 2 O solid-liquid heterojunction-based self-powered ultraviolet detector. *Nanoscale Research Letters*, 2013. **8**: p. 1–7.
108. Gao, C., et al., Titanium dioxide coated zinc oxide nanostrawberry aggregates for dye-sensitized solar cell and self-powered UV-photodetector. *Journal of Power Sources*, 2013. **239**: p. 458–465.
109. Zhan, Z., et al., Self-powered, visible-light photodetector based on thermally reduced graphene oxide–ZnO (rGO–ZnO) hybrid nanostructure. *Journal of Materials Chemistry*, 2012. **22**(6): p. 2589–2595.
110. Lin, P., et al., Self-powered UV photosensor based on PEDOT: PSS/ZnO micro/nanowire with strain-modulated photoresponse. *ACS Applied Materials & Interfaces*, 2013. **5**(9): p. 3671–3676.
111. Shen, Y., et al., A self-powered ultraviolet photodetector based on solution-processed p-NiO/n-ZnO nanorod array heterojunction. *RSC Advances*, 2015. **5**(8): p. 5976–5981.
112. Zeng, Y., et al., The enhancement of a self-powered UV photodetector based on vertically aligned Ag-modified ZnO nanowires. *RSC Advances*, 2015. **5**(82): p. 66738–66741.
113. Boruah, B.D., et al., Doping controlled pyro-phototronic effect in self-powered zinc oxide photodetector for enhancement of photoresponse. *Nanoscale*, 2018. **10**(7): p. 3451–3459.
114. Zhou, H., et al., Self-powered, visible-blind ultraviolet photodetector based on n-ZnO nanorods/i-MgO/p-GaN structure light-emitting diodes. *Journal of Materials Chemistry C*, 2015. **3**(5): p. 990–994.
115. Hatch, S.M., J. Briscoe, and S. Dunn, A self-powered ZnO-nanorod/CuSCN UV photodetector exhibiting rapid response. *Advanced Materials*, 2013. **6**(25): p. 867–871.
116. Bai, Z., et al., Self-powered ultraviolet photodetectors based on selectively grown ZnO nanowire arrays with thermal tuning performance. *Physical Chemistry Chemical Physics*, 2014. **16**(20): p. 9525–9529.
117. Yang, L., et al., A self-powered, visible-blind ultraviolet photodetector based on n-Ga: ZnO nanorods/p-GaN heterojunction. *Sensors and Actuators A: Physical*, 2017. **267**: p. 76–81.
118. Shen, Y., et al., Improved photoresponse performance of self-powered ZnO/Spiro-MeOTAD heterojunction ultraviolet photodetector by piezo-phototronic effect. *ACS Applied Materials & Interfaces*, 2016. **8**(9): p. 6137–6143.
119. Zhao, B., et al., An ultrahigh responsivity (9.7 mA W– 1) self-powered solar-blind photodetector based on individual ZnO–Ga2O3 heterostructures. *Advanced Functional Materials*, 2017. **27**(17): p. 1700264.

120. Peng, X., et al., High-performance of self-powered UV photodetector with long-term stability based on ZnO nanorods and an iodine-free quasi solid-state electrolyte. *RSC Advances*, 2017. **7**(47): p. 29440–29445.
121. Peng, W., et al., Temperature dependence of pyro-phototronic effect on self-powered ZnO/perovskite heterostructured photodetectors. *Nano Research*, 2016. **9**(12): p. 3695–3704.
122. Yang, S., J. Gong, and Y. Deng, A sandwich-structured ultraviolet photodetector driven only by opposite heterojunctions. *Journal of Materials Chemistry*, 2012. **22**(28): p. 13899–13902.
123. Li, Y., et al., An environmentally friendly and highly transparent ZnO/cellulose nanocomposite membrane for UV sensing and shielding. *Cellulose*, 2022. **29**(8): p. 4439–4453.
124. Inamdar, S., V. Ganbavle, and K. Rajpure, ZnO based visible–blind UV photodetector by spray pyrolysis. *Superlattices and Microstructures*, 2014. **76**: p. 253–263.
125. Rise, M.S., et al., Synthesis and characterization of ZnO nanorods-Zn2SiO4 nanoparticles-PMMA nanocomposites for UV-C protection. *Optical Materials*, 2022. **123**: p. 111922.
126. Yi, F., et al., Simple fabrication of a ZnO nanorod array UV detector with a high performance. *Physica E: Low-dimensional Systems and Nanostructures*, 2014. **61**: p. 180–184.
127. Gomez, J.L. and O. Tigli, Zinc oxide nanostructures: from growth to application. *Journal of Materials Science*, 2013. **48**: p. 612–624.
128. Hu, J., et al., Research advances in ZnO nanomaterials-based UV photodetectors: a review. *Nanotechnology*, 2023. **34**(23): p. 232002.
129. Liu, H. and J. Yang. *Preparation and properties of ZnO-based ultraviolet photodetectors*. in *International Conference on Agri-Photonics and Smart Agricultural Sensing Technologies (ICASAST 2022)*. 2022. SPIE.
130. Huang, H., et al., Performance enhancement of ZnO ultraviolet detector by localized surface plasmon resonance of Al nanoparticles. *Applied Physics A*, 2021. **127**: p. 1–7.
131. Liu, K., M. Sakurai, and M. Aono, ZnO-based ultraviolet photodetectors. *Sensors (Basel)*, 2010. **10**(9): p. 8604–8634.
132. Anitha, R., et al., Large area ultraviolet photodetector on surface modified Si: GaN layers. *Applied Surface Science*, 2018. **435**: p. 1057–1064.

7 ZnO Nanostructures-Based Electrochemical Sensors for Medical Applications

Kanchan Bala

7.1 INTRODUCTION

Nanotechnology and nanoscience encompass the creation of fine materials and structures with applications in sciences, chemistry, physics, engineering, and material sciences. These breakthroughs in nanotechnology advancements have given rise to innovative techniques in diverse fields, including pharmaceutical analysis, medical diagnostics, and environmental monitoring. These interdisciplinary contributions have significantly opened the path for novel and enhanced methodologies in the advancement of electrochemical sensors with exceptional selectivity and sensitivity [1–3]. Nanostructural materials (NSMs) possess distinctive physiochemical properties such as enhanced adsorption capacity, high reactivity, and significant charge-to-volume ratio enhancing their effectiveness in analytical and medical fields. The expanded surface area allows high interaction with the target and can selectively capture specific molecules of interest in analytical detection. Additionally, nanostructures functionalized with various surface modifications with functional groups enhance the sensitivity and selectivity of NSMs to the desired target which further provides a valuable platform in the advancement of electrochemical sensors and their utilization in the field of pharmaceutical, environmental, industrial, agricultural, biomedical diagnosis, and analytical applications [4–6].

Semiconducting metal oxide has been widely used as efficient NSMs in environmental remediation, high-performance electronics, and sensing applications such as remarkable catalytic, optical, physicochemical, and electronic properties [7–9]. Among various metal oxides, ZnO has been recognized as a viable, promising platform and attracted paramount attention due to its commendable and intriguing properties including its affordability, nontoxicity, good transparency, abundance, long-term stability, and wide band gap [10]. The remarkable and outstanding popularity of ZnO and its nanostructures is consistently surging. ZnO is classified as an n-type semiconductor material and possesses a bandgap of 3.37 eV. The wide band gap ensures good transparency to visible light making it suitable for optoelectronics.

DOI: 10.1201/9781003464907-7

A significant exciton binding energy of 60 meV indicates strong light and matter interactions, which play a major role in the photovoltaic and light-emitting devices.

The various routes are employed by researchers in the production of ZnO NS. The sol–gel, hydrothermal, and biological methods are highly adaptable techniques in the nanofabrication of ZnO NSs. In the sol–gel method, the hydrolysis of precursors results in the creation of a colloidal suspension known as a sol. Subsequent polymerization transforms the liquid sols into gels, ultimately yielding homogeneous and ultrafine powder [11]. J.N. Hasnidawani *et al.* significantly contributed to the metal oxide synthesis by sol–gel pathway for various applications [12]. The researchers communicated the transformation of sols from liquid phase to a gel form, wherein the viscosity reaches an infinitely high level. In a separate study, Alias and his group presented the ZnO synthesis under varying pH conditions by sol–gel method [13].

The other important method of ZnO synthesis is the hydrothermal method. It is employed under ambient and precisely controlled conditions, wherein the temperature is more than 100°C. This method is typically utilized for synthesizing particles of smaller sizes. In a study by Marlinda *et a*l., they successfully and effectively synthesized a hybrid nanocomposite of ZnO nanorods (ZnO NRs) and reduced graphene oxide (rGO) by employing hydrothermal methods. They noticed that variations in the density distribution and morphology of ZnO NRs and optical properties were influenced by the concentration of graphene oxide [14]. Zak *et al.* have contributed extensively to the research on the synthesis of ZnO NSs, particularly focusing on nano ZnO in various dimensions and employing the hydro-gel method [15].

The preparation of ZnO NSs with plant materials, such as roots, fruits, peels, flowers, stems, and leaves, has earned significant concern in all fields of science and is commonly referred to as the green or biological synthesis method. While chemical and physical methods have long been used for synthesizing nanostructures, the latest advancements in green chemistry have emerged as a non-toxic approach for generating NSs. The green synthesis approach holds promise as an alternative to traditional methods, offering numerous advantages to control the morphology without the use of noxious and expensive solvents. The scientists and researchers from various areas have utilized the green synthesis to prepare ZnO NSs [16, 17]. Mirzaei and Darroudi, for instance, highlighted the use of green and economical sources such as algae, fungi, and polymers (gelatin and starch) for synthesizing ZnO NSs and emphasized their significance in biomedical applications [18].

Continuous research and development endeavors focused on a rich variety of ZnO NSs including 0D (nanoparticles/quantum dots), 1D (nanowires/nanotubes/nanorods), 2D (nanodiscs/nanowalls), and 3D (hollow spheres and nanoflowers) are unlocking new possibilities and paving the way for advancements in fields ranging from electronics, environmental remediation, electrochemical sensing to medical applications. The utilization of 0D nanostructures enables the creation of a greater number of active sites, resulting in a significantly larger surface area. On the other side, 1D nanostructure establishes an enduring and direct pathway, facilitating electron transport rate. 2D possesses a distinctive polar surface that helps with the immobilization process. Ultimately, 3D nanostructures generate additional surface area as a result of their geometric volume, thereby enhancing their potential for various applications. The sensing performance of all these morphologies coupled with its binding ability toward

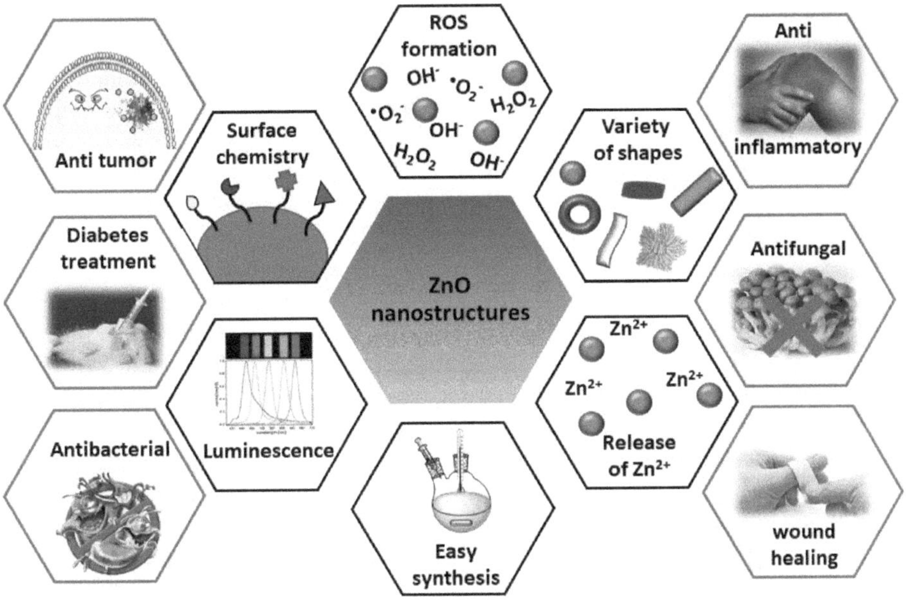

FIGURE 7.1 Main characteristics of ZnO NSs and their principal applications in biomedicine.

analyte enables ZnO suitability for the applications of electrochemical sensors in the biomedical arena as shown in Figure 7.1.

The aim and primary focus of this chapter is to provide concise and compact knowledge to the researchers about the utilization of ZnO NSs as electrochemical sensors for the sensing of glucose, uric acid (UA), cholesterol, hemoglobin (Hb), and urea. We will cover the developments and achievements made in this area in the past five years which will help the readers for future developments and research in related fields.

7.2 ZnO AS AN ELECTROCHEMICAL SENSOR AND THEIR APPLICATIONS IN MEDICAL DIAGNOSIS

In the field of medical diagnosis, the need for quick and dependable analytical techniques and equipment is paramount. Electrochemical sensors and biosensors are considered optimal candidates and have emerged as powerful tools due to their properties such as reliable and rapid analysis, constant monitoring, portability, and reliability. These sensors have been extensively employed for clinical diagnosis purposes. In this section, we will delve into the progress and evolution of ZnO NSs-based electrochemical sensors specifically designed for clinical diagnosis, focusing on the detection of glucose, UA, cholesterol, Hb, and urea.

7.2.1 Glucose

Glucose is a vital component in our bodies, serving as a primary source of energy for various physiological processes. It is a simple sugar that is derived from the digestion of carbohydrates in our diet. Glucose plays a vital role in providing energy to cells, regulating metabolism, supporting brain function, glycogen storage, regulation of hormones, and contributing to the proper functioning of various physiological processes in the body. The efficient detection of glucose levels is necessary for maintaining optimal health, managing diabetes, and facilitating scientific advancements in understanding metabolic processes.

ZnO nanostructured-based electrochemical sensors have indeed made a significant impact in the detection of glucose levels in bodily fluids, particularly in the context of diabetes management (Figure 7.2). These sensors functionalized with enzymes like glucose oxidase (GOx) help in the oxidation of glucose and lead to the production of electrical signals that correlate directly with the concentration of glucose in the sample. The rapid and accurate measurement of glucose concentrations using ZnO NSs-based electrochemical sensors is crucial for monitoring and managing diabetes. The glucose measurements are often performed in conjunction with electrolyte and blood gas analysis. A perusal literature survey of the past five years reveals that various researchers have developed a variety of ZnO NSs for glucose sensing.

Ridhun *et al.* fabricated highly oriented ZnO NRs arrays by employing a hydrothermal route. These ZnO NRs presented a suitable matrix for GOx immobilization. The prepared Nafion/GO/ZnO/ITO electrode exhibited a linear response to glucose, with a sensitivity of 48.75 μA/mM and a Michaelis–Menten constant of

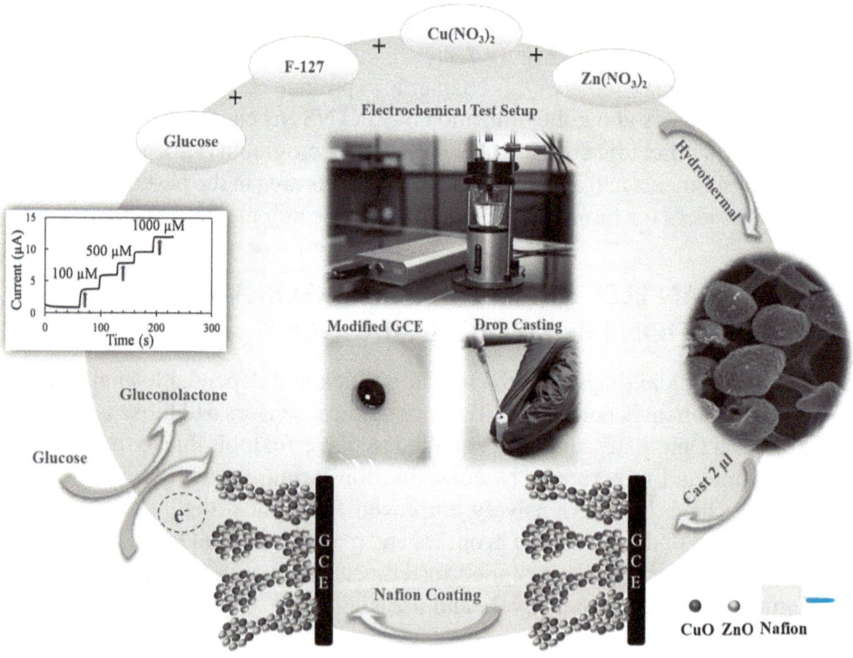

FIGURE 7.2 A nonenzymatic CuO/ZnO biosensor for glucose determination in human blood samples.

Nafion/GOx/ZnO/ITO electrode displayed a linear response to glucose with a sensitivity of 48.75 μA/mM and Michaelis–Menten constant of K_m = 0.34 mM that indicates the strong affinity of immobilized enzymes for glucose. The limit of detection (LOD) was found to be 0.06 mM [19].

One research group synthesized a cost-effective nonenzymatic Cu-doped ZnO nanoparticles (NPs) glucose sensor with sol–gel pathway. The Cu ions substitute zinc sites and there is no change in the hexagonal wurtzite structure doping of copper increases the electrochemical properties of ZnO NPs, provides more catalytic active sites, and more electron transfer rate which further enhances the glucose oxidation. It exhibits an OD of 10^{-9} M in a linear detection range of glucose from 10^{-9} M to 10^{-5} M [20]. The same research group reported that varying concentrations of copper improved the physicochemical properties of the developed sensor. After optimizing the experimental conditions, a linear response between glucose concentration and current density was made in the range from 1 nM to 100 μM. The LOD of the prepared sensor was found to be 0.7 nM. The developed sensor was effectively tested for detecting glucose in human serum samples [21].

An electrochemical sensor with remarkable sensitivity for glucose with ZnO–CeO_2 nanocomposite-modified glassy carbon electrode (GCE) was presented by Qian *et al.* The electrochemical analysis of the glucose sensor involves cyclic voltammetry (CV) and differential pulse voltammetry (DPV) within a potential range of 0–1 V in potassium buffer sulfate (PBS) with a pH = 7.2. The sensor exhibited a response to glucose within a range of 0.5– 300 μM with LOD of 0.224 μM. The modified electrochemical sensor showed sensitivity, stability, simplicity, nontoxicity, and high reproducibility [22].

Vertically grown ZnO NRs functionalized with Fe_2O_3 were investigated for the *in vivo* oxidation of glucose by Mohammed Marie *et al.* The surface of the as-grown ZnO NRs was modified with Fe_2O_3 using a dip coating technique. The fabricated sensor was characterized amperometrically using three electrodes: ZnO NRs/Fe_2O_3/Nafion, Pt electrode, and silver electrode as the working, counter, and reference electrodes, respectively. The modified glucose sensor demonstrated a high sensitivity for glucose of 0.052 μA cm^{-2} $(mg/dl)^{-1}$, a LOD of 0.95 mM, and a rapid response time of about 1 s. The device continues to function accurately for multiple measurements with good reproducibility [23].

A ZnO nanowires-based electrochemical sensor decorated with graphene nanoplates with high proficiency for glucose detection was reported. The combined effects of ZnO nanowires and graphene nanoplates have resulted in strong electron transfer, adsorption, stability, and high surface volume ratio of the proposed sensor. The sensor exhibited a response time of 5 s and a wide detection range of 0.003–30000 mg/dL for glucose measurements. The repeatability, reproducibility, and excellent stability are the magnificent characteristics of the sensor [24].

A nonenzymatic glucose sensor decorated with copper oxide nanocrystals on the surface of ZnO nanopowder was developed by Cheng *et al.* for the glucose concentration. A 1.5–10.0 mM concentration and LOD of 0.0139 mM were found for linear fitting of the modified GCE in the amperometric response [25].

A summary of different ZnO NSs for glucose sensing is given in Table 7.1.

TABLE 7.1
Summary of ZnO NSs for Glucose Sensing

S. No	Morphology	Doping/Hybrid	Technique	Analyte	Detection Limit	References
1	ZnO NRs	NiO Quantum dots	Flexible field effect transistor	Glucose	10 mM	[26]
2	ZnO NRs	Au NPs	Amperometric	Glucose	0.2 μM	[27]
3	ZnO NPs	CeO_2 Whiskers	Voltammetry	Glucose	0.224 μM	[28]
4	ZnO NRs	Graphene oxide (GO)	Voltammetry	Glucose	12 mM	[29]
5	ZnO NRs	Immobilization of GO on chitosan film	Voltammetry	Glucose	—	[30]
6	ZnO NPs	Fe	Chronoamperometry	Glucose	0.30 μM	[31]
7	ZnO NRs	GO	Field effect transistor	Glucose	1 μM	[32]
8	ZnO NSs	Mg	Voltammetry	Glucose	—	[33]
9	ZnO thin films	Al	Fluorometric	Glucose	20 μM	[34]
10	Vertically aligned ZnO nanotubes (NTs)	—	Fluorometric	Glucose	70 μM	[35]
11	ZnO hollow spheres	CuO	Voltammetry	Glucose	357.5 nM	[36]
12	ZnO NPs	Nitrogen-doped carbon sheet	Chronoamperometry, DPV	Glucose	6.3 μM	[37]
13	ZnO bird nest structure	GO	Impedimetric	Glucose	0.004 mM	[38]

7.2.2 Uric Acid (UA)

UA is an important natural antioxidant, an indicator of kidney function and the main purine metabolites present within the bloodstream and urine samples of both humans and various animal species throughout the body. UA produced within the human body is excreted by the kidneys and intestinal tract without undergoing any additional metabolic processes. The production of UA is regulated by several enzymes, including xanthine oxidase and purine nucleoside phosphorylase. Once produced, UA is released into the bloodstream. The kidneys play a significant role in regulating the concentration of UA in the body. However, in some cases, the production of UA may exceed its excretion or the kidneys may have difficulty removing it efficiently. A condition characterized by elevated levels of UA in the blood is known as hyperuricemia. Imbalances in purine metabolism within the human body can result in the excessive production of UA. Deviations in UA levels can contribute to the development of common ailments such as gout, arthritis, insulin resistance, hypertension, Parkinson's disease, hematic disease, and neurological diseases [39]. Hence, the precise measurement of UA in various biological fluids holds significant clinical value. ZnO NSs-based electrochemical sensors have been developed to measure UA levels in biological fluids, detect and quantify UA and these sensors offer fast and accurate results.

Wang *et al.* prepared a hybrid of ZnO NRs coated with graphene sheets by ultrasonic treatment. CV and DPV were employed to assess the electrochemical characteristics of the hybrid. The diameter and length of ZnO NRs were found ~50 nm and 5 μm. The sensitivity of the sensor for UA was found to be 0.3 μMA^{-1} cm^{-2} in the range of 5–80 μM and it showed good repeatability and stability. The UA concentration in urine samples was about 0.5 μM [40]. In one research work, ZnO quantum dots (QDs) were synthesized through precipitation methods. Afterward, the synthesized ZnO QDs were functionalized with uricase enzymes to fabricate UA sensors. To create a selective UA sensor, an uricase was employed to functionalize ZnO QDs. The resulting sensor exhibited a linear response spanning from 1 mM to 10 mM, demonstrating a high sensitivity of 4.0 $\mu A/mMcm^{-2}$. The LOD of UA was found at 22.97 μM [41].

Dutta *et al.* have successfully developed an electrochemical sensor specifically for the detection of UA in human blood serum. The biosensor utilizes ZnO nanoflowers (ZnO NFs) prepared through a hydrothermal process. Impressive results were obtained, with the fabricated sensor demonstrating a high sensitivity of 10.38 μA/ $mMcm^{-}$ and a LOD of 0.13 mM [42]. The detection range of the biosensor spans from 0.005 mM to 1.0 mM. Recently, one author has synthesized ZnO NPs electrochemical sensors using crude black pepper seed extract as a stabilizing, capping, and bio-reducing agent using microwave irradiation techniques. The ZnO NPs synthesized by this technique appeared to be a flake-shaped morphology. The electrocatalytic activity of fabricated ZnO NPs against UA was studied using CV and DPV. The sensor exhibited a sensitive sensitivity of 40.485 $\mu A/mMcm^{-2}$, with a LOD of 1.65 μM. It also demonstrated a linear dynamic range spanning from 50 to 500 μM [43]. The various ZnO NSs for UA sensing are summarized in Table 7.2.

7.2.3 Cholesterol

Cholesterol plays a critical role as a biomarker for various diseases and serves as a precursor for the production of essential biological compounds, including hormones and vitamin D. Foods of animal origin, such as eggs, meat, milk, and seafood, are sources of cholesterol. Elevated levels of cholesterol in the blood serve as an indicator for various clinical conditions, including cerebral thrombosis, arteriosclerosis, heart disease, coronary artery disease, and hypertension. Therefore, it is crucial to accurately and selectively measure cholesterol levels in biological samples to ensure sensitive and precise determination of cholesterol. Electrochemical sensors or biosensors have been developed to measure cholesterol levels in biological samples with high sensitivity and specificity. In this review, recent developments in ZnO nanostructure-based electrochemical cholesterol sensors are reviewed.

A ZnO NR with silver NPs-based highly sensitive nonenzymatic biosensor for cholesterol concentration monitoring was developed by one research group [54]. The sensor response was examined using CV and amperometric response was examined using PBS solution. The nonenzymatic cholesterol sensor demonstrated a linear response in the range of 1–9 mM when operating under optimal conditions. The developed

TABLE 7.2
Summary of ZnO NSs for UA Sensing

S. No.	Morphology	Doping/Hybrid	Technique	Analyte	Detection Limit	References
1	Vertically arranged ZnO NRs	Fluorine-doped tin oxide	DPV	UA	2.5 μM	[44]
2	ZnO NPs	GCE	Voltammetry	UA	0.11 μM	[45]
3	ZnO NPs	Zirconium	CV	UA	10 μM	[46]
4	ZnO	GO	Amperometry	UA	0.45 μM	[47]
5	ZnO NPs	Carboxylic acid functionalized carbon nanotubes	Chronoamperometry	UA	2.3 μM	[48]
6	ZnO NSs	KCl	DPV	UA	0.5 μM	[49]
7	ZnO NPs	Urea	Voltammetry	UA	8.49 nM	[50]
8	ZnO nanotubes	Nickel	Voltammetry	UA	5.72 nM	[51]
9	ZnO thin films	Lithium tantalate	Amperometric	UA	5 μM	[52]
10	ZnO	GO	Voltammetry	UA	0.33 μM	[53]

Ag-NPs/ ZnO-NRs-based biosensor sensitivity was 4.2 μA mM^{-1} cm^{2}. The detection limit was 1.84 mM. The sensors exhibited excellent reproducibility, stability, and reusability.

An enzyme-based electrochemical cholesterol biosensor was developed by utilizing a nanocomposite composed of gold NPs, ZnO NPs, and multi-wall carbon nanotubes (MWCNTs). The fabricated biosensor demonstrated promising performance. The Au/ZnO/MWCNTs electrode has displayed excellent analytical linear performance. It exhibited a detection range of cholesterol in human serum of 0.1–100 μM, LOD of 0.1 μM, and high sensitivity of 25.89 μA/μM from DPV data [55].

7.2.4 Hemoglobin (HB)

Hb plays an important role in diagnosing and monitoring various blood disorders. Any change in the levels or structure of Hb can lead to various diseases, including anemia and potentially life-threatening conditions. Measuring Hb levels accurately is essential for diagnosing and monitoring these blood disorders. Nowadays, the development of portable, rapid, real-time measurement electrochemical sensors has provided a promising alternative for Hb detection. Their application in diagnostic medicine can help in the early detection, monitoring, and management of blood disorders, thereby improving patient care and outcomes. Hb serves as a crucial hematological marker in diagnostic medicine, making it essential to accurately measure the levels of this hemoprotein in various human pathologies.

A thermal responsive protein imprinted polymers (TPIPs) were electrochemically prepared on the surface of ZnO NFs. The TPIPs-ZnO sensor was successfully utilized for the detection of Hb using DPV. Under ambient conditions, the biosensor exhibited a wide detection range for Hb from 1×10^{-13} to 1×10^{-1} mg/L, along with a LOD of 3.1×10^{-14} mg/L (S/N = 3). The TPIPs demonstrated excellent temperature response, selectivity, and repeatability, making them highly valuable for polymer-modified electrodes and sensitive Hb detection in the real sample [56].

Chauhan *et al.* conducted a study where they synthesized properly aligned ZnO NRs on an indium-tin oxide-coated glass plate (ITO) electrode. These ZnO NRs were then utilized for immobilizing fructose amino acid oxidase (FAO) enzymes through the physical adsorption method. The resulting biosensor demonstrated excellent sensitivity and a LOD of 0.1 μM. It exhibited a rapid response time of 4 s and a linear range from 0.1 to 2000 μM. The enzyme/working electrode remained stable for approximately 4 months when stored at 4°C. This biosensor method holds great promise for accurately detecting HbA1c in the blood samples obtained from both healthy individuals and diabetic patients [57].

7.2.5 Urea

Urea is a waste product that is produced when proteins are broken down in the body. The level of urea in biological samples can provide important information about the functioning of the kidneys. The kidneys play a pivotal role in filtering urea from the blood and excreting it in urine. Physicians use tests such as urine urea nitrogen (UUN) and blood urea nitrogen (BUN) to evaluate kidney function. The reduced urea levels may be seen in cases of low protein intake or dehydration. However, it is important to note that high and low urea levels are not always indicative of kidney failure alone. Other factors such as high protein intake or dehydration can also affect urea levels. Therefore, to accurately assess kidney function, it is crucial to have a direct, rapid, and sensitive measurement of urea levels in biological fluids.

Babitha *et al.* described a green and sustainable ZnO@rGO nanocomposite. These nanoarchitectures were constructed on a GCE, and their electrochemical performance is evaluated for the detection of urea with CV. This developed urea sensor exhibits sensitivity in the concentration range 0.02×10^{-3} mM to 7.0×10^{-3} mM with a LOD of 0.012 μM. These sensors demonstrate high sensitivity, with 682.8 μA mM^{-1} sensitivity toward urea [58].

A cost-effective nanocomposite of Fe_2O_3/ZnO and CuO/ZnO as an electrochemical sensor modified on a pencil graphite was proposed for the effective and accurate detection of urea. The results indicate that Fe_2O_3/ZnO coated graphite electrode exhibits a good response and improved sensitivity to urea as compared to CuO/ZnO. CV studies further confirm the sensing capability of Fe_2O_3/ZnO for urea detection, in a range of 0.8 μg/mL to 4 μg/mL and a LOD of 2.5 μg/mL [59].

Mikani *et al.* developed a unique (Fe_3O_4-ZnO/SnO_2:F) nanocomposite immobilized with the urease enzyme for sensitive detection of urea. The improvement in urea biosensing through the electrostatic adsorption of urease onto ZnO NPs was studied. The performance of the biosensor for urea quantification is evaluated using electrochemical techniques such as CV and electrochemical impedance spectroscopy. The results demonstrate exceptional sensitivity for urea in the concentration range of

5–150 mg/dL, with LOD of 5 mg/dL. This suggests that this method has the potential to produce disposable sensors with fast response time and repeatability [60].

Yoon *et al.* proposed a method for synthesizing silver-coated ZnO NRs and nanoflakes with varying crystallographic orientations. This involved a combination of sputter deposition and solution growth processes. Their study found that Ag/ZnO NRs, when used on carbon electrodes, exhibited higher catalytic activity and improved efficiency compared to Ag/ZnO nanoflakes on carbon electrodes. The superior performance of Ag/ZnO NRs can be attributed to their larger electrochemically active surface area of 169 cm^2 mg^{-1} on the carbon electrode, which facilitated easier electron transfer and promoted urea electrooxidation. Additionally, the Ag/ZnO NRs catalysts showed a significant reduction in the onset voltage (0.41 V vs Ag/AgCl) and an increase in current density (12.0 mA cm^{-2} mg^{-1}) at 0.55 V [61].

Ahmad *et al.* presented a novel urea biosensor that is constructed using high aspect ratio ZnO NRs vertically grown on an electrode surface through a simple one-step low-temperature solution process. The resulting electrodes demonstrate improved performance and exhibited a highly sensitive response with a sensitivity of 239.67 in a linear range of 0.01–4.56 mM. The biosensor also exhibits a rapid response time of approximately 3 s, a low detection limit of 5 nM, and a low Michaelis–Menten constant (Km) value of 0.025 mM. Furthermore, the biosensor demonstrates reproducibility, long-term stability, and high selectivity [62]. A few of the ZnO NSs used in urea sensing are given in Table 7.3.

TABLE 7.3
Summary of ZnO NSs for Urea Sensing

S. No.	Morphology	Doping/ Hybrid	Technique	Analyte	Detection Limit	References
1	ZnO NRs	Fe_2O_3	Electrolyte gate transistor	Cholesterol	60 mM	[63]
2	ZnO nanoporous	Tin oxide	Impedimetric	Urea	0.40 mM	[64]
3	ZnO NRs	Ag	Chronoamperometry	Urea	13.98 μM	[65]
4	ZnO NPs	Electrospun nanofibers	Linear sweep voltammetry	Urea	0.011 mg dL^{-1}.	[66]
5	ZnO	Polyaniline	Voltammetry	Urea	10 nM	[67]
6	ZnO NSs	TiO_2 thin film	Impedimetric	Urea	2 mg dL^{-1}	[68]
7	ZnO nanocrystals	Carbon nanotubes	Amperometric	Urea	50 mM	[69]
8	ZnO NPs	Polyaniline-Chitosan	Potentiometric	Urea	29.84 ppm	[70]
9	ZnO NPs	Pt/ZnO/AChE/Chitosan	Voltammetry	Urea	1 PM	[71]
10	ZnO	Multi-walled carbon nanotubes	—	Urea	10 ppm	[72]

7.3 CONCLUSION

Although summarizing all the research findings is a challenging task, this effort focuses on the ZnO NSs as an electrochemical sensor and their applications in the medical field for sensing glucose, UA, cholesterol, Hb, and urea in the past five years. Continuous research and development in this field on the different morphologies of ZnO NSs, and their interactions with bioanalytes contribute to the innovation and utilization of ZnO NSMs as biosensors, enabling the detection of various biomolecules. In conclusion, several advantages, various morphology, and unique properties of ZnO NSMs such as good conductivity, nontoxicity, surface area, sensitivity, selectivity, and biocompatibility have significantly explored its applications in sensing. These advancements have opened up new possibilities for detecting smaller molecules, infectious diseases, and pharmaceutical compounds in scientific research and medical fields.

REFERENCES

1. Medawar-Aguilar, V., Jofre, C. F., Fernández-Baldo, M. A., Alonso, A., Angel, S., Raba, J. & Messina, G. A. (2019). Serological diagnosis of Toxoplasmosis disease using a fluorescent immunosensor with chitosan-ZnO-nanoparticles. *Analytical Biochemistry*, *564*, 116–122.
2. Shetti, N. P., Nayak, D. S., Kuchinad, G. T., & Naik, R. R. (2018). Electrochemical behavior of thiosalicylic acid at γ-Fe2O3 nanoparticles and clay composite carbon electrode. *Electrochimica Acta*, *269*, 204–211.
3. Devarushi, U. S., Shetti, N. P., Bukkitgar, S. D., & Tuwar, S. M. (2018). Electrochemical Behavior of an Anti-Viral Drug Valacyclovir at Carbon Paste Electrode and Its Analytical Application. *Russian Journal of Electrochemistry*, *54*, 760–768.
4. Naik, E. I., Naik, H. B., Viswanath, R., Kirthan, B. R., & Prabhakara, M. C. (2020). Effect of zirconium doping on the structural, optical, electrochemical and antibacterial properties of ZnO nanoparticles prepared by sol-gel method. *Chemical Data Collections*, *29*, 100505.
5. Haque, E., Kim, J., Malgras, V., Reddy, K. R., Ward, A. C., You, J.,… & Yamauchi, Y. (2018). Recent advances in graphene quantum dots: synthesis, properties, and applications. *Small Methods*, *2*(10), 1800050.
6. Mikani, M., Talaei, S., Rahmanian, R., Ahmadi, P., & Mahmoudi, A. (2019). Sensitive electrochemical sensor for urea determination based on F-doped SnO2 electrode modified with ZnO-Fe3O4 nanoparticles transducer: application in biological fluids. *Journal of Electroanalytical Chemistry*, *840*, 285–294.
7. . Hu, Y. Y., Liu, Z., Nam, K. W., Borkiewicz, O. J., Cheng, J., Hua, X.,& Grey, C. P. (2013). Origin of additional capacities in metal oxide lithium-ion battery electrodes. *Nature Materials*, *12*(12), 1130–1136.
8. Gao, J., Wu, B., Cao, C., Zhan, Z., Ma, W., & Wang, X. (2021). Unraveling the dynamic evolution of Pd species on Pd-loaded ZnO nanorods for different hydrogen sensing behaviors. *ACS Sustainable Chemistry & Engineering*, *9*(18), 6370–6379.
9. Ortiz-Casas, B., Galdámez-Martínez, A., Gutiérrez-Flores, J., Ibañez, A. B., Panda, P. K., Santana, G., & Dutt, A. (2021). Bio-acceptable 0D and 1D ZnO nanostructures for cancer diagnostics and treatment. *Materials Today*, *50*, 533–569.
10. Zhang, Q., Dandeneau, C. S., Zhou, X., & Cao, G. (2009). ZnO nanostructures for dye-sensitized solar cells. *Advanced Materials*, *21*(41), 4087–4108.

11. Haque, F. Z., Nandanwar, R., Singh, P., Dharavath, K., & Syed, F. F. (2018). Effect of different acids and solvents on optical properties of SiO 2 nanoparticles prepared by the sol-gel process. *Silicon*, *10*, 413–419.
12. Hasnidawani, J. N., Azlina, H. N., Norita, H., Bonnia, N. N., Ratim, S., & Ali, E. S. (2016). Synthesis of ZnO nanostructures using sol-gel method. *Procedia Chemistry*, *19*, 211–216.
13. Alias, S. S., Ismail, A. B., & Mohamad, A. A. (2010). Effect of pH on ZnO nanoparticle properties synthesized by sol–gel centrifugation. *Journal of Alloys and Compounds*, *499*(2), 231–237.
14. Marlinda, A. R., Huang, N. M., Muhamad, M. R., An'Amt, M. N., Chang, B. Y. S., Yusoff, N.,... & Kumar, S. V. (2012). Highly efficient preparation of ZnO nanorods decorated reduced graphene oxide nanocomposites. *Materials Letters*, *80*, 9–12.
15. Zak, A. K., Razali, R., Majid, W. A., & Darroudi, M. (2011). Synthesis and characterization of a narrow size distribution of zinc oxide nanoparticles. *International journal of nanomedicine*, *6* 1399–1403.
16. Lingaraju, K., Raja Naika, H., Manjunath, K., Basavaraj, R. B., Nagabhushana, H., Nagaraju, G., & Suresh, D. (2016). Biogenic synthesis of zinc oxide nanoparticles using Ruta graveolens (L.) and their antibacterial and antioxidant activities. *Applied Nanoscience*, *6*, 703–710.
17. Rafique, M., Sadaf, I., Rafique, M. S., & Tahir, M. B. (2017). A review on green synthesis of silver nanoparticles and their applications. *Artificial Cells, Nanomedicine, and Biotechnology*, *45*(7), 1272–1291.
18. Mirzaei, H., & Darroudi, M. (2017). Zinc oxide nanoparticles: Biological synthesis and biomedical applications. *Ceramics International*, *43*(1), 907–914.
19. Ridhuan, N. S., Abdul Razak, K., & Lockman, Z. (2018). Fabrication and characterization of glucose biosensors by using hydrothermally grown ZnO nanorods. *Scientific Reports*, *8*(1), 13722.
20. Mahmoud, A., Echabaane, M., Omri, K., El Mir, L., & Chaabane, R. B. (2019). Development of an impedimetric non enzymatic sensor based on ZnO and Cu doped ZnO nanoparticles for the detection of glucose. *Journal of Alloys and Compounds*, *786*, 960–968.
21. Mahmoud, A., Echabaane, M., Omri, K., Boudon, J., Saviot, L., Millot, N., & Chaabane, R. B. (2021). Cu-doped ZnO nanoparticles for non-enzymatic glucose sensing. *Molecules*, *26*(4), 929.
22. Qian, J., Wang, Y., Pan, J., Chen, Z., Wang, C., Chen, J., & Wu, Z. (2020). Non-enzymatic glucose sensor based on ZnO–CeO2 whiskers. *Materials Chemistry and Physics*, *239*, 122051.
23. Marie, M., Manoharan, A., Kuchuk, A., Ang, S., & Manasreh, M. O. (2018). Vertically grown zinc oxide nanorods functionalized with ferric oxide for in vivo and non-enzymatic glucose detection. *Nanotechnology*, *29*(11), 115501.
24. Rafiee, Z., Mosahebfard, A., & Sheikhi, M. H. (2020). High-performance ZnO nanowires-based glucose biosensor modified by graphene nanoplates. *Materials Science in Semiconductor Processing*, *115*, 105116.
25. Cheng, C. E., Tangsuwanjinda, S., Cheng, H. M., & Lee, P. H. (2021). Copper oxide decorated zinc oxide nanostructures for the production of a non-enzymatic glucose Sensor, *Coatings*, 11, 936.
26. Ahmad, R.,& Hahn, Y. B. (2018). Nonenzymatic flexible field-effect transistor based glucose sensor fabricated using NiO quantum dots modified ZnO nanorods. *Journal of Colloid and Interface Science*, *512*, 21–28.
27. Zhou, F., Jing, W., Liu, S., Mao, Q., Xu, Y., Han, F., & Jiang, Z. (2020). Electrodeposition of gold nanoparticles on ZnO nanorods for improved performance of enzymatic glucose sensors. *Materials Science in Semiconductor Processing*, *105*, 104708.

28. Qian, J., Wang, Y., Pan, J., Chen, Z., Wang, C., Chen, J., & Wu, Z. (2020). Non-enzymatic glucose sensor based on ZnO–CeO2 whiskers. *Materials Chemistry and Physics*, *239*, 122051.
29. Mao, Q., Jing, W., Zhou, F., Liu, S., Gao, W., Wei, Z., & Jiang, Z. (2021). Depositing reduced graphene oxide on ZnO nanorods to improve the performance of enzymatic glucose sensors. *Materials Science in Semiconductor Processing*, *121*, 105391.
30. Bagyalakshmi, S., Sivakami, A., & Balamurugan, K. S. (2020). A Zno nanorods based enzymatic glucose biosensor by immobilization of glucose oxidase on a chitosan film. *Obesity Medicine*, *18*, 100229.
31. Raza, W., & Ahmad, K. (2018). A highly selective Fe@ ZnO modified disposable screen printed electrode based non-enzymatic glucose sensor (SPE/Fe@ ZnO). *Materials Letters*, *212*, 231–234.
32. Zong, X., & Zhu, R. (2018). ZnO nanorod-based FET biosensor for continuous glucose monitoring. *Sensors and Actuators B: Chemical*, *255*, 2448–2453.
33. Lin, C. F., Kao, C. H., Lin, C. Y., Liu, Y. W., & Wang, C. H. (2020). The electrical and physical characteristics of Mg-doped ZnO sensing membrane in EIS (electrolyte–insulator–semiconductor) for glucose sensing applications. *Results in Physics*, *16*, 102976.
34. Ghosh, J., Ghosh, R., & Giri, P. K. (2018). Tuning the visible photoluminescence in Al doped ZnO thin film and its application in label-free glucose detection. *Sensors and Actuators B: Chemical*, *254*, 681–689.
35. Mai, H. H., Tran, D. H., & Janssens, E. (2019). Non-enzymatic fluorescent glucose sensor using vertically aligned ZnO nanotubes grown by a one-step, seedless hydrothermal method. *Microchimica Acta*, *186*, 1–10.
36. Haghparas, Z., Kordrostami, Z., Sorouri, M., Rajabzadeh, M., & Khalifeh, R. (2021). Highly sensitive non-enzymatic electrochemical glucose sensor based on dumbbell-shaped double-shelled hollow nanoporous CuO/ZnO microstructures. *Scientific Reports*, *11*(1), 1–12.
37. Muthuchamy, N., Atchudan, R., Edison, T. N. J. I., Perumal, S., & Lee, Y. R. (2018). High-performance glucose biosensor based on green synthesized zinc oxide nanoparticle embedded nitrogen-doped carbon sheet. *Journal of Electroanalytical Chemistry*, *816*, 195–204.
38. Shi, F., Xu, J., Hu, Z., Ren, C., Xue, Y., Zhang, Y.,... & Yang, Z. (2021). Bird nest-like zinc oxide nanostructures for sensitive electrochemical glucose biosensor. *Chinese Chemical Letters*, *32*(10), 3185–3188.
39. Shoaie, N., Daneshpour, M., Azimzadeh, M., Mahshid, S., Khoshfetrat, S. M., Jahanpeyma, F., & Foruzandeh, M. (2019). Electrochemical sensors and biosensors based on the use of polyaniline and its nanocomposites: *A review on recent advances*. *Microchimica Acta*, *186*, 1–29.
40. Wang, Q., Yue, H., Zhang, J., Gao, X., Zhang, H., Lin, X., Bychanok, D. (2017). Electrochemical determination of uric acid in the presence of ascorbic acid by hybrid of ZnO nanorods and graphene nanosheets. *Ionics*, *24*(8), 2499–2507.
41. Ali, M., Shah, I., Kim, S. W., Sajid, M., Lim, J. H., & Choi, K. H. (2018). Quantitative detection of uric acid through ZnO quantum dots based highly sensitive electrochemical biosensor. *Sensors and Actuators A: Physical*, *283*, 282–290.
42. Dutta, P., Sharma, V., Bhardwaj, H., Agrawal, V. V., Rajesh, & Sumana, G. (2022). Fabrication of electrochemical biosensor using zinc oxide nanoflowers for the detection of uric acid. *MAPAN*, *37*(3), 585–595.
43. Ramya, B., & Priya, P. G. (2021). Rapid phytochemical microwave-assisted synthesis of zinc oxide nano flakes with excellent electrocatalytic activity for non-enzymatic electrochemical sensing of uric acid. *Journal of Materials Science: Materials in Electronics*, *32*, 21406–21424.

44. Nagal, V., Kumar, V., Khan, M., AlOmar, S. Y., Tripathy, N., Singh, K., & Ahmad, R. (2021). A highly sensitive uric acid biosensor based on vertically arranged ZnO nanorods on a ZnO nanoparticle-seeded electrode. *New Journal of Chemistry*, *45*(40), 18863–18870.
45. Pan, Y., Zuo, J., Hou, Z., Huang, Y., & Huang, C. (2020). Preparation of electrochemical sensor based on zinc oxide nanoparticles for simultaneous determination of AA, DA, and UA. *Frontiers in Chemistry*, *8*, 592538.
46. Naik, E. I., Naik, H. B., Viswanath, R., Kirthan, B. R., & Prabhakara, M. C. (2020). Effect of zirconium doping on the structural, optical, electrochemical and antibacterial properties of ZnO nanoparticles prepared by sol-gel method. *Chemical Data Collections*, *29*, 100505.
47. Eryiğit, M., Urhan, B. K., Doğan, H. Ö., Özer, T. Ö., & Demir, Ü. (2022). ZnO nanosheets-decorated ERGO layers: An efficient electrochemical sensor for non-enzymatic uric acid detection. *IEEE Sensors Journal*, *22*(6), 5555–5561.
48. Das, S. C., Pandey, R. R., Devkota, T., & Chusuei, C. C. (2018). Raman spectroscopy as an assay to disentangle zinc oxide carbon nanotube composites for optimized uric acid detection. *Chemosensors*, *6*(4), 65.
49. De Penning, R., Monzon, N., & Padalkar, S. (2022). Flexible zinc oxide-based biosensors for detection of multiple analytes. *Journal of Materials Research*, *37*(18), 2942–2950.
50. Kogularasu, S., Akilarasan, M., Chen, S. M., Chen, T. W., & Lou, B. S. (2019). Urea-based morphological engineering of ZnO; for the biosensing enhancement towards dopamine and uric acid in food and biological samples. *Materials Chemistry and Physics*, *227*, 5–11.
51. Mullani, S. B., Tawade, A. K., Tayade, S. N., Sharma, K. K. K., Deshmukh, S. P., Mullani, N. B.,… & Delekar, S. D. (2020). Synthesis of Ni 2+ ion doped ZnO–MWCNTs nanocomposites using an in situ sol–gel method: An ultra sensitive non-enzymatic uric acid sensing electrode material. *RSC Advances*, *10*(61), 36949–36961.
52. Rana, L., Gupta, R., Tomar, M., & Gupta, V. (2018). Highly sensitive Love wave acoustic biosensor for uric acid. *Sensors and Actuators B: Chemical*, *261*, 169–177.
53. Zhang, X., Zhang, Y. C., & Ma, L. X. (2016). One-pot facile fabrication of graphene-zinc oxide composite and its enhanced sensitivity for simultaneous electrochemical detection of ascorbic acid, dopamine and uric acid. *Sensors and Actuators B: Chemical*, *227*, 488–496.
54. Anh, T. T. N., Lan, H., Tam, L. T., Pham, V. H., & Tam, P. D. (2018). Highly sensitive nonenzymatic cholesterol sensor based on zinc oxide nanorods. *Journal of Electronic Materials*, *47*, 6701–6708.
55. Ghanei Agh Kaariz, D., Darabi, E., & Elahi, S. M. (2020). Fabrication of Au/ZnO/MWCNTs electrode and its characterization for electrochemical cholesterol biosensor. *Journal of Theoretical and Applied Physics*, *14*, 339–348.
56. Sun, Y., Li, S., Yang, Y., Feng, X., Wang, W., Liu, Y., & Zhang, Z. (2019). Fabrication of a thermal responsive hemoglobin (Hb) biosensor via Hb-catalyzed eATRP on the surface of ZnO nanoflowers. *Journal of Electroanalytical Chemistry*, *848*, 113346.
57. Chauhan, N., & Jain, U. (2016). Influence of zinc oxide nanorods on the sensitivity of a glycated hemoglobin biosensor. *Advanced Materials Letters*, *7*(8), 666–672.
58. Babitha, K. B., Soorya, P. S., Mohamed, A. P., Rakhi, R. B., & Ananthakumar, S. (2020). Development of ZnO@ rGO nanocomposites for the enzyme free electrochemical detection of urea and glucose. *Materials Advances*, *1*(6), 1939–1951.
59. Dhinasekaran, D., Soundharraj, P., Jagannathan, M., Rajendran, A. R., & Rajendran, S. (2022). Hybrid ZnO nanostructures modified graphite electrode as an efficient urea sensor for environmental pollution monitoring. *Chemosphere*, *296*, 133918.

60. Mikani, M., Talaei, S., Rahmanian, R., Ahmadi, P., & Mahmoudi, A. (2019). Sensitive electrochemical sensor for urea determination based on F-doped SnO2 electrode modified with ZnO-Fe3O4 nanoparticles transducer: application in biological fluids. *Journal of Electroanalytical Chemistry*, *840*, 285–294.
61. Yoon, J., Lee, D., Lee, E., Yoon, Y. S., & Kim, D. J. (2019). Ag/ZnO catalysts with different ZnO nanostructures for non-enzymatic detection of urea. *Electroanalysis*, *31*(1), 17–21.
62. Ahmad, R., Tripathy, N., Ahn, M. S., & Hahn, Y. B. (2017). Solution process synthesis of high aspect ratio ZnO nanorods on electrode surface for sensitive electrochemical detection of uric acid. *Scientific Reports*, *7*(1), 46475.
63. Khan, M., Nagal, V., Masrat, S., Tuba, T., Tripathy, N., Parvez, M. K., & Ahmad, R. (2022). Wide-linear range cholesterol detection using Fe2O3 nanoparticles decorated ZnO nanorods based electrolyte-gated transistor. *Journal of The Electrochemical Society*, *169*(2), 027512.
64. Mozaffari, S. A., Rahmanian, R., Abedi, M., & Amoli, H. S. (2014). Urea impedimetric biosensor based on reactive RF magnetron sputtered zinc oxide nanoporous transducer. *Electrochimica Acta*, *146*, 538–547.
65. Yoon, J., Lee, E., Lee, D., Oh, T. S., Yoon, Y. S., & Kim, D. J. (2017). Communication—highly sensitive Ag/ZnO nanorods composite electrode for non-enzymatic urea detection. *Journal of The Electrochemical Society*, *164*(12), B558.
66. Migliorini, F. L., Sanfelice, R. C., Mercante, L. A., Andre, R. S., Mattoso, L. H., & Correa, D. S. (2018). Urea impedimetric biosensing using electrospun nanofibers modified with zinc oxide nanoparticles. *Applied Surface Science*, *443*, 18–23.
67. Botewad, S. N., Pahurkar, V. G., Muley, G. G., Gaikwad, D. K., Bodkhe, G. A., Shirsat, M. D., & Pawar, P. P. (2020). PANI-ZnO cladding-modified optical fiber biosensor for urea sensing based on evanescent wave absorption. *Frontiers in Materials*, *7*, 184.
68. Rahmanian, R., Mozaffari, S. A., Amoli, H. S., & Abedi, M. (2018). Development of sensitive impedimetric urea biosensor using DC sputtered Nano-ZnO on TiO2 thin film as a novel hierarchical nanostructure transducer. *Sensors and Actuators B: Chemical*, *256*, 760–774.
69. Morais, P. V., Gomes, V. F., Silva, A. C., Dantas, N. O., Schöning, M. J., & Siqueira, J. R. (2017). Nanofilm of ZnO nanocrystals/carbon nanotubes as biocompatible layer for enzymatic biosensors in capacitive field-effect devices. *Journal of Materials Science*, *52*, 12314–12325.
70. Kushwaha, C. S., Singh, P., Abbas, N. S., & Shukla, S. K. (2020). Self-activating zinc oxide encapsulated polyaniline-grafted chitosan composite for potentiometric urea sensor. *Journal of Materials Science: Materials in Electronics*, *31*(14), 11887–11896.
71. Ezhilan, M., Gumpu, M. B., Ramachandra, B. L., Nesakumar, N., Babu, K. J., Krishnan, U. M., & Rayappan, J. B. B. (2017). Design and development of electrochemical biosensor for the simultaneous detection of melamine and urea in adulterated milk samples. *Sensors and Actuators B: Chemical*, *238*, 1283–1292.
72. Barthwal, S., & Singh, N. B. (2020). Urea detection by ZnO-MWCNT nanocomposite sensor. *Materials Today: Proceedings*, *29*, 749–752.

8 Photocatalytic Activity of ZnO Nanostructures

Anu Kapoor, Naveen Kumar, and Suresh Kumar

8.1 INTRODUCTION

Zinc oxide (ZnO) is a multifunctional semiconductor owing to its special characteristics, such as strong chemical stability, radiative absorption across a greater range, increased photostability, and strong electrochemical coupling coefficient [1]. Lying on the margin between ionic and covalent semiconductors, ZnO is regarded as a semiconductor with ionic properties. With a large energy gap of 3.4 eV, high binding energy of 60 meV, good optical gain of 300 cm^{-1}, and strong mechanical and thermal stability at ambient temperature, it is also very stable mechanically and thermally [1–3]. Having a stable wurzite structure, a high free carrier concentration (1021 cm^{-3}), and low Hall mobility (125 $cm^2V^{-1}s^{-1}$), ZnO is a direct band gap semiconductor. Its rigidity, radiation hardness, and piezoelectric qualities make it a crucial component in the ceramics industry [2–4]. Due to its non-toxicity, biocompatibility, and biodegradability, it is a significant biomaterial with applications in biomedicine [4]. Due to its pyroelectric and piezoelectric capabilities, ZnO is used as a sensor, energy converter, and photocatalyst. Because it is so simple to create different ZnO nanostructures and modify their properties, this sophisticated material is a top candidate for applications in the field of nanotechnology in the future [5–7]. ZnO mixed with organic materials has been found to improve the performance of solar devices in solar applications. Due to its higher electron mobility compared to TiO_2, which can also be employed, ZnO thin film is constructed as the electron-carrying layer in solar cells of the fourth generation [8]. Chemical sensors based on ZnO nanostructures have a wide band gap, high sensitivity, high electron mobility, promising response, low cost, quick recovery time, and are simple to fabricate [5, 7–9]. ZnO, one of the most often used non-titanium metal oxides, is a well-known photocatalyst having all the characteristics of photocatalytic degradation. Because of its numerous intriguing features, ZnO is a substance with multiple uses [10]. Figure 8.1 shows the mechanism of photocatalytic degradation of dyes using ZnO NPs.

8.2 GREEN SYNTHESIS OF ZnO NANOPARTICLES

'Biosynthesis of nanoparticles' is the term for the process of producing nanoparticles (NPs) with potential use in biomedicine using microorganisms and plants [11]. These techniques do not entail extreme conditions, such as toxic ingredients, high

 DOI: 10.1201/9781003464907-8

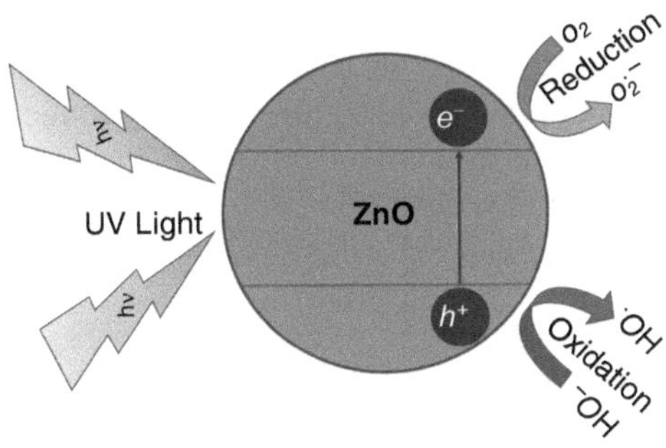

FIGURE 8.1 Photocatalytic degradation of dye using ZnO NPs under exposure to UV light.

temperature, energy, and pressure, as do chemical and physical synthesis processes and as a result, they permit environmental sustainability. Plants, bacteria, fungi, algae, and other living things can all contribute to green synthesis [12]. They make it possible to produce pure ZnO NPs in large quantities on an industrial basis. NPs made by a biomimetic process have increased catalytic activity and need fewer costly and hazardous ingredients [13, 14]. The production of uniformly sized ZnO nanoflowers from *B. licheniformis* cell-soluble proteins showed improved photocatalytic activity and photostability, as evidenced by the 83% degradation of the pollutant dye methylene blue (MB) in the presence of ZnO nanoflowers, even though MB's self-degradation was null (it was observed that MB's self-degradation was 0%). Certain phytochemicals, which function as capping or stabilizing agents as well as reducing agents, are secreted by these natural strains and plant extracts [15].

8.2.1 Plant Source and Herbs

Plant components, such as fruits, seeds, leaves, stems, and roots, have also been used to create NPs because their extract contains a wealth of phytochemicals with stabilizing and antioxidant qualities. Spherical ZnO NPs have been made using leaf extract from a range of plants, such as *A. indica, O. basilicum, S. multiflorus, C. alata, E. crassipes, V. trifolia, S. chirayita, L. var. purpurascens Benth, and C. roseus* [16–20]. Using *S. rebaudiana* leaves, Seo et al. created spherical and flower-shaped ZnO nanoparticles [21]. ZnO nanoparticles, even the smallest ones, gather to create large clusters when the annealing temperature is increased. Leaf extracts from P. pinnata and S. Album were formed into nanorod-like structures [22, 23]. Plants were used in the biosynthesis of ZnO nanomaterials which includes their fruit peel, endosperm, leaves, flowers, and rhizomes. Three citrus fruit peel extracts—*Citrus sinensis* (orange), *Citrus paradis* (grapefruit), and *Citrus aurantifolia* (lemon)—were used to create uniform, monodispersed ZnO nanoparticles for their reducing characteristics [24]. Flowers' many different color emissions give them a high aesthetic value. They are essential to the creation of nanomaterials. Ogunyemi et al. used flower extract from *Matricaria chamomilla* to make green ZnO NPs [25]. Using the rhizome of *C. Chinensis*, L. Nagajyothi et al. created ZnO NPs that were almost spherical and rod-shaped [26]. Coconut water, which is generated from the *Cocos Nucifera* tree's endosperm, was used to create ZnO NPs (20–80 nm) [27]. Plant-mediated synthesis in combination with the aloe vera plant allowed Kavyashree et al. to create ZnO superstructures made of various micro/nanoentities, such as platelets and bowls [28]. Green synthesis was used by Jayachandran et al. to create ZnO nanoparticles using leaf extract from *Cayratia pedata* [29].

8.2.2 Microorganism as Source

Due to their relative ease of handling and genetic manipulation, microorganisms like bacteria and fungi have become popular for the evolution of nanoparticles [30]. Bacteria typically lack several organic functional groups that can decrease Zn^{2+} ions [31]. Gram-negative bacteria, *Serratiaureilytica*, and Gram-positive bacteria, *Rhodococcuspyridinivorans*, are two examples of microorganisms that were used as capping and reducing agents in the biosynthesis of ZnO NPs [32, 33]. The rigorous monitoring of culture and the screening of bacteria, however, takes time. To create ZnO NPs, a harmful bacterium called *B. licheniformis* was employed. These ZnO nanoflowers demonstrated high MB dye breakdown efficiency [34]. Fungi are a popular choice for the extracellular synthesis of ZnO NPs because of their high production rate, ease of downstream processing, and affordability [31]. Due to their great tolerance capacity and superior metal bioaccumulation properties, fungi have superior traits to bacteria. *Aspergillus* species and *Candida albicans* fungi were mostly used to synthesize ZnO NPs, and the majority of these NPs were found to have a spherical form [35].

8.2.3 Algae as a Source

Algae is the most important renewable resource found in the ocean environment. Algae can be categorized as brown (Pheophytes), red (Rhodophytes), or green (Chlorophytes) seaweed depending on the color of their pigments [36]. Because algal cells have a negative charge on their surface, nanoparticle nucleation and growth occur more quickly. Important bioactive compounds, such as amines, polysaccharides, carboxyl, sulfates, and hydroxyl functional groups are also present in seaweed macroalgae. Through adsorption, these functional groups form a chelate with Zn^{2+}. The resulting Zn-seaweed chelate is subsequently broken down to yield ZnO NPs. ZnO nanoparticles (NPs) encapsulated with brown seaweed of *Padinate strastromatica* were synthesized by Pandimurugan and Thambidurai (2016) using four distinct zinc sources: zinc nitrate, zinc acetate, zinc chloride, and zinc sulfate via precipitation [37].

8.3 PHOTOCATALYTIC MECHANISM

The system of nanoparticle photocatalyst activity is mainly due to the combined effects of multiple elements, such as charge stimulation/uncoupling and light immersion. Several essential elements of the photocatalytic system are as follows: four processes comprise photocatalysis: (i) light absorption to produce e^-–h^+ pairs; (ii) charge carrier separation; (iii) transfer of e^- and h^+ to the surface of semiconductor; and (iv) use charges on the surface in redox reactions. Ultimately, a large number of e^-–h^+ recombine with one another or rise to the surface leading to the release of heat or light energy as a result of this recombination. The sorbed kinds' acceptor/donor properties dictate the particular redox pathways that are triggered by charges on the surface which are photo-induced and persistent [38–41]. The absorbed photon must possess energy equal to or greater than the photocatalyst's forbidden band gap to excite e^- from VB to CB resulting in charge generation. It is commonly known that ZnO has strong conductivity/acceptors and can generate electrons when exposed to light. Owing to its remarkable e-conductivity, it has the potential to significantly enhance the charge separation caused by light, hence promoting quicker interfacial charge transfer and prolonging the lifetime of the photogenerated e^-–h^+ pair. Consequently, this greatly increases the semiconductor-ZnO heterojunction and site activation [42, 43].

8.4 PHOTOCATALYTIC ACTIVITY OF GREEN SYNTHESIZED ZnO

8.4.1 Leaf Extract

The most important and visible component of a plant is its leaves, which through photosynthesis eventually provide food and support for all terrestrial animals. The literature describes several leaf extracts for the creation of ZnO NPs as well as their photocatalytic capabilities. Using Zn acetate precursor, Hassan et al. produced ZnO NPs with *Coriandrum sativum* leaf extract and investigated their catalytic properties on a polycyclic aromatic hydrocarbon, anthracene, commonly present as a contaminant in wastewater [44]. Standard studies revealed the presence of several phytochemicals in *Coriandrum sativum* that promote the development of ZnO NPs with sizes ranging from 9 to 18 nm. According to the conducted study, in a 240-min reaction time, the degradation percentage of anthracene for chemically and green synthesized ZnO was 31% and 96%, respectively. The ideal conditions for degradation of anthracene (100 μg L^{-1}) were 25°C ambient temperature and pH 7 with 1000 μg L^{-1} ZnO NP present. This finding highlights the superior efficacy of green ZnO NPs for water remediation as compared to chemically synthesized ZnO.

Additionally, Saraswathi et al. investigated the degradation of methyl orange (MO) in sunlight exposure using manufactured ZnO mediated by *Lagerstroemia speciosa* leaf extract [45]. Numerous phytochemicals, including flavonoids, glycosides, alkaloids, terpenoids, saponins, phenols, saponins, tannins, vitamins, and proteins, are present in it. The synthesized ZnO nanoparticles, with an average particle size of about 40 nm, were in a hexagonal phase and very stable. Within the reaction period of two hours, the synthesized ZnO NPs were able to degrade MO up to 93.5%. Following the photocatalytic experiment, mineralization research was conducted to estimate the chemical oxygen demand (COD) of MO solution during pre-photocatalytic and post-photocatalytic stages, thereby examining the efficiency of mineralization. After 100 min of sun exposure, COD values considerably dropped from 5600 to 374 mg/L, suggesting that there was enough MO decomposition.

In another study, Gawade et al. used *Calotropis procera* to prepare aqueous leaf extract to create spherical-shaped ZnO nanoparticles with sizes varying from 15 to 25 nm [46]. Numerous phytochemicals that serve as stabilizing and reducing factors are present in this leaf extract, including amines, carboxylic acid, ketones, aldehydes, hydroxyl groups, and phenolic and terpenoid compounds. Therefore, employing these biogenic ZnO NPs, it was discovered that the photocatalytic degradation was around 81% in 100 min for MO dye when irradiated with UV light. The ideal loading of the catalyst was 1.5 g/dm^3 and was determined to yield the highest photodegradation efficiency of MO.

Rathnasamy et al. used hexagonal wurtzite ZnO NPs made from leaf extract of *Carica papaya* to investigate the photodegradation of MB dye [47]. Under UV light irradiation, the MB dye completely degraded (approximately 100%) in 180 min of reaction time. Significant phytochemicals found in the *Carica papaya* leaf extract, including alkaloids, cardiac glycoside, saponins, tannins, hydroxyl groups, and ketones were essential for the creation of prospective ZnO NPs that showed improved photocatalytic activity.

8.4.2 Flower Extract

A flower is thought to be the most beautiful feature of a plant and is also what allows the plant to reproduce. ZnO NPs were synthesized using a variety of flower extracts, some of which are briefly covered here.

The green synthesis of ZnO NP preparation was carried out, using pale flower powder of *Butea monosperma*, which serves as an appropriate factor for stabilization and reduction [48]. The green synthesized nanoparticles, according to various characterization investigations, are irregularly sized, having an average crystallite size of around 17.6 nm. The photocatalytic degradation rate was determined to be approximately 62% for phenol at 4 hours into the experiment. After 6 hours, the phenol concentration dropped from 100 to 35.8 mg/L. Using *Canna indica* flower extract, Nguyen et al. created green ZnO NPs, which they then employed for CBBG dye from wastewater. In addition to increasing the root and shoot lengths of red and green bean seedlings, photocatalytically degrading MB dye (94.23%) in the presence of sunlight, the nanoparticles demonstrated good recyclability [49].

In a more environmentally friendly method, Soto-Robles et al. assessed ZnO NPs for their photocatalytic activity while synthesizing them using flower extract with *Hibiscus sabdariffa* [50]. The generated nanoparticles possessed a hexagonal crystalline structure (Wurtzite), and their sizes ranged from 8 to 30 nm, according to X-ray diffraction research. The flower extract's characterization investigations also revealed the existence of various phytochemicals, such as flavonoids and phenolic, which may cause the decrease and stabilization of zinc salts and control the size of ZnO NPs. Using these green-produced ZnO NPs, it was discovered that the photocatalytic breakdown percentage of MB dye was 97% in 150 min.

8.4.3 Fruit Extract

In botanical terms, a fruit is a structure on a flower-bearing plant that holds seeds that develop from the ovary after flowering. To this day, a growing number of research reports on the manufacture of ZnO NPs from fruit extracts have been witnessed. This section discusses a few of them. Using star fruit (*Averroe carambola*) extract, Chakraborty et al. examined the biogenesis of ZnO NPs and used it to investigate the photocatalytic degradation of a carcinogenic dye primarily used in the textile sector named Congo Red (CR) [51]. The phytochemical oxalate group found in star fruit facilitates the effective reduction of ZnO size to nanoflakes with an approximate size of ~20 nm. The photodegradation investigation conducted on CR dye showed good photocatalysis efficiency of green ZnO, which had an adsorption capacity of 285 mg g^{-1} with a percentage degradation of 93% during 150 min under UV light irradiation. ZnO NPs were made using jujube fruit extract by Golmohammadi et al., who also investigated the degradation efficiency for MB and eriochromeblack-T (ECBT), two organic dyes, in direct sunshine [52]. Fruit extract of *Jujube* has a greater reduction factor for the green ZnO NPs due to the presence of flavonoid biomolecules. For MB and ECBT, the ZnO NPs-assisted degradation efficiencies were approximately 92% and 86% within 5 hours, respectively. Fowsiya et al. used ZnO (flower-shaped), created by microwave-accelerated extraction from the fruit of *Carissa edulis* plant, to study the photodegradation of CR dye [53]. ZnO NPs were made possible by the terpenoid, phenolic, and flavonoid components found in the fruit extract of *Carissa edulis*. Consequently, after 135 min of reaction, 97% of the CR dye had reached its maximum degradation in the presence of ZnONPs. Rana et al. examined the green creation of ZnO spherical-shaped nanoparticles using aqueous *Terminalia chebula* plant fruit extract and further photocatalytic degradation of rhodamine B. After five hours of exposure, almost 70% photodegradation was attained [54].

8.4.4 Peel Extract

Fruit peels have also been used in conjunction with actual fruits to generate ZnO nanoparticles. The rambutan (*Nepheliumlappaceum L.*) fruit peel extract was used by Karnan and Selvakumar to generate ZnO NPs, which were then applied to study the photocatalytic degradation of MO dye [55]. Numerous phytochemicals, including ellagitannins, geranin, corilagin, and ellagic acid among others, aid in stabilizing and limiting the size of ZnO NPs over a particular threshold. Furthermore, at pH 7.01, 83.99% of elimination effectiveness was achieved after 120 min of UV light irradiation. In addition to the photocatalytic activity, a COD activity was conducted to ascertain the mineralization efficiency. Following 120 min of UV irradiation, the mineralization efficiency significantly dropped to 481 from 616.2 mg/l. This demonstrated that the artificial ZnO NPs had satisfactorily broken down the MO. In another study, ZnO NPs with a particle size of 23.56 nm were generated by Aminuzzaman et al. via a biogenic method using *Garcinia mangostana* (*G. mangostana*) fruit's pericarp [56]. Many phytochemicals, such as phenol, flavonoids, anthocyanin, and xanthone compounds, which have the ability to function as reducing and capping agents, were found to be abundant in the pericarp of *G. mangostana*. Under solar irradiation, malachite green dye—which is primarily used in the aquaculture industry as a fungicide and parasiticide—was degraded to test the photocatalytic properties of ZnO nanoparticles. Ultimately, after 180 min of radiation, about 99% of the MG had been degraded.

8.4.5 Seed Extract

Although plant seed extract has been used quite successfully to create metal nanoparticles, there are not many studies in this area of metal oxide nanoparticle synthesis. Using seed extract of Kalonji (*Nigella sativa*), Kavyashree et al. manufactured ZnO NPs with diversified morphologies and demonstrated outstanding performance in photocatalytic degradation of MB dye under both sunlight and UV irradiation [57]. The whole process was done in an easy, green approach, where many phytochemicals in the seed extract, including tannins, phenolic compounds, alkaloids, and flavonoids, among others, aid in capping and stabilizing the ZnO NPs, which have an average particle size of ~34 nm. Using an ideal catalyst dose of 500 mg/L, photocatalytic activity was found to be approximately 90% for flower-shaped ZnO. ZnO NPs were also synthesized by Moghaddas et al. using *Cydonia oblonga* as a stabilizing agent [58]. Numerous analysis findings attested to the existence of pristine ZnO NPs, which have rather homogeneous crystals of an average size of 25 nm. Here, the growth of ZnO NPs and subsequent adsorption of additional Zn ions on the nanosurface are caused by the mucilage obtained from quince seeds. The extracts contained biomolecules such as galactose, mannose, xylose, and arabinose. The degradation of over 80% of MB dye using green-produced ZnO NPs has been observed, with a 2-hour irradiation UV light.

8.4.6 Latex Extract

Generally speaking, latex constitutes milk-like material that is present in 10% of all angiosperms, or plants that can bear flowers. In general, sugars, starches, alkaloids, proteins, oils, tannins, gums, and resins are the main ingredients of latex, which is a complex emulsion that coagulates when exposed to air. Kumar et al. used latex of *E. tirucalli* plant and prepared effective ZnO NPs to study the photocatalytic degradation of MB dye under UV light exposure where the source of MB is from different tannery and textile industries [59]. Many phytochemicals that stabilized the latex extract, such as flavonoids, alkaloids, and tannins phenols, were able to generate ZnO NPs with precise morphology. Consequently, the synthesized NPs demonstrated a 96% degradation efficiency within 120 min of reaction time while photocatalyzing the breakdown of MB dye at pH 12. Using *Carica papaya* milk (CPM) latex, Sharma conducted a bio-route synthesis of ZnO with various morphologies and used it to degrade alizarin red-S (ARS) dye [60]. Papaya latex is extracted as a milk-like fluid that is thixotropic and composed of nutritious elements including proteins and vitamins as well as phytochemicals such as flavones, alkaloids, and phenolic compounds. In ZnO NP evolution, these phytochemicals serve as potential stabilizing agents. By adjusting the latex concentration, the ZnO shape was optimized, with the ZnO nano-flower exhibiting maximum photocatalytic activity. Under UV light, the ZnO nano-flower broke down the ARS dye solution to about 99% during 90 min. To confirm the ARS degradation, the total oxygen carbon (TOC) was also assessed as a mineralization index. Also highest TOC elimination percentage in a 90-min irradiation was discovered to be approximately 81%. In the event of its commercial use to stop pollution, it was highly crucial that most of the amount of ARS dye must have calcified during photocatalysis.

8.4.7 Stem Extract

The part of the plant that is above ground is called the stem, and it serves several purposes that are critical to many different biochemical processes, including photosynthesis. This section highlights some of the available reports on the photoactivity of stem extract and the green production of ZnO NPs. In one of the reports, Nahi et al. prepared ZnO-assimilated nanocellulose from sugar bagasse and *Tinospora cordifolia* plant's stem extract to photodegrade the antibiotic enrofloxacin (EF), which is commonly used in veterinary medicine to treat bacterial infections in animals [61]. Numerous characterization analyses revealed that the ZnO NPs having flower shapes were enhanced with significant phytochemicals that function as capping and reducing agents when ZnO nanoparticles are formed. As a result, 97% of the degrading efficiency was achieved in 120 min of reaction time under sun radiation.

8.4.8 Root Extract

Subterranean roots are an essential component of various vascular plants. The said portion of the plant is essentially responsible for securing it firmly in the ground and allowing it to absorb water, nutrients, and essential minerals from the surrounding soil. Food is also stored in roots. Numerous researchers used root extract to create green ZnO, as mentioned in this section. Chen et al. used a root extract from *Scutellaria baicalensis* (*S. baicalensis*) to approach the biogenic production of a spherical-shaped and stable ZnO NPs composite [62]. The ZnO NPs are stabilized because flavonoid phytochemicals are present. ZnO NPs have shown an efficient ability to destroy MB dye. Purportedly, this led to the achievement of over 98.6% deterioration in 210 min under UV light exposure.

8.4.9 Plant Gum Extract

Polysaccharides, principally found in plant gums, have been gaining significant attention in sustainable nanofabrication due to their affordability, abundance, and environmentally friendly characteristics. They are also used for surface modification of nanostructured materials, as a matrix for nanocomposites, and as stabilizing and dispersing agents. Suganya et al. investigated the photocatalytic decomposition of TB dye in the presence of UV light using ZnO NPs made via the gum-assisted sol–gel method of neem (*Azadirachta indica*) [63]. In this case, neem gum is essential because it is a powerful chelating and cross-linking mediator for Zn^{2+} ions, allowing the metal ions to be distributed evenly all over the gum matrix. In the photocatalytic degradation experiment of tuberculosis, the maximum degradation percentage of 97% was achieved at 180 min. Taghavi Fardood et al. demonstrated the synthesis of wurtzite ZnO NPs through the use of Arabic gum which acts as a biopolymeric agent [64]. The sol–gel process was employed to assess the ZnO NPs' photocatalytic activity in the degradation of direct blue 129 dye (DB129) when exposed to visible light. In 105 min, the catalyst eliminated about 95% of the DB129.

8.4.10 Marine Algae Extract

Because they are abundant and follow environmentally friendly protocols, biomanufacturing of NPs using marine materials has been accomplished and has gained notable significance. To this end, a wide range of seaweeds have been examined for the creation of nanoparticles. However, it is important to note that few studies have been conducted and described in the literature in this particular instance of the synthesis of ZnO nanoparticles. *Ulva lactuca* seaweed extract was used by Ishwaryaa et al. to demonstrate the biosynthesis of ZnO NPs with an average crystal size of 10–50 nm [65]. Sea lettuce or *U. lactuca* is an edible seaweed that is mostly found in marine habitats across the world. It contains phenol and alcohol groups, having prospects of being used as capping and stabilizing agents. In 120 min at optimal pH 7, ZnO NPs, which were asymmetrically shaped and sponge-like, demonstrated photocatalytic degradation of MB dye by approximately 90.4% under sunlight. Rajaboopath and Thambidurai studied the simple synthesis of silver-doped ZnO nanoparticles with an average particle size ranging from 20 to 40 nm mediated by *Padinagymnospora* algae extract using the chemical route [66]. Therefore, the water-soluble polyphenols, proteins, minerals, fibers, and biomolecule polysaccharides found in the algae extract serve as a capping agent during the nanoparticle manufacturing process. Furthermore, under direct sunlight irradiations, the catalyst nanoparticles' photodegradation and decolorization performances were demonstrated by both MB and Reactive Blue (RB) 198 dyes. As a result, after 30 min of radiation exposure, MB and RB 198 dyes, respectively, showed 85% and 95% degradation. Finally, the mineralization of 69% and 73% obtained with MB and RB198 dyes is supported by the COD values at various time intervals.

Some of the studies related to biosynthesized ZnO NPs and their photocatalytic efficiency are listed in Table 8.1.

TABLE 8.1
Studies Related to Biosynthesized ZnO NPs and Their Photocatalytic Efficiency

S. No.	Precursor for ZnO NPs	Plant Source	Plant Extract	Size of NPs (nm)	Light Source	Type of Dye	Degradation Efficiency % (time)	References
1	Zinc acetate	*Coriandrum sativum*	Leaf	24	Direct sunlight	Yellow 186 dye	93.3–130 min	[67]
2	Zinc nitrate	*Mussaenda frondosa*	Leaf/stem	5–20	UV	MB	80–100 min	[68]
3	Zinc nitrate	*Carica papaya*	Latex	11–26	UV	Alizarin red-S dye	99–120 min	[69]
4	Zinc nitrate	*Sea buckthorn*	Fruit	17	UV	Malachite green Congo red	99–80 min 99–70 min	[70]
5	Zinc nitrate	*Azadirachta indica*	Leaves	9–38	UV	MB	85–120 min	[71]
5	Zinc acetate	*Oak*	Fruit Hull	34	Visible	Basic violet 3	93–2:30 h	[72]
6	Zinc nitrate	*Nephelium lappaceumL*	Peel Extract	25–40	UV	MB	120 min	[73]
7	Zinc sulfate	*Serratia nematodiphila*	Bacteria	10–50	UV	MO	80 min	[74]
8	Zinc acetate	*Acremonium potronii*	Fungus	13–15		MB		[75]

8.5 CONCLUSIONS

ZnO nanoparticles' greater photocatalytic activity over other photocatalysts was caused by their special qualities, which included a large surface area, a potent adsorption capacity, and effective charge transfer. The produced ZnO nanoparticles exhibited minimal photocatalytic activity loss after many reuses after being effortlessly extracted from the reaction fluid. The latest advancements in the photocatalytic applications of ZnO nanoparticles for environmental remediation, as well as their production using plant derivatives, have been extensively reviewed.

REFERENCES

1. Kołodziejczak-Radzimska, A., & Jesionowski, T. (2014). Zinc oxidefrom synthesis to application: a review. *Materials*, *7*(4), 2833–2881.
2. Mishra, P. K., Mishra, H., Ekielski, A., Talegaonkar, S., & Vaidya, B. (2017). Zinc oxide nanoparticles: a promising nanomaterial for biomedical applications. *Drug Discovery Today*, *22*(12), 1825–1834.
3. Chaari, M., & Matoussi, A. (2014). Structural and dielectric properties of undoped ZnO pellets prepared by solid state route. *Applied Physics A*, *116*, 1149–1160.
4. Murugan, K., Subramaniam, J., Rajaganesh, R., Panneerselvam, C., Amuthavalli, P., Vasanthakumaran, M., ... & Aziz, A. T. (2021). Efficacy and side effects of biofabricated sardine fish scale silver nanoparticles against malarial vector Anopheles stephensi. *Scientific Reports*, *11*(1), 19567.
5. Noman, M. T., Amor, N., & Petru, M. (2022). Synthesis and applications of ZnO nanostructures (ZONSs): A review. *Critical Reviews in Solid State and Materials Sciences*, *47*(2), 99–141.
6. Rohani, R., Dzulkharnien, N. S. F., Harun, N. H., & Ilias, I. A. (2022). Green approaches, potentials, and applications of zinc oxide nanoparticles in surface coatings and films. *Bioinorganic Chemistry and Applications*, *2022*, (1), 3077747(1-22).
7. Prasad, A. R., Williams, L., Garvasis, J., Shamsheera, K. O., Basheer, S. M., Kuruvilla, M., & Joseph, A. (2021). Applications of phytogenic ZnO nanoparticles: A review on recent advancements. *Journal of Molecular Liquids*, *331*, 115805.
8. Miao, Y., Du, P., Wang, Z., Chen, Q., & Eslamian, M. (2018). Ultrasonic vibration imposed on nanoparticle-based ZnO film improves the performance of the ensuing perovskite solar cell. *Materials Research Express*, *5*(2), 026404.
9. Sadiq, H., Sher, F., Sehar, S., Lima, E. C., Zhang, S., Iqbal, H. M.,... & Nuhanović, M. (2021). Green synthesis of ZnO nanoparticles from Syzygium Cumini leaves extract with robust photocatalysis applications. *Journal of Molecular Liquids*, *335*, 116567.
10. Weldegebrieal, G. K. (2020). Synthesis method, antibacterial and photocatalytic activity of ZnO nanoparticles for azo dyes in wastewater treatment: A review. *Inorganic Chemistry Communications*, *120*, 108140.
11. Duan, H., Wang, D., & Li, Y. (2015). Green chemistry for nanoparticle synthesis. *Chemical Society Reviews*, *44*(16), 5778–5792.
12. Pathania, D., Kumar, S., Thakur, P., Chaudhary, V., Kaushik, A., Varma, R. S.,... & Khosla, A. (2022). Essential oil-mediated biocompatible magnesium nanoparticles with enhanced antibacterial, antifungal, and photocatalytic efficacies. *Scientific Reports*, *12*(1), 11431.
13. Bhuyan, T., Mishra, K., Khanuja, M., Prasad, R., & Varma, A. (2015). Biosynthesis of zinc oxide nanoparticles from Azadirachta indica for antibacterial and photocatalytic applications. *Materials Science in Semiconductor Processing*, *32*, 55–61.

14. Akhter, S. M. H., Mahmood, Z., Ahmad, S., & Mohammad, F. (2018). Plant-mediated green synthesis of zinc oxide nanoparticles using Swertia chirayita leaf extract, characterization and its antibacterial efficacy against some common pathogenic bacteria. *Bionanoscience*, *8*, 811–817.
15. Raliya, R., & Tarafdar, J. C. (2013). ZnO nanoparticle biosynthesis and its effect on phosphorous-mobilizing enzyme secretion and gum contents in Clusterbean (Cyamopsis tetragonoloba L.). *Agricultural Research*, *2*, 48–57.
16. Al-Dhabi, N. A., & Valan Arasu, M. (2018). Environmentally-friendly green approach for the production of zinc oxide nanoparticles and their anti-fungal, ovicidal, and larvicidal properties. *Nanomaterials*, *8*(7), 500.
17. Happy, A., Soumya, M., Kumar, S. V., Rajeshkumar, S., Sheba, R. D., Lakshmi, T., & Nallaswamy, V. D. (2019). Phyto-assisted synthesis of zinc oxide nanoparticles using Cassia alata and its antibacterial activity against Escherichia coli. *Biochemistry and Biophysics Reports*, *17*, 208–211.
18. Elumalai, K., Velmurugan, S., Ravi, S., Kathiravan, V., & Raj, G. A. (2015). Bio-approach: Plant mediated synthesis of ZnO nanoparticles and their catalytic reduction of methylene blue and antimicrobial activity. *Advanced Powder Technology*, *26*(6), 1639–1651.
19. Vanathi, P., Rajiv, P., Narendhran, S., Rajeshwari, S., Rahman, P. K., & Venckatesh, R. (2014). Biosynthesis and characterization of phyto mediated zinc oxide nanoparticles: a green chemistry approach. *Materials Letters*, *134*, 13–15.
20. Bhuyan, T., Mishra, K., Khanuja, M., Prasad, R., & Varma, A. (2015). Biosynthesis of zinc oxide nanoparticles from Azadirachta indica for antibacterial and photocatalytic applications. *Materials Science in Semiconductor Processing*, *32*, 55–61.
21. Hoon Seo, K., Markus, J., Soshnikova, V., Oh, K. H., Anandapadmanaban, G., Elizabeth Jimenez Perez, Z.,... & Yang, D. C. (2019). Facile and green synthesis of zinc oxide particles by Stevia Rebaudiana and its in vitro photocatalytic activity. *Inorganic and Nano-Metal Chemistry*, *49*(1), 1–6.
22. Sundrarajan, M., Ambika, S., & Bharathi, K. (2015). Plant-extract mediated synthesis of ZnO nanoparticles using Pongamia pinnata and their activity against pathogenic bacteria. *Advanced Powder Technology*, *26*(5), 1294–1299.
23. Madan, H. R., Sharma, S. C., Suresh, D., Vidya, Y. S., Nagabhushana, H., Rajanaik, H.,... & Maiya, P. S. (2016). Facile green fabrication of nanostructure ZnO plates, bullets, flower, prismatic tip, closed pine cone: their antibacterial, antioxidant, photoluminescent and photocatalytic properties. *Spectrochimica Acta Part A: Molecular and Biomolecular Spectroscopy*, *152*, 404–416.
24. Nava, O. J., Soto-Robles, C. A., Gómez-Gutiérrez, C. M., Vilchis-Nestor, A. R., Castro-Beltrán, A., Olivas, A., & Luque, P. A. (2017). Fruit peel extract mediated green synthesis of zinc oxide nanoparticles. *Journal of Molecular Structure*, *1147*, 1–6.
25. Ogunyemi, S. O., Abdallah, Y., Zhang, M., Fouad, H., Hong, X., Ibrahim, E.,... & Li, B. (2019). Green synthesis of zinc oxide nanoparticles using different plant extracts and their antibacterial activity against Xanthomonas oryzae pv. oryzae. *Artificial Cells, Nanomedicine, and Biotechnology*, *47*(1), 341–352.
26. Nagajyothi, P. C., Sreekanth, T. V. M., Tettey, C. O., Jun, Y. I., & Mook, S. H. (2014). Characterization, antibacterial, antioxidant, and cytotoxic activities of ZnO nanoparticles using Coptidis Rhizoma. *Bioorganic & Medicinal Chemistry Letters*, *24*(17), 4298–4303.
27. Krupa, A. N. D., & Vimala, R. (2016). Evaluation of tetraethoxysilane (TEOS) sol–gel coatings, modified with green synthesized zinc oxide nanoparticles for combating microfouling. *Materials Science and Engineering: C*, *61*, 728–735.

28. Kavyashree, D., Shilpa, C. J., Nagabhushana, H., Daruka Prasad, B., Sreelatha, G. L., Sharma, S. C.,… & Premkumar, H. B. (2015). ZnO superstructures as an antifungal for effective control of Malassezia furfur, dermatologically prevalent yeast: prepared by aloe vera assisted combustion method. *ACS Sustainable Chemistry & Engineering*, *3*(6), 1066–1080.
29. Jayachandran, A., Aswathy, T. R., & Nair, A. S. (2021). Green synthesis and characterization of zinc oxide nanoparticles using Cayratia pedata leaf extract. *Biochemistry and Biophysics Reports*, *26*, 100995.
30. Li, X., Xu, H., Chen, Z. S., & Chen, G. (2011). Biosynthesis of nanoparticles by microorganisms and their applications. *Journal of Nanomaterials*, *2011*(1), 270974.
31. Chandra, H., Kumari, P., Bontempi, E., & Yadav, S. (2020). Medicinal plants: Treasure trove for green synthesis of metallic nanoparticles and their biomedical applications. *Biocatalysis and Agricultural Biotechnology*, *24*, 101518.
32. Kundu, T. K., Karak, N., Barik, P., & Saha, S. (2011). Optical properties of ZnO nanoparticles prepared by chemical method using poly (vinyl alcohol)(PVA) as capping agent. *International Journal of Soft Computing and Engineering*, *1*, 19–24.
33. Dhandapani, P., Siddarth, A. S., Kamalasekaran, S., Maruthamuthu, S., & Rajagopal, G. (2014). Bio-approach: ureolytic bacteria mediated synthesis of ZnO nanocrystals on cotton fabric and evaluation of their antibacterial properties. *Carbohydrate Polymers*, *103*, 448–455.
34. Raliya, R., & Tarafdar, J. C. (2013). ZnO nanoparticle biosynthesis and its effect on phosphorous-mobilizing enzyme secretion and gum contents in Clusterbean (Cyamopsis tetragonoloba L.). *Agricultural Research*, *2*, 48–57.
35. Mashrai, A., Khanam, H., & Aljawfi, R. N. (2017). Biological synthesis of ZnO nanoparticles using C. albicans and studying their catalytic performance in the synthesis of steroidal pyrazolines. *Arabian Journal of Chemistry*, *10*, S1530–S1536.
36. Baker, S., Harini, B. P., Rakshith, D., & Satish, S. (2013). Marine microbes: invisible nanofactories. *Journal of Pharmacy Research*, *6*(3), 383–388.
37. Pandimurugan, R., & Thambidurai, S. (2016). Novel seaweed capped ZnO nanoparticles for effective dye photodegradation and antibacterial activity. *Advanced Powder Technology*, *27*(4), 1062–1072.
38. Adhikari, S., Sarkar, D., & Madras, G. (2015). Highly efficient WO 3–ZnO mixed oxides for photocatalysis. *RSC Advances*, *5*(16), 11895–11904.
39. Sun, C., Fu, Y., Wang, Q., Xing, L., Liu, B., & Xue, X. (2016). Ultrafast piezo-photocatalytic degradation of organic pollutions by Ag 2 O/tetrapod-ZnO nanostructures under ultrasonic/UV exposure. *RSC Advances*, *6*(90), 87446–87453.
40. Qi, K., Cheng, B., Yu, J., & Ho, W. (2017). Review on the improvement of the photocatalytic and antibacterial activities of ZnO. *Journal of Alloys and Compounds*, *727*, 792–820.
41. Li, Z., Jin, D., & Wang, Z. (2020). ZnO/CdSe-diethylenetriamine nanocomposite as a step-scheme photocatalyst for photocatalytic hydrogen evolution. *Applied Surface Science*, *529*, 147071.
42. Shekofteh-Gohari, M., Habibi-Yangjeh, A., Abitorabi, M., & Rouhi, A. (2018). Magnetically separable nanocomposites based on ZnO and their applications in photocatalytic processes: a review. *Critical Reviews in Environmental Science and Technology*, *48*(10–12), 806–857.
43. Goktas, A., Mutlu, I. H., & Yamada, Y. (2013). Influence of Fe-doping on the structural, optical, and magnetic properties of ZnO thin films prepared by sol–gel method. *Superlattices and Microstructures*, *57*, 139–149.

44. Hassan, S. S., El Azab, W. I., Ali, H. R., & Mansour, M. S. (2015). Green synthesis and characterization of ZnO nanoparticles for photocatalytic degradation of anthracene. *Advances in Natural Sciences: Nanoscience and Nanotechnology*, *6*(4), 045012.
45. Saraswathi, V. S., Tatsugi, J., Shin, P. K., & Santhakumar, K. (2017). Facile biosynthesis, characterization, and solar assisted photocatalytic effect of ZnO nanoparticles mediated by leaves of L. speciosa. *Journal of Photochemistry and Photobiology B: Biology*, *167*, 89–98.
46. Gawade, V. V., Gavade, N. L., Shinde, H. M., Babar, S. B., Kadam, A. N., & Garadkar, K. M. (2017). Green synthesis of ZnO nanoparticles by using Calotropis procera leaves for the photodegradation of methyl orange. *Journal of Materials Science: Materials in Electronics*, *28*, 14033–14039.
47. Rathnasamy, R., Thangasamy, P., Thangamuthu, R., Sampath, S., & Alagan, V. (2017). Green synthesis of ZnO nanoparticles using Carica papaya leaf extracts for photocatalytic and photovoltaic applications. *Journal of Materials Science: Materials in Electronics*, *28*, 10374–10381.
48. Bharati, R., & Suresh, S. (2016). Synthesis of ZnO Based Nanocatalyst with Palash Flower Powder for Degradation of Phenol. In *Recent Advances in Chemical Engineering: Select Proceedings of ICACE 2015* (pp. 177–184). Springer, Singapore.
49. Nguyen, D. T. C., Le, H. T., Nguyen, T. T., Nguyen, T. T. T., Bach, L. G., Nguyen, T. D., & Van Tran, T. (2021). Multifunctional ZnO nanoparticles bio-fabricated from Canna indica L. flowers for seed germination, adsorption, and photocatalytic degradation of organic dyes. *Journal of Hazardous Materials*, *420*, 126586.
50. Soto-Robles, C. A., Luque, P. A., Gómez-Gutiérrez, C. M., Nava, O., Vilchis-Nestor, A. R., Lugo-Medina, E.,… & Castro-Beltrán, A. (2019). Study on the effect of the concentration of Hibiscus sabdariffa extract on the green synthesis of ZnO nanoparticles. *Results in Physics*, *15*, 102807.
51. Chakraborty, S., Farida, J. J., Simon, R., Kasthuri, S., & Mary, N. L. (2020). Averrhoe carrambola fruit extract assisted green synthesis of zno nanoparticles for the photodegradation of congo red dye. *Surfaces and Interfaces*, *19*, 100488.
52. Golmohammadi, M., Honarmand, M., & Ghanbari, S. (2020). A green approach to synthesis of ZnO nanoparticles using jujube fruit extract and their application in photocatalytic degradation of organic dyes. *Spectrochimica Acta Part A: Molecular and Biomolecular Spectroscopy*, *229*, 117961.
53. Fowsiya, J., Madhumitha, G., Al-Dhabi, N. A., & Arasu, M. V. (2016). Photocatalytic degradation of Congo red using Carissa edulis extract capped zinc oxide nanoparticles. *Journal of Photochemistry and Photobiology B: Biology*, *162*, 395–401.
54. Rana, N., Chand, S., & Gathania, A. K. (2016). Green synthesis of zinc oxide nano-sized spherical particles using Terminalia chebula fruits extract for their photocatalytic applications. *International Nano Letters*, *6*(2), 91–98.
55. Karnan, T., & Selvakumar, S. A. S. (2016). Biosynthesis of ZnO nanoparticles using rambutan (Nephelium lappaceumL.) peel extract and their photocatalytic activity on methyl orange dye. *Journal of Molecular Structure*, *1125*, 358–365.
56. Aminuzzaman, M., Ying, L. P., Goh, W. S., & Watanabe, A. (2018). Green synthesis of zinc oxide nanoparticles using aqueous extract of Garcinia mangostana fruit pericarp and their photocatalytic activity. *Bulletin of Materials Science*, *41*, 1–10.
57. Kavyashree, D., Anandakumari, R., Nagabhushana, H., Basavaraj, R. B., Raja Naik, H., & Lingaraju, K.(2015, October).Kalonji seed extract mediated green synthesis of ZnO nanopowders: photocatalytic, antibacterial and antioxidant activities. In *Materials Science Forum* (Vol. 830, pp. 533–536). Trans Tech Publications Ltd.

58. Moghaddas, S. M. T. H., Elahi, B., & Javanbakht, V. (2020). Biosynthesis of pure zinc oxide nanoparticles using Quince seed mucilage for photocatalytic dye degradation. *Journal of Alloys and Compounds*, *821*, 153519.
59. Kumar, K. S., Dhananjaya, N., & Yadav, L. R. (2018). E. tirucalli plant latex mediated green combustion synthesis of ZnO nanoparticles: Structure, photoluminescence and photo-catalytic activities. *Journal of Science: Advanced Materials and Devices*, *3*(3), 303–309.
60. Sharma, S. C. (2016). ZnO nano-flowers from Carica papaya milk: degradation of Alizarin Red-S dye and antibacterial activity against Pseudomonas aeruginosa and Staphylococcus aureus. *Optik*, *127*(16), 6498–6512.
61. Nahi, J., Radhakrishnan, A., & Beena, B. J. M. T. P. (2021). Green synthesis of zinc oxide incorporated nanocellulose with visible light photocatalytic activity and application for the removal of antibiotic enrofloxacin from aqueousmedia. *Materials Today: Proceedings*, *41*, 583–589.
62. Chen, L., Batjikh, I., Hurh, J., Han, Y., Huo, Y., Ali, H.,... & Yang, D. C. (2019). Green synthesis of zinc oxide nanoparticles from root extract of Scutellaria baicalensis and its photocatalytic degradation activity using methylene blue. *Optik*, *184*, 324–329.
63. Suganya, S., & Vivekanandhan, S. (2019). Neem (Azadirachta indica) gum assisted sol–gel synthesis and characterization of ZnO nanoparticles for photocatalytic application. *Journal of the Australian Ceramic Society*, *55*, 433–442.
64. Taghavi Fardood, S., Ramazani, A., Moradi, S., & Azimzadeh Asiabi, P. (2017). Green synthesis of zinc oxide nanoparticles using arabic gum and photocatalytic degradation of direct blue 129 dye under visible light. *Journal of Materials Science: Materials in Electronics*, *28*, 13596–13601.
65. Ishwarya, R., Vaseeharan, B., Kalyani, S., Banumathi, B., Govindarajan, M., Alharbi, N. S.,... & Benelli, G. (2018). Facile green synthesis of zinc oxide nanoparticles using Ulva lactuca seaweed extract and evaluation of their photocatalytic, antibiofilm and insecticidal activity. *Journal of Photochemistry and Photobiology B: Biology*, *178*, 249–258.
66. Rajaboopathi, S., & Thambidurai, S.(2018). Enhanced photocatalytic activity of Ag-ZnO nanoparticles synthesized by using Padina gymnospora seaweed extract. *Journal of Molecular Liquids*, *262*, 148–160.
67. Singh, J., Kaur, S., Kaur, G., Basu, S., & Rawat, M. (2019). Biogenic ZnO nanoparticles: A study of blueshift of optical band gap and photocatalytic degradation of reactive yellow 186 dye under direct sunlight. *Green Processing and Synthesis*, *8*(1), 272–280.
68. Jayappa, M. D., Ramaiah, C. K., Kumar, M. A. P., Suresh, D., Prabhu, A., Devasya, R. P., & Sheikh, S. (2020). Green synthesis of zinc oxide nanoparticles from the leaf, stem and in vitro grown callus of Mussaenda frondosa L.: characterization and their applications. *Applied nanoscience*, *10*, 3057–3074.
69. Sharma, S. C. (2016). ZnO nano-flowers from Carica papaya milk: degradation of Alizarin Red-S dye and antibacterial activity against Pseudomonas aeruginosa and Staphylococcus aureus. *Optik*, *127*(16), 6498–6512.
70. Rupa, E. J., Kaliraj, L., Abid, S., Yang, D. C., & Jung, S. K. (2019). Synthesis of a zinc oxide nanoflower photocatalyst from sea buckthorn fruit for degradation of industrial dyes in wastewater treatment. *Nanomaterials*, *9*(12), 1692.
71. Madan, H. R., Sharma, S. C., Suresh, D., Vidya, Y. S., Nagabhushana, H., Rajanaik, H.,... & Maiya, P. S. (2016). Facile green fabrication of nanostructure ZnO plates, bullets, flower, prismatic tip, closed pine cone: their antibacterial, antioxidant, photoluminescent and photocatalytic properties. *Spectrochimica Acta Part A: Molecular and Biomolecular Spectroscopy*, *152*, 404–416.

72. Sorbiun, M., Shayegan Mehr, E., Ramazani, A., & Taghavi Fardood, S. (2018). Green synthesis of zinc oxide and copper oxide nanoparticles using aqueous extract of oak fruit hull (jaft) and comparing their photocatalytic degradation of basic violet 3. *International Journal of Environmental Research*, *12*, 29–37.
73. Karnan, T., & Selvakumar, S. A. S. (2016). Biosynthesis of ZnO nanoparticles using rambutan (Nephelium lappaceumL.) peel extract and their photocatalytic activity on methyl orange dye. *Journal of Molecular Structure*, *1125*, 358–365.
74. Jain, D., Shivani, Bhojiya A. A., Singh, H., Daima, H. K., Singh, M.,… & Singh, A. (2020). Microbial fabrication of zinc oxide nanoparticles and evaluation of their antimicrobial and photocatalytic properties. *Frontiers in Chemistry*, *8*, 778.
75. Ameen, F., Dawoud, T., & AlNadhari, S. (2021). Ecofriendly and low-cost synthesis of ZnO nanoparticles from Acremonium potronii for the photocatalytic degradation of azo dyes. *Environmental Research*, *202*, 111700.

9 ZnO Nanostructures
A Potential Candidate for Gas Sensing

Payal Patial

9.1 INTRODUCTION

Wide-bandgap semiconductors, particularly semiconducting metal oxides (SMOs), attract considerable interest in the field of gas sensing. Their unique ability to detect different gases, combined with easy production, cost-effectiveness, and being compatible with other components, makes them highly desirable. Among these metal oxides, zinc oxide (ZnO) stands out due to its simple synthesis, lower temperature for crystal formation, and the ability to create large, low-temperature coatings. It also stands out for its unique combination of a wide bandgap (around 3.37 electron volts) and a robust exciton binding energy (60 meV at room temperature). This makes it a particularly attractive material for developing gas sensors. Researchers have actively explored various techniques for synthesizing ZnO nanostructures, with a strong emphasis on creating one-dimensional (1D) forms. These 1D nanostructures hold significant promise in the fields of nanoscience and nanotechnology [1, 2]. Several well-established methods exist for creating ZnO nanostructures. These techniques primarily fall under the category of gaseous methods, including chemical solution deposition (CSD), metal-organic chemical vapor deposition (MOCVD), pulsed laser deposition (PLD), magnetron sputtering technique (MST), and hydrothermal processes [3–6]. While effective, these methods present challenges for large-scale productivity due to their complexity, need for advanced equipment, and high operating temperatures. Additionally, these approaches may face limitations in achieving uniform samples, working with specific substrates, and producing high yields. Solution-based approaches offer a more attractive alternative. These methods are advantageous due to lower growth temperatures, potential for scalability, and simpler operation for creating high-density ZnO arrays [7]. Vapor methods, especially those that operate at room temperature, also hold promise for the cost-effective ZnO nanostructure synthesis. Both solution-based and vapor techniques offer significant cost benefits compared to traditional methods. These approaches can generate a broad range of nanostructures, consisting of nanorods, nanowires, nanocombs, nanoflakes, nanoflowers, and many others [3, 4, 7–9]. ZnO, a wide-bandgap semiconductor, holds significant promise for various applications due to its unique properties and affordability. Being nontoxic, readily available, and easy to produce, ZnO is widely used in paints, protective coatings, sunscreens, and more

DOI: 10.1201/9781003464907-9

[10–13]. These properties endow them with remarkable versatility, making them applicable in diverse fields [14–17]. Traditionally, gas sensor fabrication has been expensive and complex, hindering its widespread use. Recent research focuses on achieving selectivity for specific gases, rapid response/recovery times, and lower costs. Fast recovery times ensure sensors can quickly return to a baseline state for subsequent detections. ZnO gas sensors rely on the chemical sensitivity of their surface to various adsorbed gases, which alters the sensor's resistance. Porosity plays a key role in both sensitivity and response time [18, 19]. Studies by S. Basu et al. indicate that larger grain sizes in ZnO sensors can decrease gas sensitivity [14]. Researchers can enhance the capability of ZnO gas sensors via introducing dopants into base material. The type and concentration of these dopants significantly influence the sensor's response to various gases. Doping creates electronic imperfections within the material that can amplify the influence of oxygen molecules on the sensor's conductivity. This amplified response allows for more sensitive gas detection [19]. Nanostructured metal oxides, particularly those with exceptionally large surface areas, are especially promising for gas sensors. This is because a large surface area provides more efficient adsorption–desorption for gas molecules, enhancing sensitivity [20]. In comparison to bulk materials, nanostructured gas sensors can better distinguish between various molecules of gas at lower concentrations due to their increased surface area. This translates to superior sensing characteristics.

This mini-review delves into recent advancements in ZnO nanostructure engineering for enhanced gas sensor performance. We explore various strategies, including metal incorporation, noble metal nanoparticle decoration, carbonaceous nanomaterial integration, metal oxide nanocomposite formation, UV activation, and electron beam irradiation. By dissecting these modification techniques, we aim to elucidate their impact on sensor behavior. Additionally, we will provide a concise overview of gas-sensing mechanisms in ZnO and future directions for ZnO-based gas sensor development.

9.2 LITERATURE REVIEW

Air pollution, a consequence of rapid industrialization and urbanization, primarily stems from vehicle and industrial emissions, threatening the environment and human health. Clean air is vital for a healthy planet, and leaks of toxic and combustible gases pose significant safety risks. This highlights the critical need for accurate, real-time gas monitoring equipment and sensors. Gas sensors play an increasingly crucial role due to their diverse applications across various sectors. ZnO gas sensors possess a broad area of applications in industrial settings. They play a crucial role in fuel cells for detecting hydrogen gas for safety and efficiency purposes; in emission control for monitoring nitrogen dioxide (NO_2) emissions from vehicles, particularly relevant in mining and automotive industries; in oil refineries for identifying hydrocarbons, which are essential components in the refining process; and in fertilizer production for detecting ammonia leaks, ensuring safety and proper functioning of production facilities [22–26]. An ideal gas sensor for these tasks should be compact, affordable, reliable, energy-efficient, and highly sensitive, effectively preventing accidents caused by hazardous and flammable gases [27, 28].

Recent advancements in nanomaterial fabrication hold promise for improved gas-sensing efficiency compared to bulk materials. Nanostructured materials offer advantages like small size and high surface area, leading to enhanced sensing capabilities [29, 30]. To improve gas-sensing performance, researchers are investigating metal oxide (MOx) materials with various morphologies, including CuO, Fe_2O_3, SnO_2, TiO_2, WO_3, and ZnO [31–37]. The characteristic length scale of these nanostructures has significant impact on the sensor's response [38, 39]. Consequently, various synthesis techniques have been developed to create diverse nanostructures. These include electrospinning, hydrothermal synthesis, MOCVD, molecular beam epitaxy, radio frequency sputtering (RF), sol–gel processing, and synthesis through template [40–46]. These fabrication techniques offer remarkable versatility, allowing researchers to create ZnO nanostructures in a vast array of shapes and configurations. Some prominent examples include nanofibers, nanoflowers, nanoneedles, nanoplates, nanoribbons, nanorods, nanospheres, nanotubes, nanowires, and thin films [47]. As a result, MOx-based gas sensors can detect a broad spectrum of aimed gases, including the gases that can both oxidize and reduce, for example acetone, ethanol, CH_4, CO, CO_2, SO_2, H_2, H_2S, NH_3, NO_2, and toluene [48].

ZnO stands out among MOx sensors for its versatility in diverse applications. This material offers a compelling combination of properties: inherent non-toxicity, exceptional resilience to heat and chemicals, straightforward fabrication, remarkable sensitivity to target gases, and the ability to distinguish between them with high accuracy. Additionally, ZnO classifies it as an intrinsic n-type semiconductor [49, 50]. ZnO nanostructures have transcended their traditional role in gas sensing and are now attracting attention in photovoltaic devices. Venkatesan et al. showcased a hybrid solar cell combining a polymer with ZnO, achieving an extended lifespan by modifying ZnO film interfaces using a cationic polymer (PEIE) [51]. Their research suggests that modified ZnO offers improved device longevity compared to the original material. Xu et al. explored the development and characterization of polymer-inorganic hybrid solar cells, evaluating their efficiency and potential improvement methods [52]. Similarly, in their study, Kumar et al. explored how grain boundaries affect electrical, optical, and structural electrical characteristics of thin films made from Al-doped ZnO (AZO). They achieved this by depositing the AZO films using a technique called RF-superimposed DC sputtering. [53]. The research revealed a standout AZO film formulation with a unique ratio of components. This formulation exhibited several key advantages: significantly lower residual stress, substantially higher electron mobility, and a notable increase in crystallite size when compared to the rest of the thin films of AZO.

ZnO is an environmentally beneficial gas sensor due to its inexpensiveness and non-toxicity, which sets it apart from other MOx-based gas sensors. As a chemiresistive sensor, ZnO's response relies heavily on the chemisorbed oxygen ions on its surface. These oxygen ions play a critical role in the sensor's electrical resistance, and their interaction with target gases directly affects how the resistance changes. Additionally, when atmospheric air is present, molecules of oxygen accumulate on the outermost layer of ZnO. The formation of these molecules is the result of electrons being taken out of ZnO's conduction band, a process that raises ZnO's resistance. The decrease in resistance occurs when reducing gases react by interacting with chemisorbed oxygen ions on ZnO's surfaces as these oxygen ions provide free electrons that enter the ZnO conduction band. Moreover, operating temperatures and concentration of gases reflect an important effect on the ZnO-based gas sensor's sensing ability [54]. According to Ranwa et al. (2014) and Xu Ho et al., ZnO-based gas sensors that rely on a working temperature typically need enough light energy on their surface to reduce the barrier to activation energy and boost the redox process, which in turn improves the sensing response [55, 56]. Nevertheless, there are certain requirements that must be met in order to have an improved sensing response at high operating temperatures. Furthermore, at high operating temperatures, ZnO nanostructures' surface morphology is modified. Additionally, it was discovered that the sensor's detecting response declines as a result of chemisorbed ions of oxygen being less present on the surface above its ideal working temperature [57].

Several techniques are explored to enhance the gas-sensing capabilities of ZnO nanostructures. These strategies incorporate doping ZnO with metals, incorporating noble metal nanoparticles onto its surface, creating composites by combining ZnO with other metal oxides, utilizing ultraviolet (UV) light activation, subjecting ZnO nanostructures to high-energy irradiation, and integrating carbon-based nanomaterials into ZnO [58]. While ZnO-based gas sensors have been widely researched, there's a gap in knowledge regarding the comprehensive exploration of these various enhancement methods [59]. Fewer studies delve into all the available options for improving ZnO sensor performance and elucidate the underlying gas-sensing mechanisms [60].

9.3 SYNTHESIS OF ZnO NANOSTRUCTURES

ZnO boasts several advantages from a manufacturing standpoint. It is a cost-effective material with well-established fabrication techniques, leading to consistent and reliable sensor production. Additionally, ZnO exhibits excellent stability, ensuring long-term sensor performance. Importantly, ZnO's nontoxic nature makes it safe. Researchers introduced a comprehensive suite of synthesis methods to tailor ZnO nanostructures into various 1D, 2D, and 3D morphologies. These methodologies include CVD, RF sputtering, hydrothermal synthesis, electrospinning, sol–gel processing, among others. When it comes to ZnO nanostructures for gas sensors, researchers have a toolbox filled with various shapes and sizes. These nanostructures are classified into three classes based on their dimensions: 1D, 2D, and 3D. 1D Menagerie: This group boasts nanowires, nanorods, nanofibers, nanoneedles, nanoribbons, and nanotubes – a diverse range of elongated structures. 2D Sheets and Plates: Imagine atomically thin, flat sheets and slightly thicker platelets – that's the world of 2D ZnO nanostructures. 3D Complexity: Here, intricate structures like nanoflowers, dendrimers, and urchins showcase a fascinating arrangement of ZnO atoms in three dimensions.

CVD is a method for growing tiny structures (nanostructures and microstructures) on a heated surface. During CVD, vapors containing the desired material react on the hot surface, building up the structures layer by layer. Researchers have identified several factors that influence the final shape and size of these ZnO structures grown with CVD. These factors include the way the substrate is prepared, the conditions inside the deposition chamber (temperature, pressure, gas flow), and how long the deposition process lasts [61]. By carefully adjusting these parameters, scientists can create ZnO nanostructures in various shapes and sizes. One interesting approach to control nanostructure growth involves placing metal catalysts on the substrate's surface. This technique, known as vapor–liquid–solid (VLS) growth, can promote the formation of 1D nanostructures like nanorods [61]. For instance, researchers have used VLS to grow ZnO nanorods by first depositing a thin layer of ZnO seed material on a silicon substrate [62, 63]. In a separate study, Tuayjaroen and Jutarosaga investigated how to create both 1D and 2D ZnO microstructures using a CVD method with additional oxygen. They found that increasing the amount of oxygen vapor in the chamber led to a shift from 1D to 2D structures, due to a change in the dominant growth mechanism [64].

Another effective method to create ZnO nanostructures is RF sputtering. It is the best method for depositing ZnO nanostructures because of its excellent adherence and homogeneity across the substrate. Atoms are ejected from the material of source and deposited on the substrate, which is the fundamental idea of sputtering. With argon serving as a supporting gas, the process of deposition takes place in plasma. The primary variables that affect the sputtering method are the pressure of the gas, distance between target and substrate, argon supply, substrate temperature, and RF power. The majority of literature reviews describe the use of RF sputtering to create ZnO thin films. Nevertheless, a small number of researchers have also looked into the RF sputtering method of ZnO nanorod synthesis. Employing an RF MST, Ranwa et al. were able to fabricate vertically aligned ZnO nanorods without the need for any kind of metal catalysts on the substrate. The formation characteristics were set at a temperature of 500 °C for the substrate, $2*10^{-2}$ mbar pressure of chamber, 60 sccm supply of Ar gas, 150 W RF power, and 14 cm distance between substrate and target correspondingly [55].

Moreover, the electrospinning method is frequently employed to create 1D nanostructures. This method, which is quite easy to use and inexpensive, enables the synthesis of a wide range of materials at both the nano and microscales [65]. The basic idea behind electrospinning is to develop as-spun nanofibers by implementing an electric field between the collector, an aluminum plate, and the capillary tube, which is typically used as a syringe. The end product is produced following the procedure of calcination, and these nanofibers must be gathered from the collector surface. In this section, the suggested material's metal ions polymeric solution is in the syringe. Wei et al. developed hollow and solid ZnO nanofibers by employing single capillary electrospinning [66].

The precursor that has been dissolved in solvents is hydrolyzed in the sol–gel process. Hydrolysis, also known as alcoholysis, causes sol to form, and recondensation and aging causes gel to form. After that, the materials go through various processes including sintering and drying in order to create nanostructures [67]. Precursor materials are often inorganic or metal salts. ZnO thin film was created by Kim et al. utilizing the sol–gel technique [68].

ZnO nanostructures can also be crafted using a template-assisted method. In this approach, a mold with predefined shapes and sizes guides the formation of the nanostructures. These molds can be rigid or flexible. Common examples of rigid templates (hard templates) include carbon nanotubes (CNTs) and porous alumina. Soft templates, on the other hand, can be biomolecules, single molecules, or flexible polymers [69]. Researchers have successfully employed various template techniques to create complex ZnO structures. For instance, Zhang et al. utilized a template derived from bacteria to synthesize hollow ZnO spheres. In another study, Zhou et al. preferred hydrothermal techniques for fabrication of core-shell structures of ZnO-ZIF. He has taken two zeolitic imidazolate (ZIFs) having pore size of 3.4 Å for ZIF-8 and 4.8 Å for ZIF-71 and deposited onto ZnO nanorods, essentially using the nanorods as a template for ZIF growth [70]. Figure 9.1(a–c) showcases these vertically aligned ZnO nanorod arrays (NRAs) and the ZnO-ZIF core-shell structures using FESEM images.

Another method for depositing ZnO nanostructures is molecular beam epitaxy (MBE). In its procedure, reactive oxygen is present when a molecule of zinc is sprayed on the substrate. This entire deposition procedure takes place in a chamber with an extremely high vacuum. On the other hand, growth is sluggish and equipment and maintenance expenditures are substantial. Reshchikov et al. used H_2O_2 as a source of reactive oxygen to deposit MBE-grown ZnO films on a sapphire substrate [71]. The production of ZnO nanostructures, an approach that is frequently employed is MOCVD. Metal-organic compounds are typically utilized as base for the process, and formation happens through the processes of evaporation, reaction, and deposition on the target substrate. Employing MOCVD technique with the operating temperature at 300 °C and pressure of 1 Torr, Mohanta et al. created ZnO nanorods on a substrate of c-sapphire [72].

Another prevalent method for creating ZnO nanostructures is MOCVD. This technique utilizes metal-containing organic molecules as precursors. These precursors

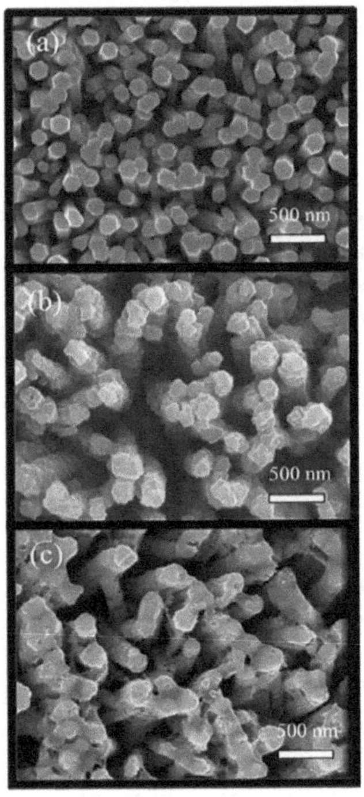

FIGURE 9.1 SEM pictures of three distinct ZnO NRAs: (a) vertically oriented ZnO NRAs, (b) ZIF-8-coated ZnO NRAs, and (c) ZIF-71-coated ZnO NRAs [70].

TABLE 9.1
Displays Several ZnO Nanostructure Types Developed Through Different Synthesis Methods

S. No.	ZnO Nanostructures	Synthesis Method	References
1.	ZnO nanorods	CVD, RF magnetron sputtering, hydrothermal	[85]
2.	ZnO nanowires	CVD	[86]
3.	ZnO nanosheets	Hydrothermal	[87]
4.	ZnO nanoflowers	Hydrothermal	[88]
5.	ZnO nanotubes	Electrospinning	[89]
6.	ZnO nanofibers	Electrospinning	[90]
7.	ZnO nanobelts	MBE	[91]
8.	ZnO nanotips	MOCVD	[92]
9.	ZnO quantum dots	Aptamer template	[93]
10.	ZnO thin film	Sol–gel, MOCVD	[94]

undergo a three-step process within the MOCVD chamber: evaporation, reaction, and deposition onto a designated substrate. Researchers like Mohanta et al. have successfully employed MOCVD to fabricate well-aligned ZnO nanorods. In their experiment, they achieved this at a relatively low temperature of 300°C and a pressure of 1 Torr on a c-sapphire substrate [72]. As shown in Table 9.1 details on synthesis methods and resulting structures used to create diverse ZnO nanostructures.

9.4 GAS MATERIAL INTERACTION FOR SENSING

ZnO nanostructures rely on a unique surface phenomenon for gas detection. In clean air, molecules of oxygen latch onto the surface of ZnO, stealing electrons and forming a layer of negatively charged oxygen ions. This electron depletion near the surface leads to the formation of a depletion layer that increases the overall resistance of ZnO. Band bending theory helps explain this effect: oxygen's high electronegativity pulls electrons toward the surface, bending the energy bands upwards and widening the depletion layer. Temperature plays a crucial role – adsorption of oxygen ions is more efficient at higher temperatures [55]. Interestingly, shining ultraviolet light on ZnO can boost reaction rates at lower temperatures. When ZnO encounters reducing gases like ammonia (NH_3), hydrogen (H_2), formaldehyde (HCHO), or methane (CH_4), these gases interact directly with the pre-adsorbed oxygen ions. This chemical reaction donates electrons to ZnO, effectively shrinking the depletion layer and lowering the resistance. In contrast, oxidizing gases like NO_2 and CO_2 have opposite effects on ZnO's resistance compared to reducing gases [28].

One-dimensional (1D) ZnO nanostructures exhibit a distinct advantage in gas-sensing applications compared to thin films. This advantage stems from the dominant role of their extensive surface area in governing electrical behavior, as opposed to grain boundaries in thin films by Xu et al. [73]. Consequently, 1D ZnO nanostructures possess a significantly larger surface area, facilitating the creation of a wider depletion region – a region near the area with diminished free electron concentration. This expanded depletion layer translates to a more pronounced variation

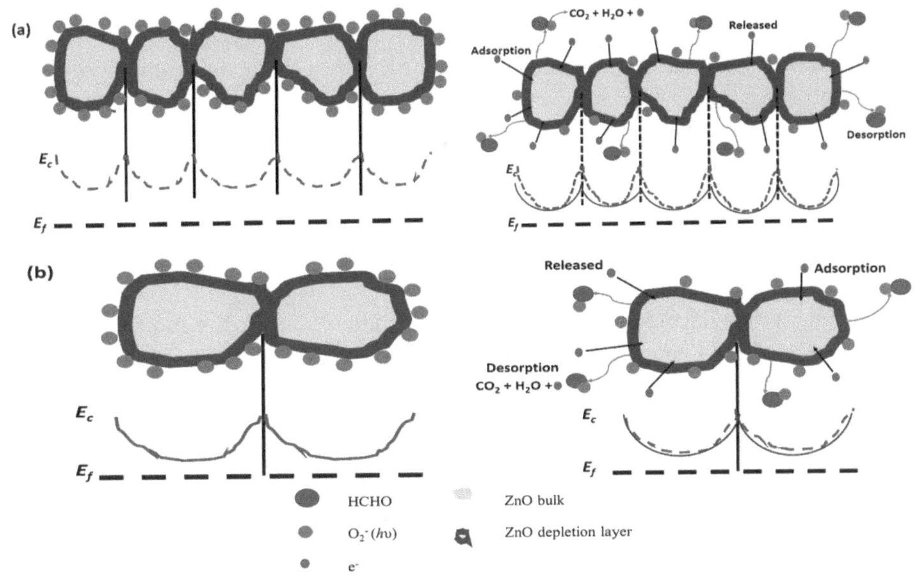

FIGURE 9.2 Probing the influence of UV light on the gas-sensing behavior of ZnO nanostructures: (a) comparison between nanoplates and (b) nanoflowers for formaldehyde (HCHO) detection.

in electrical resistance when target gas is introduced, enhancing sensor's sensitivity. Cui et al. investigated the gas sensor response for various 1D ZnO nanostructure morphologies toward formaldehyde gas under UV light, refer to Figures 9.2(a and b) [74]. Their findings revealed that nanofibers, characterized by their smaller particle size, demonstrated the highest sensitivity. The significant alteration of the depletion layer upon gas interaction leads to excellent performance, as depicted in Figure 9.2(a). Conversely, nanoplates and nanoflowers, possessing larger particle sizes, exhibited comparatively smaller changes in the depletion region and, correspondingly, the overall sensing response is decreased as shown in Figure 9.2(b). Interestingly, the study also observed a change in the barrier length for nanofibers as compared to the other morphologies, potentially contributing to their enhanced sensitivity.

9.5 OPTIMIZING ZnO NANOSTRUCTURE SENSORS FOR IMPROVED RESPONSE

ZnO nanostructures excel in gas sensing, but their sensitivity can be further amplified: metal doping, foreign metal ions enhance reactivity and response; noble metal nanoparticles, these act as catalysts for faster gas reactions; carbonaceous nanoparticles, improve conductivity and offer more adsorption sites; MOx nanostructure mixing, combines advantages of different materials for better selectivity; e-beam irradiation, creates defects for gas adsorption and potentially modifies sensor response. These strategies allow researchers to fine-tune ZnO gas sensors for specific needs, leading to superior sensitivity and detection.

9.5.1 High-Power Irradiation

High-powered irradiation methods, for example electron beam, gamma ray, ion beam, and laser irradiation, have emerged as promising posttreatment strategies to enhance the responsiveness of ZnO nanostructures gas sensors [68]. These methods alter the sensor's physicochemical properties by introducing energy into the material, leading to core-level electron excitation and the creation of electron–hole pairs [55]. Moreover, irradiation can also induce structural defects, which can impact the sensor's performance. The optimal response is achieved by carefully balancing the irradiation dose and energy [68].

Several studies have explored the effect of irradiation on hydrogen (H_2) sensing performance of ZnO nanostructures. Kim et al. investigated electron beam (e-beam) irradiation of electrospun ZnO nanofibers. They found that treating the fibers with an e-beam energy of 1 MeV at a dose of 150 kGy significantly enhanced the response to 10 ppm H_2 while maintaining excellent selectivity against interfering gases (C_7H_8, CO, C_6H_6, and C_2H_5OH). Their work suggests that joint effects of grain boundary modification, the influence of the e-beam, metallization effects, and radiation modulation itself contribute to the improved sensing mechanism as shown in Figure 9.3(a–d) [68]. Similarly, Ranwa showed the effect of gamma irradiation on ZnO nanorods for H_2 detection. Their findings revealed that a dose of 1 kGy yielded the highest response for 1% H_2 at 150 °C compared to sensors treated with different doses. Higher doses (5 and 10 kGy) resulted in a decrease in sensitivity. This optimal response at 1 kGy was attributed to the creation of a limited number of charge centers and defects due to the ionization process. Additionally, the generated electrons during ionization facilitated the enhancement of adsorbed oxygen ions on the nanorod surface [55]. Table 9.2 summarizes the impacts of different irradiation sources on response of the gas sensor of ZnO nanostructures.

9.5.2 Incorporation of Carbonaceous Nanostructures

Research by Xiao et al. suggests that combining carbon-based nanomaterials with ZnO nanostructures offers the most promising approach for achieving sensors with superior performance. This strategy has the potential to enhance sensing response, improve selectivity toward target gases, and achieve faster response times [75]. Carbon materials come in a number of forms, including graphene (2D) and its related compounds, CNTs (1D), and carbon nanofibers (1D). Since pure graphene lacks band gap and functional groups, it is not widely used in gas sensors. Therefore, adding functional groups to graphene's surface is crucial for adjusting the band gap and giving target gas molecules more active sites [76]. Reduced graphene oxide (rGO)/ZnO-based sensors have received the most research attention of all carbon nanomaterials because of their exceptional characteristics, which include great selectivity, quick response, and outstanding sensing response. Due to their distinct operations and semiconducting qualities, carbon materials and ZnO must form a heterojunction for the core principle of enhanced sensing responsiveness to occur. The nanoporous nature of carbon materials further contributes to the amplification effect. This intricate network of pores effectively disperses and presents target gas molecules, enhancing their interaction with the sensor surface. Additionally, the existence of

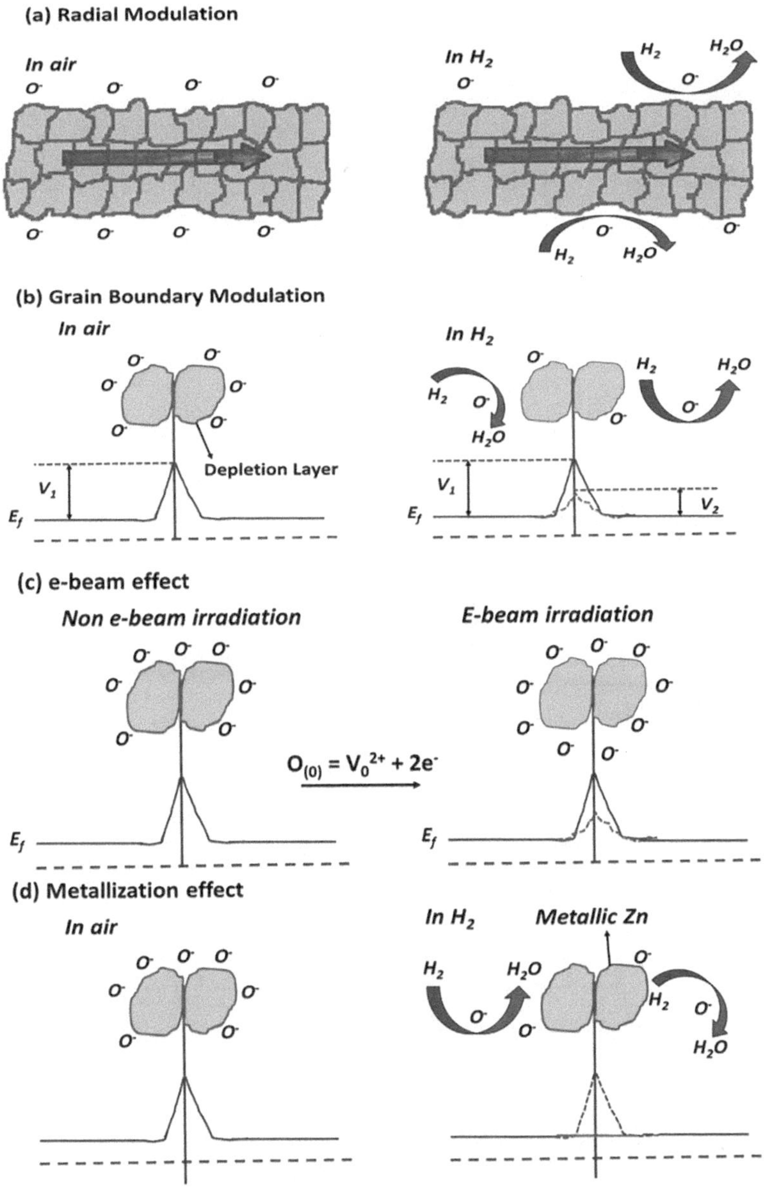

FIGURE 9.3 (a–d) Unveiling the H_2 sensing principle in e-beam modified sensors.

modified oxygen groups offers some active sites that help to boost the sensitivity of the sensor. Abideen et al. suggested using electrospinning for loading ZnO nanofibers with rGO nanosheets (0, 0.04, 0.11, 0.17, 0.44, 0.77, 1.04 wt% rGO) for NO_2 gas. At 400 °C, 5 ppm NO_2 was detected using all of the sensors [77]. The maximum sensing response was seen in 0.44 wt% rGO-loaded ZnO-based sensors and shown in Figure 9.4 (a and b).

TABLE 9.2
The Gas-Sensing Response Depending on Various Energy Sources Irradiated on ZnO Nanostructures

S. No.	Material	Source of Irradiation	Synthesis Method	Morphology	Concentration (ppm)	Target Gas	Temp (°C)	Sensitivity	Res Time (sec)	Ref
1.	--	Au ions	RF Sputtering	Thin Films	50000	H_2	175	89.94	35	[55, 95]
2.	Palladium	Electron beam	Electronspinning	Nanofibers	0.1	H_2	350	74.7	—	[96]
3.	SiO_2	Electron beam	Sol–gel	Thin Film		Acetone	300	38	—	[97]
4.	--	He ions	Nanowires	Nanowire	300	H_2S	RT	9	—	[38, 39]

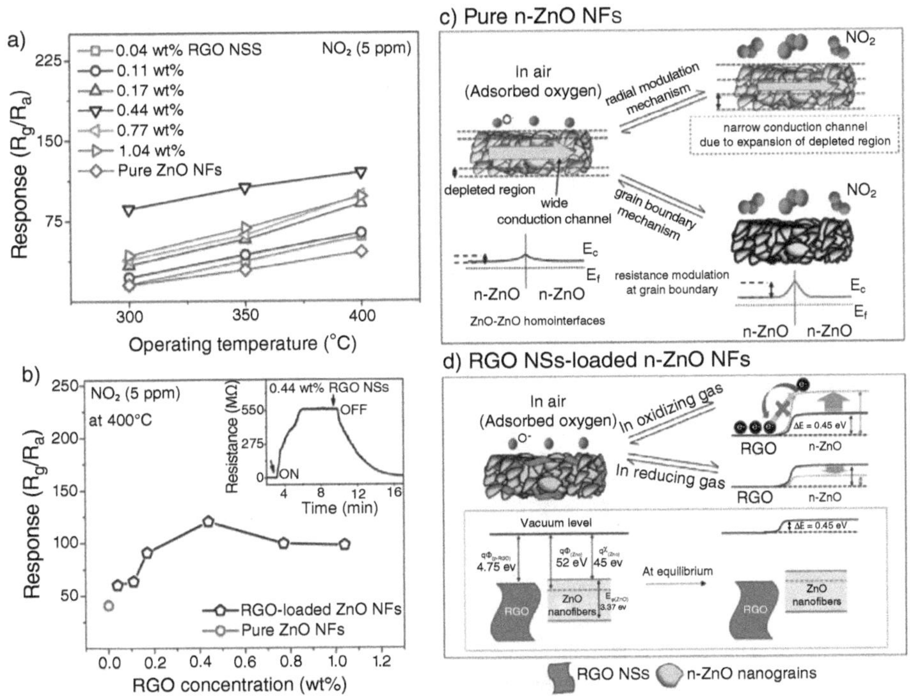

FIGURE 9.4 The study investigates the response of NO_2 gas sensors at 5 ppm concentration: (a) sensor response changes with varying temperatures. (b) Sensor response is affected by different concentrations of reduced graphene oxide (rGO) at a fixed temperature of 400°C (c and d). The underlying mechanisms behind the gas sensing in both pure n-ZnO nanofibers (NFs) and rGO-ZnO heterojunctions [77].

The following explanations were given in Figure 9.4(c and d) for the outstanding sensing response. This research explores the mechanisms behind NO_2 gas sensing in ZnO nanostructures and their composites: (i) surface interactions – the study examines the adsorption and desorption of NO_2 molecules on the sensor surface, which plays a crucial role in the sensing process as these molecules interact with the material. (ii) Depletion layer effect: – The presence of ZnO/ZnO interfaces (homointerfaces) can influence the depletion layer within the material, impacting its electrical conductivity and gas-sensing response. (iii) Heterojunction formation – The introduction of a p-type material like rGO creates heterointerfaces with n-type ZnO. These interfaces provide more adsorption sites for gas molecules of NO_2, potentially enhancing the sensitivity of the sensor. (iv) Functional groups: – The availability of functional groups on the surface of rGO can further influence interaction with NO_2 molecules, potentially improving the sensor response. Additionally, the text mentions a study by Schütt et al. [78] that employed ZnO-CNT hybrid tetrapods for NH_3 sensing. This research highlights the potential for engineering ZnO nanostructures and composites for various gas-sensing applications. The gas-sensing responses determined by different ZnO nanostructures and carbon nanomaterials at multiple temperatures are displayed in Table 9.3.

TABLE 9.3
Gas-Sensing Response at Different Temperatures Using Carbon Nanomaterials and ZnO Nanostructures

S. No.	Method of Synthesis	Carbon Nanomaterial	Type of Nanostructure	Target Gas	Concentration (ppm)	Temperature (°C)	Sensitivity	Response Time (sec)	Reference
1.	Electronspinning	Carbon	Nanofibres	H_2S	50	250	102	—	[98]
2.	Flame transport	CNT	Nanowires	NH_3	50	RT	430	—	[4, 78]
3.	In situ method	Graphene	Nanoparticles	H_2	200	150	3.5	22	[99]
4.	In situ method	MWCNT	Nanoparticles	NO_2	10	300	12	5.1 min	[100]
5.	Hydrothermal		Nanosheets	CO	50	100	12.3	4.9	[101]
6.	Grating	Nanoparticle	Nanoparticles	Acetone	300	400	52.5	—	[102]
7.	Solution process	rGO	Flower	NO_2	50 ppb	100	12	5.1	[103]
8.	In situ method		Nanoparticles	NO_2	5	RT	25.6	165	[104]

9.5.3 Integration of Noble Metal Nanoparticles on ZnO Surface

By functionalizing nanoparticles of noble metals (such as Au, Pd, and Pt) on the ZnO nanostructure surface, the gas sensor response can be improved [69]. The enhanced sensing response is typically explained by different methods. "Chemical mechanism" is the first category, while "electronic mechanism" is the subsequent one. The spillover action forms the basis of the chemical mechanism [28]. Oxygen molecule adsorption and desorption are stimulated by functionalized noble metal nanoparticles on ZnO nanostructures. These noble metal nanoparticles divide oxygen molecules into oxygen atoms, which are then dispersed over the surface upon interaction. After that, these oxygen atoms transform into oxygen ions by removing electrons from ZnO's conduction band. Moreover, the sensor's total resistance rises and the depletion layer of ZnO surface is reinforced. Reductive gases react by interacting with catalytic nanoparticles on the sensor surface and dissociate on the ZnO surface as soon as they are exposed to them. As a result, there are plenty of oxygen ions present for interaction with the atoms in the reductive gas. This leads to a significant decrease in the sensor's resistance and depletion layer, in addition to a spillover of free electrons into the ZnO's conduction band. As a result, this action lowers the ZnO-based sensor's operating temperature while improving its selectivity and detecting responsiveness. Since the noble metal nanoparticles and ZnO each possess distinct work functions, electrons are moved to the nanoparticles in the electrical mechanism, leading to the development of a nano-Schottky junction between both of them. Consequently, band bending occurs at the interface of ZnO/noble metal nanoparticles, raising the sensor's resistance and its depletion layers substantially.

Kim and colleagues produced ZnO nanowires modified with Pt and Pd. To lower the operating temperature, a self-heating state was introduced for sensors in their working mode [69]. Pt-modified nanowires have a maximum sensing response of 2.86 for 50 parts per million toluene at ambient temperature, whereas Pd-modified nanowires demonstrated a highest sensing response of 2.20 for 50 parts per million benzene. The outstanding catalytic nature of Pt and Pd was shown to be responsible for the creation of the Schottky barrier, which increases the adsorption sites on the surface of the sensor and serves as the sensing function. To detect NO_2, Chen et al. used ZnO nanowires, which they grew using a hydrothermal approach and doped with Pd nanoparticles (0, 1, 2, and 5% Pd) [79]. Response time of 1 ppm at 100 °C, the sensor demonstrated an excellent sensing response of around 13.5 (Ra/Rg). The synergistic effect of chemical and electrical systems led to an increased sensing mechanism, as illustrated in Figure 9.5. Table 9.4 compares the gas-sensing response at multiple temperatures using ZnO nanostructures and modified noble metal nanoparticles.

9.5.4 Mixture of MOx Nanostructures

The gas-sensing response of ZnO nanostructures can be attributed to the formation of junctions between the particles. In pure ZnO, exposure to target gas creates a homojunction, where the sensing response arises from changes in the barrier height at the ZnO–ZnO interface. However, combining ZnO with another metal oxide material introduces a different mechanism. Due to varying work functions, electrons flow

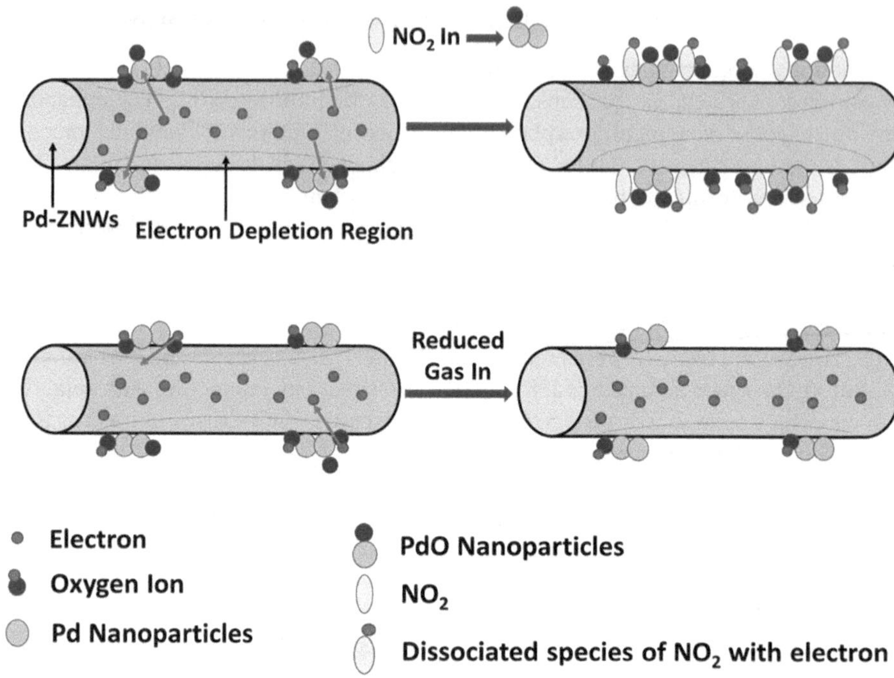

FIGURE 9.5 Sensing mechanism for Pd-ZNWs under NO_2 and reducing gases.

from the lower work function of MOx to the higher one until their Fermi levels align. This creates a "band bending" phenomenon at the contact point, forming a heterojunction. This heterojunction typically leads to a higher overall resistance and barrier height compared to a homojunction. In p-n-type nanocomposites, the significant difference in work functions leads to a particularly pronounced barrier height. When exposed to reducing or oxidizing gases, these heterojunctions experience a significant change in resistance and depletion layer width [80]. This change in the electrical properties is the basis for enhanced gas-sensing response observed in ZnO nanostructures with engineered heterojunctions.

Hydrothermal method was applied by Zhang et al. to synthesize flower-like p-CuO (0, 0.125, 0.25, and 0.5)/n-ZnO heterojunction nanorods, and their crystalline structures can be seen in the FESEM images as shown in Figure 9.6(a–h). They discovered that a 0.25:1 CuO/ZnO nanorod-based sensor with a sensing response of 100 ppm ethanol at 300 °C had the best sensing response [69]. This sensor's detecting response was 2.5 times that of a ZnO-based sensor, as shown in Figure 9.6(i). As demonstrated in Figure 9.6(j–k), the improved sensing response was attributable to the creation of a heterojunction between the p-type of CuO and the n-type of ZnO. Furthermore, ethanol vapor caused a significant drop in barrier height. Katoch et al. demonstrated the SnO_2-ZnO composite nanofibers with 10 ppm H_2 at 300 °C [81]. The improved sensing response for H_2 was due to the development of SnO_2-SnO_2 homointerfaces and the metallic impact at SnO_2-ZnO

TABLE 9.4
Comparison Table for Gas-Sensing Response of ZnO Nanostructures with Noble Metal Modification Temperature Dependence

S.No.	Method of Synthesis	Noble Metal	Type of Nanostructure	Target Gas	Concentration (ppm)	Temperature (°C)	Sensitivity	Response Time (sec)	Reference
1.	Hydrothermal and dipping	Bimetallic Au/Pd	Nanowires	NO_2	1	100	94.2	35	[105]
2.	Precipitation	Ag	Hollow nanocages	Ethane	100	250	84.6	5	[106]
3.	Precipitation		3D microspheres	Triethylamine		183.5	6043	1	[107]
4.	Hydrothermal	Au	Nanorods	CO	1000	150	12	—	[108]
5.	Hydrothermal		Nanowires	Ethanol	100	380	33.6	3	[109]
6.	Photochemical	Pd	Nanorod arrays	H_2	500	260	1106	—	[110]
7.	VLS Growth		Nanowires	CO	0.1	RT	1.02	210	[111]
8.	VLS Growth		Nanowires	H_2	100	350	87.17	—	[112]
9.	Hydrothermal	Pt-Au bimetallic	Nanorods	H_2	250	RT	25	115	[113]

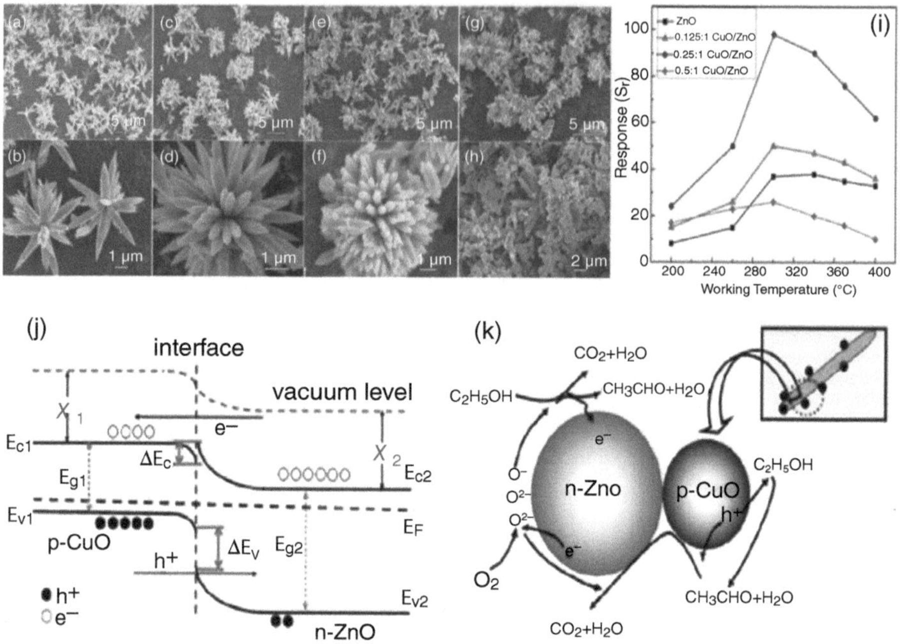

FIGURE 9.6 High-resolution FESEM images reveal the microstructure of the sensor materials: (a) and (b) show pure ZnO, (c) and (d) depict ZnO doped with CuO at a ratio of 0.125:1, (e) and (f) present ZnO doped with CuO at a ratio of 0.25:1, and (g) and (h) illustrate ZnO doped with CuO at a ratio of 0.5:1 [69].

heterointerfaces. Table 9.5 contains an exhaustive list of additional documented works on heterostructures based on ZnO nanostructures.

Researchers have explored fabricating flower-like nanorods with a p-type copper oxide and n-type ZnO heterojunction using a hydrothermal method. The crystal structures of these nanorods were visualized using FESEM imaging as shown in Figure 9.6(a–h). Studies revealed that a sensor based on these 0.25:1 CuO/ZnO nanorods exhibited a significant enhanced response to 100 ppm ethanol at 300°C compared to a pure ZnO sensor [69]. The improved response, 2.5 times higher, is attributed to the formation of the p-n heterojunction between CuO and ZnO, as shown in Figure 9.6(j–k). This heterojunction facilitates a significant decrease in the barrier height upon exposure to ethanol vapor. Another study by Katoch et al. demonstrated improved hydrogen sensing using SnO2-ZnO composite nanofibers at 300°C for a concentration of 10 ppm H_2 [81]. The enhanced response is explained by the formation of SnO_2–SnO_2 homointerfaces and the influence of metallic character at the SnO_2–ZnO heterointerfaces.

9.5.5 Metal Doping into ZnO

One of the key elements that enhances ZnO nanostructures' sensing responsiveness is doping [42]. A range of transition metals, including nickel (Ni), cobalt (Co), iron

TABLE 9.5
An Exhaustive List of Additional Documented Works on Heterostructures Based on ZnO Nanostructures

S.No.	Method of Synthesis	Metal Oxide	Type of Nanostructure	Target Gas	Concentration (ppm)	Temperature (°C)	Sensitivity	Response Time (sec)	Reference
1.	ALD	CuO	Nanofibers	H_2S	100	250	60	26	[114]
2.	Spray pyrolysis		Nanotubular array	H_2S	5	50	25	37	[115]
3.	Template	Fe_2O_3	3D ordered inverse opal	Acetone	50	485	47	5	[116]
4.	Chemical route	SnO_2	Film	H_2	10000	150	90	60	[117]
5.	Electrospinning		Nanofibers	H_2	0.1	350	78	74	[118]
6.	Hydrothermal		Nanorods	Ethanol	100	250	60	26	[119]
7.	Hydrothermal		Nanorods	Triethylamine	50	40	6.8	20	[120]

(Fe), and copper (Cu), are widely explored as dopants in the field of ZnO nanostructures [31]. Moreover, dopants lead to lower the working temperature of the sensor and improve its uniformity, specificity, and rapid reaction to the target gas [59]. It was recently noted that the introduction of metal dopants modifies the surface morphology of ZnO nanostructures [82]. Based on the additional constraint on crystallite mobility that occurs when edges separating the dopant and the host material touch with each other, Maciel et al noticed that the ZnO nanostructures having dopants have smaller particle size when compared to pure ZnO nanostructures. The dopant infusion leads to crystal growth to stop. Although smaller particles possess a high surface area and thus raise the amount of chemisorbed oxygen ions and barrier height, the gas-sensing response can be improved. A series of Ni-doped ZnO nanostructures with varying Ni content (0%, 2%, 4%, and 6%) were deposited onto silicon substrates [59]. Table 9.6 displays different ZnO nanostructures based on various metal-doped ZnO nanostructures at various operating temperatures.

Figure 9.7(a–h), shows the shape of tightly packed undoped ZnO nanorods was changed into Ni-doped ZnO nanoplates. Among all samples, 4% Ni-doped ZnO nanostructures exhibited the highest relative response at 150°C for 100 ppm hydrogen gas in Figure 9.7(i). This enhanced response is attributed to two factors: an abundance of active sites for H_2 gas adsorption and the lowest activation energy for the reaction on the 4% Ni-doped ZnO nanostructures. In addition, they demonstrated the good selectivity and great stability of 4% Ni-doped ZnO. Furthermore, since dopant and zinc have comparable ionic radii, donor defects are created when Zn ions are substituted in the ZnO framework by the dopant. Co-doped ZnO (0, 2, 4, 6, 8% Co) nanorods were created on a glass substrate using a hydrothermal technique [83].

The response for the gas sensor at 300 °C was examined using various H_2 doses. As illustrated in Figure 9.8(a–b), 8% Co-doped ZnO nanorods had maximum response. In Figure 9.8(c), hike in surface defect and high surface-to-volume ratio of nanorods were the cause of enhanced sensor response. According to other research, the p-n junction that is developed between the n-type ZnO nanostructures and p-type dopants (Ni and Cr) can also enhance the sensing response [84].

9.6 CONCLUSION

ZnO finds extensive use in sensing applications, primarily because of the ease of producing ZnO in the shape of nanostructures. While several ZnO sensors have been documented for detecting various gases, there is still ample scope for enhancement. Further endeavors are required to pave the path for heightened selectivity in ZnO gas sensors and to prolong the operational lifespan of these devices. To develop a gas sensor, ZnO nanostructures are considered and proved to be a versatile platform. Lowering the operating temperature of ZnO nanostructure gas sensors still remained as a challenging situation, despite their potential for low-power operation. A variety of synthesis techniques are being employed to produce ZnO nanostructures with tailored morphologies for optimized gas-sensing applications. Despite their promise, ZnO nanostructure gas sensors still face challenges in achieving the trifecta of high sensitivity, exceptional selectivity for target gases, and long-term stability – all while operating at lower temperatures. To overcome these hurdles, researchers

TABLE 9.6
Displays Various Metal-doped ZnO Nanostructure-based Gas Sensors at Various Operating Temperatures

S.No.	Method of Synthesis	Dopant	Type of Nanostructure	Target Gas	Concentration (ppm)	Temperature (°C)	Sensitivity	Response Time (sec)	Reference
1.	FSP	Al	Nanoparticles	Acetone	10	450	245	3	[121]
2.	Precipitation	Ce	Nanoflower	Ethanol	100	300	72.6	3	[42]
3.	Hydrothermal	Co	Hierarchical microspheres	Dichloroethane	100	440	23.7	30	[122]
4.	Thermal evaporation		Nanowires	Xylene	5	400	19.55	—	[123]
5.	Spray pyrolysis		Thin film	NH_3	100	RT	3.48	100	[124]
6.	Spray pyrolysis	Cr	Nano-whiskers array	H_2	400	RT	133	—	[125]
7.	Template method	In	3D macroporous	Ethanol	100	250	88	25	[126]
8.	Electrospinning	La	Nanobeads	Acetone	10	340	64	16	[127]
9.	Hydrothermal	Ni	Nanorods	H_2S	100	200	23.7	30	[128]
10.	Co-precipitation	Tb	Nanoparticles	Ethanol	50	280	49	—	[129]

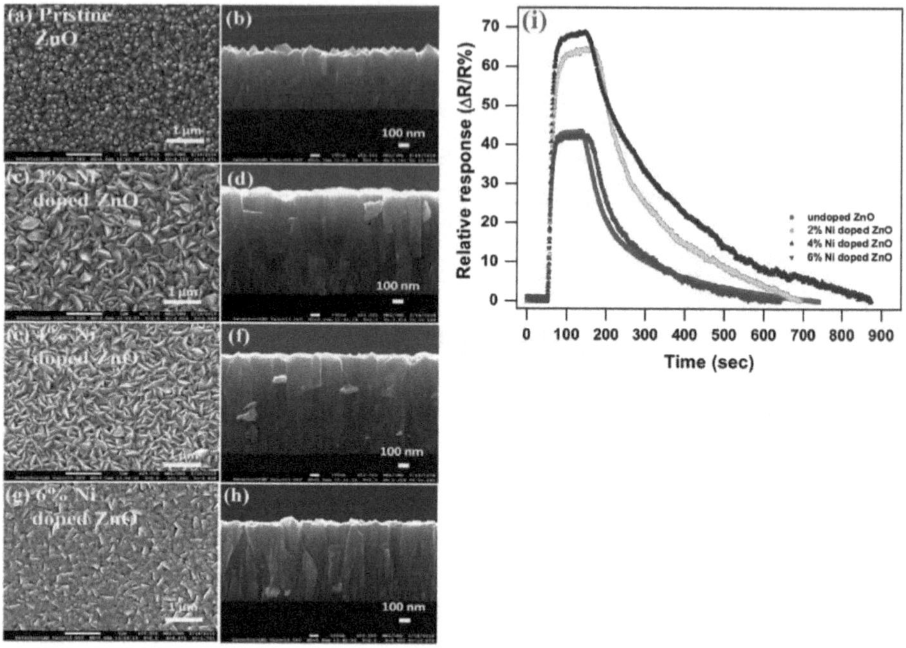

FIGURE 9.7 FESEM images: (a) and (b) depict undoped ZnO nanostructures. (c) and (d) show ZnO nanostructures doped with 2% nickel (Ni). (e) and (f) illustrate ZnO nanostructures doped with 4% Ni. (g) and (h) present ZnO nanostructures doped with 6% Ni [83].

are actively exploring techniques to improve gas-sensing capability of the material. These involve decorating them with nanoparticles of noble metals, incorporating elements from the transition metal group, introducing carbonaceous nanomaterials, creating composites, using ultraviolet light activation, and high-energy radiation posttreatment. Researchers have explored various approaches to lower the working temperature of gas sensors based on ZnO nanostructures without compromising sensitivity and selectivity. These strategies often involve manipulating the interaction of molecules of target gas with the surface of the sensor. When the molecules of target gas are adsorbed on the upper layer surface, they may influence various adsorbed oxygen ions. This, in turn, significantly alters the depletion layer within the material, ultimately leading to the electrical response measured by the sensor. By fine-tuning this interaction, scientists aim to achieve reliable gas detection at lower temperatures. The critical next step is to develop ZnO nanostructure gas sensors that are not only highly sensitive and cost-effective but also demonstrably selective for specific target gases. Current research has not conclusively shown methods for achieving strong selectivity in ZnO-based sensors. We believe these findings will offer a springboard for upcoming research in ZnO nanostructure gas sensors. The goal is to create sensors that can reflect high sensitivity and selectivity and can come at an accessible cost. By achieving low-temperature operation, these sensors could become even more practical for real-world applications. This research could also shed light on mechanisms for achieving superior selectivity in these sensors.

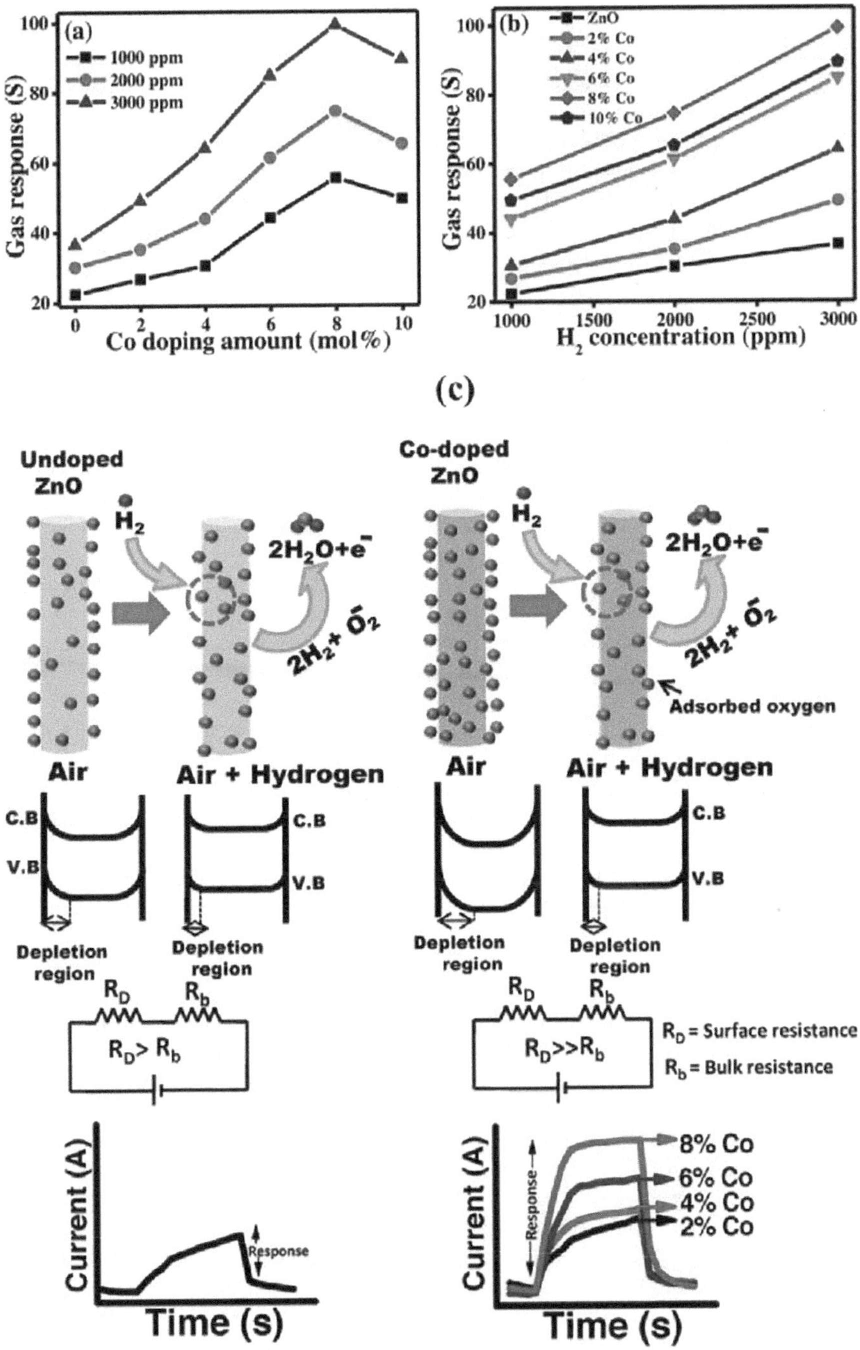

FIGURE 9.8 (a) The impact of varying cobalt doping levels on the gas response at 300°C. (b) The response of all sensors to different H_2 concentrations, also measured at 300°C. (c) The underlying mechanism by which both undoped and cobalt-doped ZnO NRs detect H_2 gas [84].

REFERENCES

1. Le, V. T., Nguyen, D. H., Dang, T. T. L., Do, T. V., Phuong, D. T., Le, A. T., & Nguyen, V. H. (2010). On-chip fabrication of SnO2-nanowire gas sensor: The effect of growth time on sensor performance. *Sensors and Actuators B*, *146*, 361–367.
2. Arafat, M. M., Dinan, B., Akbar, S. A., & Haseeb, A. S. M. A. (2012). Gas sensors based on one dimensional nanostructured metal-oxides: a review. *Sensors*, *12*(6), 7207–7258.
3. ShaoMin, Z., HongLei, Y., LiSheng, L., XiLiang, C., ShiYun, L., YaoMing, H.,... & Ning, L. (2010). Magnetic properties of Ni-doped ZnO nanocombs by CVD approach. *Nanoscale Research Letters*, *5*, 1284–1288.
4. Lupan, O., Chow, L., Ono, L. K., Cuenya, B. R., Chai, G., Khallaf, H.,... & Schulte, A. (2010). Synthesis and characterization of Ag-or Sb-doped ZnO nanorods by a facile hydrothermal route. *The Journal of Physical Chemistry C*, *114*(29), 12401–12408.
5. Yousif, A. A., Habubi, N. F., & Haidar, A. A. (2012). Nanostructure zinc oxide with cobalt dopant by PLD for gas sensor applications. *Journal of Nano-and Electronic Physics*, *4*(2), 02007.
6. Barnabé, A., Lalanne, M., Presmanes, L., Soon, J. M., Tailhades, P., Dumas, C.,... & Normand, P. (2009). Structured ZnO-based contacts deposited by non-reactive rf magnetron sputtering on ultra-thin SiO2/Si through a stencil mask. *Thin Solid Films*, *518*(4), 1044–1047.
7. Greene, L. E., Yuhas, B. D., Law, M., Zitoun, D., & Yang, P. (2006). Solution-grown zinc oxide nanowires. *Inorganic Chemistry*, *45*(19), 7535–7543.
8. Fan, X., Fang, G., Guo, S., Liu, N., Gao, H., Qin, P.,... & Zhao, X. (2011). Controllable synthesis of flake-like Al-doped ZnO nanostructures and its application in inverted organic solar cells. *Nanoscale Research Letters*, *6*, 1–6.
9. Ibrahim, A. A., Dar, G. N., Zaidi, S. A., Umar, A., Abaker, M., Bouzid, H., & Baskoutas, S. (2012). Growth and properties of Ag-doped ZnO nanoflowers for highly sensitive phenyl hydrazine chemical sensor application. *Talanta*, *93*, 257–263.
10. Abdullah, N. A., Khusaimi, Z., & Rusop, M. (2013). A review on zinc oxide nanostructures: Doping and gas sensing. *Advanced Materials Research*, *667*, 329–332.
11. Mathiazhagan, A., & Joseph, R. (2011). Nanotechnology-a New prospective in organic coating-review. *International Journal of Chemical Engineering and Applications*, *2*(4), 225.
12. Abdullah, N. A., Khusaimi, Z., & Rusop, M. (2013). A review on zinc oxide nanostructures: Doping and gas sensing. *Advanced Materials Research*, *667*, 329–332.
13. Choi, Y. S., Kang, J. W., Hwang, D. K., & Park, S. J. (2009). Recent advances in ZnO-based light-emitting diodes. *IEEE Transactions on electron devices*, *57*(1), 26–41.
14. Patil, D. R., Patil, L. A., & Amalnerkar, D. P. (2007). Ethanol gas sensing properties of Al 2 O 3-doped ZnO thick film resistors. *Bulletin of Materials Science*, *30*, 553–559.
15. Abdullah, N. A., Khusaimi, Z., & Rusop, M. (2013). A review on zinc oxide nanostructures: Doping and gas sensing. *Advanced Materials Research*, *667*, 329–332.
16. Wang, X., Li, R., & Fan, D. (2011). Nanostructured Al-ZnO/CdSe/Cu 2 O ETA solar cells on Al-ZnO film/quartz glass templates. *Nanoscale Research Letters*, *6*, 1–5.
17. Goncalves, A. S., Goes, M. S., Fabregat-Santiago, F., Moehl, T., Davolos, M. R., Bisquert, J.,... & Bueno, P. R. (2011). Doping saturation in dye-sensitized solar cells based on ZnO: Ga nanostructured photoanodes. *Electrochimica Acta*, *56*(18), 6503–6509.
18. Roy, S., & Basu, S. (2002). Improved zinc oxide film for gas sensor applications. *Bulletin of Materials Science*, *25*, 513–515.
19. Navale, S. C., Ravi, V., Mulla, I. S., Gosavi, S. W., & Kulkarni, S. K. (2007). Low temperature synthesis and NOx sensing properties of nanostructured Al-doped ZnO. *Sensors and Actuators B: Chemical*, *126*(2), 382–386.

20. Siriwong, C., Wetchakun, K., Wisitsoraat, A., & Phanichphant, S. (2009, October). Gas sensing properties of WO 3-doped ZnO nanoparticles synthesized by flame spray pyrolysis. In *SENSORS, 2009 IEEE* (pp. 118–123). IEEE.
21. Hübert, T., Boon-Brett, L., Black, G., & Banach, U. (2011). Hydrogen sensors–a review. *Sensors and Actuators B: Chemical, 157*(2), 329–352.
22. Sansone, F. J. (1992). Fuel cell hydrogen sensor for marine applications. *Marine Chemistry, 37*(1-2), 3–14.
23. Bhattacharyya, P., Basu, P. K., Saha, H., & Basu, S. (2007). Fast response methane sensor using nanocrystalline zinc oxide thin films derived by sol–gel method. *Sensors and Actuators B: Chemical, 124*(1), 62–67.
24. Wales, D. J., Grand, J., Ting, V. P., Burke, R. D., Edler, K. J., Bowen, C. R.,... & Burrows, A. D. (2015). Gas sensing using porous materials for automotive applications. *Chemical Society Reviews, 44*(13), 4290–4321.
25. Wetchakun, K., Samerjai, T., Tamaekong, N., Liewhiran, C., Siriwong, C., Kruefu, V.,... & Phanichphant, S. (2011). Semiconducting metal oxides as sensors for environmentally hazardous gases. *Sensors and Actuators B: Chemical, 160*(1), 580–591.
26. Ghosh, R., Midya, A., Santra, S., Ray, S. K., & Guha, P. K. (2013). Chemically reduced graphene oxide for ammonia detection at room temperature. *ACS Applied Materials & Interfaces, 5*(15), 7599–7603.
27. Patial, P., & Deshwal, M. (2022). Selectivity and sensitivity property of metal oxide semiconductor based gas sensor with dopants variation: A review. *Transactions on Electrical and Electronic Materials, 23*(1), 6–18.
28. Özgür, Ü., Alivov, Y. I., Liu, C., Teke, A., Reshchikov, M. A., Doğan, S.,... & Morkoç, A. H. (2005). A comprehensive review of ZnO materials and devices. *Journal of Applied Physics, 98*(4), 041301.
29. Wang, Z. L. (2008). Splendid one-dimensional nanostructures of zinc oxide: a new nanomaterial family for nanotechnology. *ACS Nano, 2*(10), 1987–1992.
30. Li, C., Du, Z., Yu, H., & Wang, T. (2009). Low-temperature sensing and high sensitivity of ZnO nanoneedles due to small size effect. *Thin Solid Films, 517*(20), 5931–5934.
31. Zhang, J., Zhu, Z., Chen, C., Chen, Z., Cai, M., Qu, B.,... & Zhang, M. (2018). ZnO-carbon nanofibers for stable, high response, and selective H2S sensors. *Nanotechnology, 29*(27), 275501.
32. Dwivedi, P., Chauhan, N., Vivekanandan, P., Das, S., Kumar, D. S., & Dhanekar, S. (2017). Scalable fabrication of prototype sensor for selective and sub-ppm level ethanol sensing based on TiO2 nanotubes decorated porous silicon. *Sensors and Actuators B: Chemical, 249*, 602–610.
33. Vallejos, S., Gràcia, I., Lednický, T., Vojkuvka, L., Figueras, E., Hubálek, J., & Cané, C. (2018). Highly hydrogen sensitive micromachined sensors based on aerosol-assisted chemical vapor deposited ZnO rods. *Sensors and Actuators B: Chemical, 268*, 15–21.
34. Urasinska-Wojcik, B., Vincent, T. A., Chowdhury, M. F., & Gardner, J. W. (2017). Ultrasensitive WO3 gas sensors for NO2 detection in air and low oxygen environment. *Sensors and Actuators B: Chemical, 239*, 1051–1059.
35. Umar, A., Alshahrani, A. A., Algarni, H., & Kumar, R. (2017). CuO nanosheets as potential scaffolds for gas sensing applications. *Sensors and Actuators B: Chemical, 250*, 24–31.
36. Sun, P., Wang, W., Liu, Y., Sun, Y., Ma, J., & Lu, G. (2012). Hydrothermal synthesis of 3D urchin-like α-Fe2O3 nanostructure for gas sensor. *Sensors and Actuators B: Chemical, 173*, 52–57.
37. Choi, K. J., & Jang, H. W. (2010). One-dimensional oxide nanostructures as gas-sensing materials: review and issues. *Sensors, 10*(4), 4083–4099.

38. Liao, L., Lu, H. B., Li, J. C., Liu, C., Fu, D. J., & Liu, Y. L. (2007). The sensitivity of gas sensor based on single ZnO nanowire modulated by helium ion radiation. *Applied Physics Letters*, *91*(17), 173110.
39. Liao, L., Lu, H. B., Li, J. C., He, H., Wang, D. F., Fu, D. J., & Zhang, W. F. (2007). Size dependence of gas sensitivity of ZnO nanorods. *The Journal of Physical Chemistry C*, *111*(5), 1900–1903.
40. Ameen, S., Akhtar, M. S., Song, M., & Shin, H. S. (2012). Vertically aligned ZnO nanorods on hot filament chemical vapor deposition grown graphene oxide thin film substrate: solar energy conversion. *ACS Applied Materials & Interfaces*, *4*(8), 4405–4412.
41. Baratto, C. (2018). Growth and properties of ZnO nanorods by RF-sputtering for detection of toxic gases. *RSC Advances*, *8*(56), 32038–32043.
42. Yang, Q., Lu, Z., Liu, J., Lei, X., Chang, Z., Luo, L., & Sun, X. (2013). Metal oxide and hydroxide nanoarrays: Hydrothermal synthesis and applications as supercapacitors and nanocatalysts. *Progress in Natural Science: Materials International*, *23*(4), 351–366.
43. Mondal, K., & Sharma, A. (2016). Recent advances in electrospun metal-oxide nanofiber based interfaces for electrochemical biosensing. *RSC Advances*, *6*(97), 94595–94616.
44. Sui, R., & Charpentier, P. (2012). Synthesis of metal oxide nanostructures by direct sol–gel chemistry in supercritical fluids. *Chemical Reviews*, *112*(6), 3057–3082.
45. Boyjoo, Y., Wang, M., Pareek, V. K., Liu, J., & Jaroniec, M. (2016). Synthesis and applications of porous non-silica metal oxide submicrospheres. *Chemical Society Reviews*, *45*(21), 6013–6047.
46. Ogata, K. I. O. K. I., Kawanishi, T. K. T., Maejima, K. M. K., Sakurai, K. S. K., Fujita, S. F. S., & Fujita, S. F. S. (2001). Improvements of ZnO qualities grown by metal-organic vapor phase epitaxy using a molecular beam epitaxy grown ZnO layer as a substrate. *Japanese Journal of Applied Physics*, *40*(7A), L657.
47. Niu, G., Saint-Girons, G., & Vilquin, B. (2018). Epitaxial systems combining oxides and semiconductors. *Molecular Beam Epitaxy*, 377–402.
48. Wang, W., Li, Z., Zheng, W., Huang, H., Wang, C., & Sun, J. (2010). Cr2O3-sensitized ZnO electrospun nanofibers based ethanol detectors. *Sensors and Actuators B: Chemical*, *143*(2), 754–758.
49. Rai, P., Kim, Y. S., Song, H. M., Song, M. K., & Yu, Y. T. (2012). The role of gold catalyst on the sensing behavior of ZnO nanorods for CO and NO2 gases. *Sensors and Actuators B: Chemical*, *165*(1), 133–142.
50. Zheng, J., Jiang, Z. Y., Kuang, Q., Xie, Z. X., Huang, R. B., & Zheng, L. S. (2009). Shape-controlled fabrication of porous ZnO architectures and their photocatalytic properties. *Journal of Solid State Chemistry*, *182*(1), 115–121.
51. Venkatesan, S., Ngo, E., Khatiwada, D., Zhang, C., & Qiao, Q. (2015). Enhanced lifetime of polymer solar cells by surface passivation of metal oxide buffer layers. *ACS Applied Materials & Interfaces*, *7*(29), 16093–16100.
52. Xu, T., Chen, Q., Lin, D. H., Wu, H. Y., Lin, C. F., & Qiao, Q. (2011). Self-assembled thienylsilane molecule as interfacial layer for ZnO nanowire/polymer hybrid system. *Journal of Photonics for Energy*, *1*(1), 011107.
53. Kumar, N., Dubey, A., Bahrami, B., Venkatesan, S., Qiao, Q., & Kumar, M. (2018). Origin of high carrier mobility and low residual stress in RF superimposed DC sputtered Al doped ZnO thin film for next generation flexible devices. *Applied Surface Science*, *436*, 477–485.
54. Kumar, M., Singh Bhati, V., Ranwa, S., Singh, J., & Kumar, M. (2017). Pd/ZnO nanorods based sensor for highly selective detection of extremely low concentration hydrogen. *Scientific Reports*, *7*(1), 236.

55. Ranwa, S., Barala, S. S., Fanetti, M., & Kumar, M. (2016). Effect of gamma irradiation on Schottky-contacted vertically aligned ZnO nanorod-based hydrogen sensor. *Nanotechnology*, *27*(34), 345502.
56. Xu, F., & Ho, H. P. (2017). Light-activated metal oxide gas sensors: A review. *Micromachines*, *8*(11), 333.
57. Jing, Z., & Zhan, J. (2008). Fabrication and gas-sensing properties of porous ZnO nanoplates. *Advanced Materials*, *20*(23), 4547–4551.
58. Ganesh, R. S., Durgadevi, E., Navaneethan, M., Patil, V. L., Ponnusamy, S., Muthamizhchelvan, C & Hayakawa, Y. (2017). Low temperature ammonia gas sensor based on Mn-doped ZnO nanoparticle decorated microspheres. *Journal of Alloys and Compounds*, *721*, 182–190.
59. Bhati, V. S., Ranwa, S., Fanetti, M., Valant, M., & Kumar, M. (2018). Efficient hydrogen sensor based on Ni-doped ZnO nanostructures by RF sputtering. *Sensors and Actuators B: Chemical*, *255*, 588–597.
60. Kumar, R., Al-Dossary, O., Kumar, G., & Umar, A. (2015). Zinc oxide nanostructures for NO 2 gas–sensor applications: A review. *Nano-Micro Letters*, *7*, 97–120.
61. Yang, C., Wu, P., Gan, W., Habib, M., Xu, W., Fang, Q., & Song, L. (2016). Low temperature CVD growth of ultrathin carbon films. *AIP Advances*, *6*(5), 055310.
62. Solís-Pomar, F., Martínez, E., Meléndrez, M. F., & Pérez-Tijerina, E. (2011). Growth of vertically aligned ZnO nanorods using textured ZnO films. *Nanoscale Research Letters*, *6*, 1–11.
63. Babu, E. S., & Hong, S. K. (2015). Effect of indium concentration on morphology of ZnO nanostructures grown by using CVD method and their application for H2 gas sensing. *Superlattices and Microstructures*, *82*, 349–356.
64. Tuayjaroen, R., & Jutarosaga, T. (2017). The influence of oxygen partial pressure on the shape transition of ZnO microstructure by thermal evaporation. *Thin Solid Films*, *631*, 213–218.
65. Guo, J., Song, Y., Chen, D., & Jiao, X. (2010). Fabrication of ZnO nanofibers by electrospinning and electrical properties of a single nanofiber. *Journal of Dispersion Science and Technology*, *31*(5), 684–689.
66. Wei, S., Zhou, M., & Du, W. (2011). Improved acetone sensing properties of ZnO hollow nanofibers by single capillary electrospinning. *Sensors and Actuators B: Chemical*, *160*(1), 753–759.
67. Znaidi, L. (2010). Sol–gel-deposited ZnO thin films: A review. *Materials Science and Engineering: B*, *174*(1-3), 18–30.
68. Kim, H. T., Lee, S. Y., & Park, C. (2017). Controls of surface morphology on sol-gel derived ZnO films under isothermal treatment conditions. *Vacuum*, *143*, 312–315.
69. Zhang, J., Zhu, Z., Chen, C., Chen, Z., Cai, M., Qu, B. & Zhang, M. (2018). ZnO-carbon nanofibers for stable, high response, and selective H2S sensors. *Nanotechnology*, *29*(27), 275501.
70. Zhou, T., Sang, Y., Wang, X., Wu, C., Zeng, D., & Xie, C. (2018). Pore size dependent gas-sensing selectivity based on ZnO@ ZIF nanorod arrays. *Sensors and Actuators B: Chemical*, *258*, 1099–1106.
71. Reshchikov, M. A., Avrutin, V., Izyumskaya, N., Shimada, R., & Morkoc, H. (2007). Anomalous shifts of blue and yellow luminescence bands in MBE-grown ZnO films. *Physica B: Condensed Matter*, *401*, 374–377.
72. Mohanta, S. K., Kim, D. C., Cho, H. K., Chua, S. J., & Tripathy, S. (2008). Structural and optical properties of ZnO nanorods grown by metal organic chemical vapor deposition. *Journal of Crystal Growth*, *310*(13), 3208–3213.

73. Xu, X. L., Chen, Y., Ma, S. Y., Li, W. Q., & Mao, Y. Z. (2015). Excellent acetone sensor of La-doped ZnO nanofibers with unique bead-like structures. *Sensors and Actuators B: Chemical, 213*, 222–233.
74. Cui, J., Shi, L., Xie, T., Wang, D., & Lin, Y. (2016). UV-light illumination room temperature HCHO gas-sensing mechanism of ZnO with different nanostructures. *Sensors and Actuators B: Chemical, 227*, 220–226.
75. Xiao, Z., Kong, L. B., Ruan, S., Li, X., Yu, S., Li, X.,... & Li, S. (2018). Recent development in nanocarbon materials for gas sensor applications. *Sensors and Actuators B: Chemical, 274*, 235–267.
76. Chatterjee, S. G., Chatterjee, S., Ray, A. K., & Chakraborty, A. K. (2015). Graphene–metal oxide nanohybrids for toxic gas sensor: A review. *Sensors and Actuators B: Chemical, 221*, 1170–1181.
77. Abideen, Z. U., Katoch, A., Kim, J. H., Kwon, Y. J., Kim, H. W., & Kim, S. S. (2015). Excellent gas detection of ZnO nanofibers by loading with reduced graphene oxide nanosheets. *Sensors and actuators B: chemical, 221*, 1499–1507.
78. Schütt, F., Postica, V., Adelung, R., & Anan, O. (2017). Single and networked ZnO–CNT hybrid tetrapods for selective room-temperature high-performance ammonia sensors. *ACS Applied Materials & Interfaces, 9*(27), 23107–23118.
79. Chen, Y., Li, X., Li, X., Wang, J., & Tang, Z. (2016). UV activated hollow ZnO microspheres for selective ethanol sensors at low temperatures. *Sensors and Actuators B: Chemical, 232*, 158–164.
80. Han, C., Li, X., Shao, C., Li, X., Ma, J., Zhang, X., & Liu, Y. (2019). Composition-controllable p-CuO/n-ZnO hollow nanofibers for high-performance H2S detection. *Sensors and Actuators B: Chemical, 285*, 495–503.
81. Katoch, A., Kim, J. H., Kwon, Y. J., Kim, H. W., & Kim, S. S. (2015). Bifunctional sensing mechanism of SnO2–ZnO composite nanofibers for drastically enhancing the sensing behavior in H2 gas. *ACS Applied Materials & Interfaces, 7*(21), 11351–11358.
82. Iqbal, J., Jan, T., & Ronghai, Y. (2013). Effect of Co doping on morphology, optical and magnetic properties of ZnO 1-D nanostructures. *Journal of Materials Science: Materials in Electronics, 24*, 4393–4398.
83. Bhati, V. S., Ranwa, S., Fanetti, M., Valant, M., & Kumar, M. (2018). Efficient hydrogen sensor based on Ni-doped ZnO nanostructures by RF sputtering. *Sensors and Actuators B: Chemical, 255*, 588–597.
84. Sett, D., & Basak, D. (2017). Highly enhanced H2 gas sensing characteristics of Co: ZnO nanorods and its mechanism. *Sensors and Actuators B: Chemical, 243*, 475–483.
85. Mudusu, D., Nandanapalli, K. R., Dugasani, S. R., Park, S. H., & Tu, C. W. (2016). Zinc oxide nanorods shielded with an ultrathin nickel layer: tailoring of physical properties. *Scientific Reports, 6*(1), 28561.
86. Rodwihok, C., Choopun, S., Ruankham, P., Gardchareon, A., Phadungdhitidhada, S., & Wongratanaphisan, D. (2019). UV sensing properties of ZnO nanowires/nanorods. *Applied Surface Science, 477*, 159–165.
87. Ju, D., Xu, H., Zhang, J., Guo, J., & Cao, B. (2014). Direct hydrothermal growth of ZnO nanosheets on electrode for ethanol sensing. *Sensors and Actuators B: Chemical, 201*, 444–451.
88. Song, Y., Chen, F., Zhang, Y., Zhang, S., Liu, F., Sun, P.,... & Lu, G. (2019). Fabrication of highly sensitive and selective room-temperature nitrogen dioxide sensors based on the ZnO nanoflowers. *Sensors and Actuators B: Chemical, 287*, 191–198.
89. Alali, K. T., Liu, J., Liu, Q., Li, R., Zhang, H., Aljebawi, K.,... & Wang, J. (2017). Enhanced acetone gas sensing response of ZnO/ZnCo2O4 nanotubes synthesized by single capillary electrospinning technology. *Sensors and Actuators B: Chemical, 252*, 511–522.

90. Wang, W., Li, Z., Zheng, W., Huang, H., Wang, C., & Sun, J. (2010). Cr2O3-sensitized ZnO electrospun nanofibers based ethanol detectors. *Sensors and Actuators B: Chemical, 143*(2), 754–758.
91. Kennedy, O. W., Coke, M. L., White, E. R., Shaffer, M. S., & Warburton, P. A. (2018). MBE growth and morphology control of ZnO nanobelts with polar axis perpendicular to growth direction. *Materials Letters, 212*, 51–53.
92. Biswas, P., Kundu, S., Banerji, P., & Bhunia, S. (2013). Super rapid response of humidity sensor based on MOCVD grown ZnO nanotips array. *Sensors and Actuators B: Chemical, 178*, 331–338.
93. Daniel, S. K., Kumar, A., Sivasakthi, K., & Thakur, C. S. (2019). Handheld, low-cost electronic device for rapid, real-time fluorescence-based detection of Hg2+, using aptamer-templated ZnO quantum dots. *Sensors and Actuators B: Chemical, 290*, 73–78.
94. Chia, C. H., Tsai, W. C., & Chiou, J. W. (2013). Thickness effect on luminescent properties of sol–gel derived ZnO thin films. *Journal of Luminescence, 136*, 160–164.
95. Ranwa, S., Kumar, M., Kulriya, P. K., Fanetti, M., Valant, M., & Kumar, M. (2016). Improvement in the sensing response of nano-crystalline ZnO-based hydrogen sensor: effect of swift heavy ion irradiation. *IEEE Sensors Journal, 16*(21), 7586–7592.
96. Kim, J. H., Mirzaei, A., Kim, H. W., Wu, P., & Kim, S. S. (2019). Design of supersensitive and selective ZnO-nanofiber-based sensors for H2 gas sensing by electron-beam irradiation. *Sensors and Actuators B: Chemical, 293*, 210–223.
97. Nalimova, S. S., Myakin, S. V., & Moshnikov, V. A. (2016). Controlling surface functional composition and improving the gas-sensing properties of metal oxide sensors by electron beam processing. *Glass Physics and Chemistry, 42*, 597–601.
98. Zhang, Y., Li, D., Qin, L., Zhao, P., Liu, F., Chuai, X.,… & Lu, G. (2018). Preparation and gas sensing properties of hierarchical leaf-like SnO2 materials. *Sensors and Actuators B: Chemical, 255*, 2944–2951.
99. Anand, K., Singh, O., Singh, M. P., Kaur, J., & Singh, R. C. (2014). Hydrogen sensor based on graphene/ZnO nanocomposite. *Sensors and Actuators B: Chemical, 195*, 409–415.
100. Kwon, Y. J., Mirzaei, A., Kang, S. Y., Choi, M. S., Bang, J. H., Kim, S. S., & Kim, H. W. (2017). Synthesis, characterization and gas sensing properties of ZnO-decorated MWCNTs. *Applied Surface Science, 413*, 242–252.
101. Hojati, T., Ebrahimi, M., & Afzalzadeh, R. (2018). Highly sensitive CO sensor based on ZnO/MWCNT nano sheet network grown via hydrothermal method. *Materials Chemistry and Physics, 207*, 50–57.
102. Tian, H., Fan, H., Guo, H., & Song, N. (2014). Solution-based synthesis of ZnO/carbon nanostructures by chemical coupling for high performance gas sensors. *Sensors and Actuators B: Chemical, 195*, 132–139.
103. Liu, J., Li, S., Zhang, B., Xiao, Y., Gao, Y., Yang, Q.,… & Lu, G. (2017). Ultrasensitive and low detection limit of nitrogen dioxide gas sensor based on flower-like ZnO hierarchical nanostructure modified by reduced graphene oxide. *Sensors and Actuators B: Chemical, 249*, 715–724.
104. Liu, S., Yu, B., Zhang, H., Fei, T., & Zhang, T. (2014). Enhancing NO2 gas sensing performances at room temperature based on reduced graphene oxide-ZnO nanoparticles hybrids. *Sensors and Actuators B: Chemical, 202*, 272–278.
105. Cheng, Z., Zhou, S., Chen, T., Dong, Y., Zhang, W., & Chu, X. (2012). Acetic acid gas sensors based on Ni2+ doped ZnO nanorods prepared by using the solvothermal method. *Journal of Semiconductors, 33*(11), 112003.
106. Zhang, J., Lu, H., Zhang, L., Leng, D., Zhang, Y., Wang, W.,… & Wang, C. (2019). Metal–organic framework-derived ZnO hollow nanocages functionalized with nanoscale Ag catalysts for enhanced ethanol sensing properties. *Sensors and Actuators B: Chemical, 291*, 458–469.

107. Shen, Z., Zhang, X., Mi, R., Liu, M., Chen, Y., Chen, C., & Ruan, S. (2018). On the high response towards TEA of gas sensors based on Ag-loaded 3D porous ZnO microspheres. *Sensors and Actuators B: Chemical*, *270*, 492–499.
108. Rai, P., Kim, Y. S., Song, H. M., Song, M. K., & Yu, Y. T. (2012). The role of gold catalyst on the sensing behavior of ZnO nanorods for CO and NO2 gases. *Sensors and Actuators B: Chemical*, *165*(1), 133–142.
109. Guo, J., Zhang, J., Zhu, M., Ju, D., Xu, H., & Cao, B. (2014). High-performance gas sensor based on ZnO nanowires functionalized by Au nanoparticles. *Sensors and Actuators B: Chemical*, *199*, 339–345.
110. Chang, C. M., Hon, M. H., & Leu, I. C. (2013). Outstanding H2 sensing performance of Pd nanoparticle-decorated ZnO nanorod arrays and the temperature-dependent sensing mechanisms. *ACS Applied Materials & Interfaces*, *5*(1), 135–143.
111. Choi, S. W., & Kim, S. S. (2012). Room temperature CO sensing of selectively grown networked ZnO nanowires by Pd nanodot functionalization. *Sensors and Actuators B: Chemical*, *168*, 8–13.
112. Kim, J. H., Lee, J. H., Park, Y., Kim, J. Y., Mirzaei, A., Kim, H. W., & Kim, S. S. (2019). Toluene-and benzene-selective gas sensors based on Pt-and Pd-functionalized ZnO nanowires in self-heating mode. *Sensors and Actuators B: Chemical*, *294*, 78–88.
113. Fan, F., Zhang, J., Li, J., Zhang, N., Hong, R., Deng, X.,... & Li, D. (2017). Hydrogen sensing properties of Pt-Au bimetallic nanoparticles loaded on ZnO nanorods. *Sensors and Actuators B: Chemical*, *241*, 895–903
114. Han, C., Li, X., Shao, C., Li, X., Ma, J., Zhang, X., & Liu, Y. (2019). Composition-controllable p-CuO/n-ZnO hollow nanofibers for high-performance H2S detection. *Sensors and Actuators B: Chemical*, *285*, 495–503.
115. Li, Y., Tao, Z., Luo, N., Sun, G., Zhang, B., Jin, H.,... & Wang, Y. (2019). Single-crystalline porous nanoplates-assembled ZnO hierarchical microstructure with superior TEA sensing properties. *Sensors and Actuators B: Chemical*, *290*, 607–615.
116. Zhang, X., Dong, Z., Liu, S., Shi, Y., Dong, Y., & Feng, W. (2017). Maize straw-templated hierarchical porous ZnO: Ni with enhanced acetone gas sensing properties. *Sensors and Actuators B: Chemical*, *243*, 1224–1230.
117. Mondal, B., Basumatari, B., Das, J., Roychaudhury, C., Saha, H., & Mukherjee, N. (2014). ZnO–SnO2 based composite type gas sensor for selective hydrogen sensing. *Sensors and Actuators B: Chemical*, *194*, 389–396.
118. Katoch, A., Kim, J. H., Kwon, Y. J., Kim, H. W., & Kim, S. S. (2015). Bifunctional sensing mechanism of SnO2–ZnO composite nanofibers for drastically enhancing the sensing behavior in H2 gas. *ACS Applied Materials & Interfaces*, *7*(21), 11351–11358.
119. Yang, X., Zhang, S., Zhao, L., Sun, P., Wang, T., Liu, F.,... & Lu, G. (2019). One step synthesis of branched SnO2/ZnO heterostructures and their enhanced gas-sensing properties. *Sensors and Actuators B: Chemical*, *281*, 415–423.
120. Ju, D. X., Xu, H. Y., Qiu, Z. W., Zhang, Z. C., Xu, Q., Zhang, J.,... & Cao, B. Q. (2015). Near room temperature, fast-response, and highly sensitive triethylamine sensor assembled with Au-loaded ZnO/SnO2 core–shell nanorods on flat alumina substrates. *ACS applied materials & interfaces*, *7*(34), 19163–19171.
121. Yoo, R., Güntner, A. T., Park, Y., Rim, H. J., Lee, H. S., & Lee, W. (2019). Sensing of acetone by Al-doped ZnO. *Sensors and Actuators B: Chemical*, *283*, 107–115.
122. Zhu, G., Xu, H., Liu, Y., Xu, X., Ji, Z., Shen, X., & Xu, Z. (2012). Enhanced gas sensing performance of Co-doped ZnO hierarchical microspheres to 1, 2-dichloroethane. *Sensors and Actuators B: Chemical*, *166*, 36–43.
123. Woo, H. S., Kwak, C. H., Chung, J. H., & Lee, J. H. (2014). Co-doped branched ZnO nanowires for ultraselective and sensitive detection of xylene. *ACS Applied Materials & Interfaces*, *6*(24), 22553–22560.

124. Mani, G. K., & Rayappan, J. B. B. (2015). A highly selective and wide range ammonia sensor—Nanostructured ZnO: Co thin film. *Materials Science and Engineering: B*, *191*, 41–50.
125. Renitta, A., & Vijayalakshmi, K. (2016). Highly sensitive hydrogen safety sensor based on Cr incorporated ZnO nano-whiskers array fabricated on ITO substrate. *Sensors and Actuators B: Chemical*, *237*, 912–923.
126. Wang, Z., Tian, Z., Han, D., & Gu, F. (2016). Highly sensitive and selective ethanol sensor fabricated with In-doped 3DOM ZnO. *ACS Applied Materials & Interfaces*, *8*(8), 5466–5474.
127. Xu, X. L., Chen, Y., Ma, S. Y., Li, W. Q., & Mao, Y. Z. (2015). Excellent acetone sensor of La-doped ZnO nanofibers with unique bead-like structures. *Sensors and Actuators B: Chemical*, *213*, 222–233.
128. Modaberi, M. R., Rooydell, R., Brahma, S., Akande, A. A., Mwakikunga, B. W., & Liu, C. P. (2018). Enhanced response and selectivity of H2S sensing through controlled Ni doping into ZnO nanorods by using single metal organic precursors. *Sensors and Actuators B: Chemical*, *273*, 1278–1290.
129. Hastir, A., Kohli, N., & Singh, R. C. (2016). Temperature dependent selective and sensitive terbium doped ZnO nanostructures. *Sensors and Actuators B: Chemical*, *231*, 110–119.

10 Application of ZnO Nanostructures for Plant Growth and Development

Mandeep Kaur

10.1 INTRODUCTION

Nanotechnology has brought revolution in numerous fields such as cancer treatment, drug delivery, biomedicines, wastewater treatment, electronics, and biosensors [1]. The nanoparticles (NPs) have unique physicochemical characteristics [2–6] compared to their bulk counterparts and plants can absorb NPs more efficiently than bulk particles [7]. One captivating area that may enhance crop management is the application of NPs in agriculture. For example, the use of nano-encased pesticides is very effective in releasing chemicals in a precisely aimed way, which results in simple pest management techniques [8, 9].

Zinc oxide nanoparticles (ZnO-NPs) hold significant capability in agriculture to enhance crop productivity and sustainability. Their unique properties enable efficient nutrient delivery, particularly zinc (Zn), which is crucial for plant metabolism. ZnO-NP is the third most widely used metal-containing nanomaterial, with an estimated 550–33,400 tons produced per year worldwide [10–12]. ZnO is a biosafe substance with the ability to photo-oxidize and catalyze chemicals and biota [13, 14].

10.2 NUTRIENT DELIVERY TO PLANTS BY ZnO-NP

ZnO-NP has emerged as an innovative approach to delivering essential Zn nutrients to plants, addressing one of the most common micronutrient deficiencies affecting crop productivity worldwide (Figure 10.1). Zn is crucial in enzyme activation, protein synthesis, hormone regulation, and defense against oxidative stress. However, conventional Zn fertilizers often face challenges related to low bioavailability and environmental losses, making ZnO-NPs a promising alternative.

ZnO-NP is used extensively in numerous materials, such as solar cells, medicine, cosmetics, rubber, and concrete, due to high thermal conductivity, UV protection, and antibacterial properties [15].

DOI: 10.1201/9781003464907-10

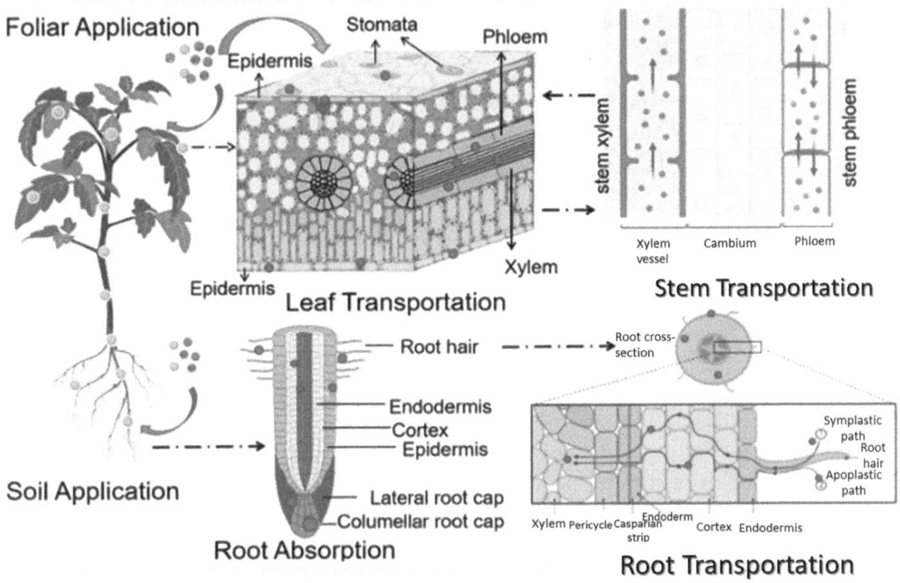

FIGURE 10.1 Mechanism of nutrient delivery to plants by zinc oxide nanoparticles [38].

For plants to grow normally, Zn is a necessary micronutrient. The effect of Zn on the metabolism of plants and crop nutrition has been the subject of research [16]. According to [17], Zn is necessary for the correct operation of several enzymes that are necessary for a variety of vital physiological processes. It was also reported that glutathione peroxidase activity rose as the deficit grew worse [18]. Reduced trunk size, leaf size, and yield are noted in cases of severe Zn shortage. Reactive oxygen species detoxification is linked to increased enzyme peroxidase activity.

ZnO-NP has shown significant promise in agricultural boosts due to its potent antimicrobial properties, which can help protect plants from a range of pathogenic microbes. These NPs act as a sustainable alternative to traditional chemical pesticides and fungicides, offering a means to reduce the reliance on conventional agrochemicals.

10.2.1 Advantages of ZnO Nanoparticles for Nutrient Delivery

- The major benefits of using ZnO-NPs instead of traditional Zn fertilizers is its higher bioavailability. The surface area to volume ratio is increased in NPs which facilitates better interaction with plant root systems. Due to this, plants uptake Zn effectively even in soils where it is typically less available such as soil with high pH.
- ZnO-NPs can be altered to provide a sustained nutrient supply to plants. This mechanism affirms that plants receive a steady amount of Zn over time and minimizes the risk of nutrient leaching.

10.3 ANTIMICROBIAL EFFECTS OF ZnO-NP AND PLANT PROTECTION

ZnO-NPs are progressively acknowledged for its strong antimicrobial properties, which make it significant in various fields such as medicine, agriculture, and environmental implications [19–23].

10.3.1 Antimicrobial Mechanisms of ZnO-NP

The primary modes of action include:

1. ZnO-NPs on exposure to light can generate reactive oxygen species such as hydrogen peroxide (H_2O_2), superoxide anions (O_2^-), as well as hydroxyl radicals (OH) shown in Figure 10.2. Reactive Oxygen Species (ROS) are extremely reactive and hold the capacity to damage essential cellular components such as lipids, proteins, nucleic acids, etc. This oxidative stress causes the destruction of microbial cell membranes, eventually leading to cell death of harmful microbiomes.

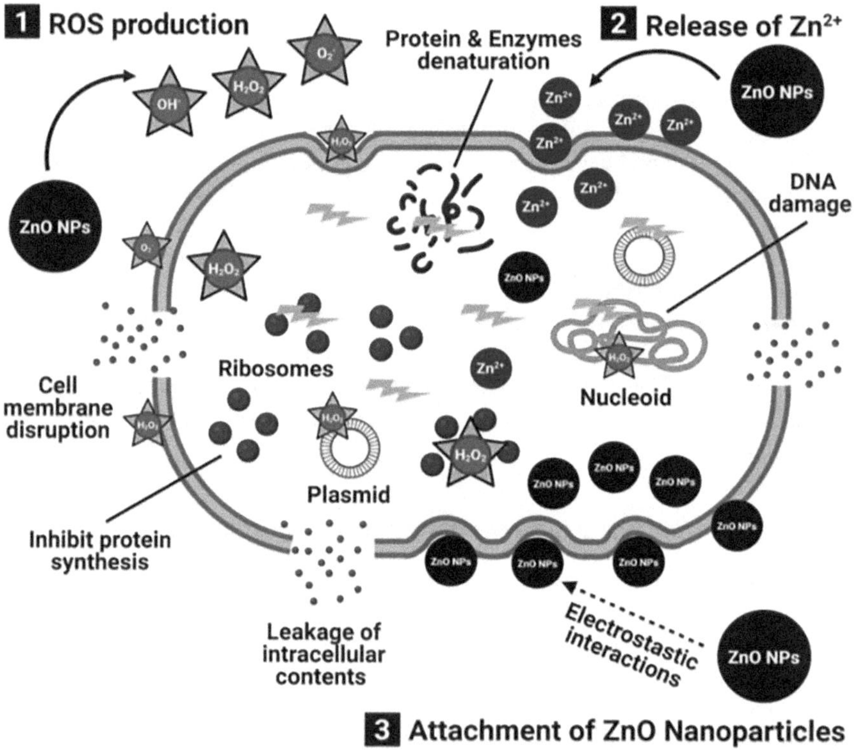

FIGURE 10.2 Antimicrobial mechanism of zinc oxide nanoparticles in plant protection [39].

2. ZnO-NPs can append to the cell membranes of bacteria and cause structural turmoil. These NPs strengthen the permeability of membrane which results in the leakage of cellular contents. This causes loss of essential cellular functions.
3. ZnO-NPs dissolve in the biological ecosystem and releases Zn^{2+} ions. These ions penetrate microbial cells and interfere with metabolic processes, including enzyme activity and DNA replication. The ions also disrupt cellular functions of microbial inhibition.

10.3.2 Applications of Antimicrobial Effects of ZnO NPs in Plant Protection

The antimicrobial properties of ZnO-NPs make them remarkably potent in safeguarding plants from different pathogens:

1. ZnO-NPs hinder the most common fungal pathogens such as Botrytis cinerea, Fusarium oxysporum, and Alternaria alternata. They hinder spore germination and disrupt the cell walls of fungi by acting as a preventive shield against fungal infections.
2. **Bacterial Disease Prevention**: Xanthomonas spp. and Pseudomonas syringae, which cause grave plant diseases, are prevented by ZnO-NP. The NPs suppress bacterial growth to reduce the spread of infections to crops.
3. **Resistance to Viral Infections**: While research is still evolving, ZnO-NP has shown potential in controlling viral infections either by interfering with viral replication or by protecting plants from viral penetration.

10.3.3 Factors Influencing Antimicrobial Activity

The antimicrobial effectiveness of ZnO-NP depends on several factors, including:

- **Particle Size and Shape**: Smaller NPs have a high surface area to volume ratio which enhances their communication with microbial cells. Additionally, specific shapes, such as rods or spherical forms, may exhibit different antimicrobial efficiencies.
- **Concentration**: A high concentration of ZnO-NP leads to greater microbial inhibition. However, optimizing the concentration is crucial to avoid toxicity to non-target organisms or human cells.

10.4 GROWTH PROMOTION AND STRESS TOLERANCE

ZnO-NPs hold significant potential and represents a cutting-edge solution for sustainable agriculture for enhancing stress tolerance in plants. This dual role is crucial in the context of the increasing need for sustainable farming practices. ZnO-NP offers a novel approach to boosting plant's resilience to drought, salinity, and pathogen attacks [24, 25].

10.4.1 Mechanisms of Growth Promotion

1. **Enzyme Activation**: Zn is a structural and functional part of many enzymes such as carbonic anhydrase, SOD, and alcohol dehydrogenase. These enzymes are essential in photosynthesis, cellular respiration, and detoxification of ROS. Providing Zn in a readily available form by ZnO-NP will increase enzyme activities, thereby enhancing the metabolic processes and overall vigor of the plant.
2. **Improved Photosynthesis**: ZnO-NP increases chlorophyll production and photosynthetic activity. The involvement of Zn in carbonic anhydrase activity catalyzes carbon dioxide assimilation during photosynthesis, hence increasing biomass productivity. Experimental results have been found that the plants exposed to ZnO-NP showed greater amounts of chlorophyll, giving greener leaves with higher strength and increased photosynthetic rate.
3. **Hormonal Regulation**: Zn increases the biosynthesis of auxins, which are vital plant hormones for root development. ZnO-NP has shown to stimulate root elongation and branching, thereby enhancing water and nutrient uptake. This root system enhancement results in stronger, healthier plants with greater access to soil resources.

10.4.2 Mechanisms of Stress Tolerance Enhancement

1. ZnO-NP helps the plants to confront stress by reducing oxidative damage. Stress causes the surplus production of ROS damaging the proteins and DNA. ZnO-NP elevates the activity of antioxidant enzymes. These enzymes help in neutralizing ROS, protecting cellular structures, and maintaining the integrity of physiological functions.
2. ZnO-NPs augment the drought stress tolerance in the plants. This boosts the absorption of water into the plant by enhancing the hydrostatic pressure. This reduces water stress in the plants. The elevation in antioxidant enzyme activity helps the plants to survive by hindering the cell damage due to dehydration.
3. ZnO-NP reduces the salt stress by improving ion consistency and membrane balance. High salt content in the soil poses a great problem in agriculture because it adversely impacts the ion balance in the plants. ZnO-NP helps in the ion accumulation balance and hence controls the intake of vital nutrients.

10.5 PHOTOCATALYTIC ACTIVITY FOR PEST MANAGEMENT

This method uses the ability of ZnO-NP to gather sunlight and produce ROS, which are very powerful in killing pests and harmful organisms that cause crop damage[26–31].

10.5.1 Photocatalytic Mechanism

ZnO-NPs possess a wide bandgap and exhibits strong photocatalytic activity when exposed to ultraviolet (UV) and visible light. This activity is driven by the following steps:

1. ZnO-NP absorbs photons and generates electron–hole pairs when exposed to light energy, particularly in the UV range. Positively charged holes remain in the valence band as a result of the electrons jumping from the valence band to the conduction band due to the energy from the light.
2. ROS such as OH, (O_2^-), and H_2O_2 are created when excited electrons and holes interact with environmental oxygen and water molecules. These reactive species have significant oxidative properties and are highly unstable.
3. The molecules present in the cellular membranes and structures of bacteria and pests can be degraded by the ROS generated by ZnO-NPs. The aimed organisms are as a result of the breakdown of cell membranes and damage to nucleic acids.

10.5.2 Application in Pest Management

1. **Controlling Insect Pests**: ZnO-NP has an opportunity to be used against the insect pests that cause crop infestation through its photocatalytic activity. While being applied to the surface of plants, ZnO-NP acts as a protective layer that activates under sunlight. The ROS created in this process can disrupt the cuticle and digestive system of insects, thus reducing populations of pests and damaging the crops.
2. **Antibacterial Effects**: Plant diseases associated with bacterial and fungal pathogenesis can be successfully managed using ZnO-NP. By interpenetrating bacterial and fungal cell walls, ROS generated by the photocatalytic process hinders the development of infection. This property helps in protecting the plant from common diseases such as bacterial wilt and powdery mildew.
3. **Degradation of Pesticide Residues**: ZnO-NP can break down pesticide residues on crop surfaces. This ensures that pesticides have a little effect on the environment and the crops.

10.6 SEED COATING FOR INCREASED GERMINATION USING ZnO-NP

With novel methods like seed coating, which improves seed germination, the agricultural landscape has been completely changed by ZnO-NP. ZnO-NP-based seed coverings are a groundbreaking agricultural technology to improve crop yield by overcoming nutritional deficiencies and low germination rates [32–35].

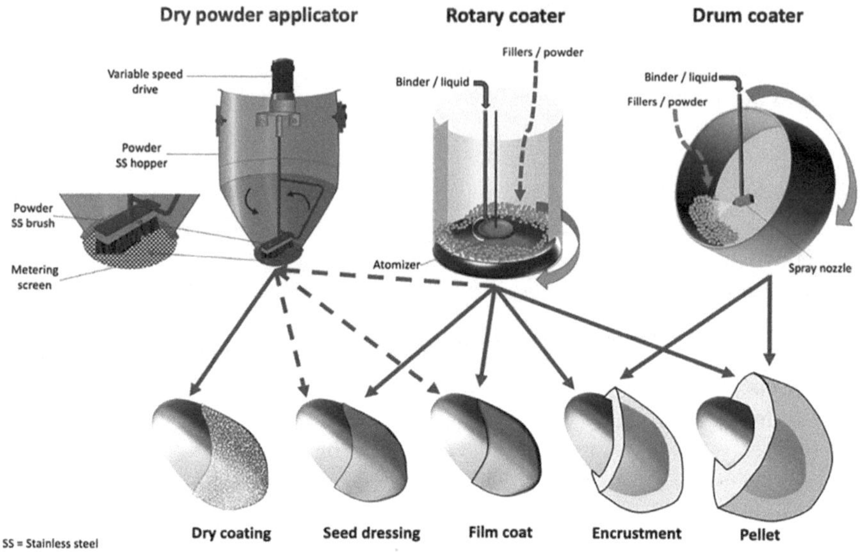

FIGURE 10.3 Seed coating using zinc oxide nanoparticles [40].

10.6.1 Seed Coating

The seed coating is a significant agricultural method for boosting germination, augmenting seed performance, and improving defense against diseases and stress (Figure 10.3). ZnO-NP has emerged as a potential material among other NPs that can successfully protect seeds and give essential nutrients. Below is a thorough description of the application procedure for seed coating using ZnO-NPs.

10.6.2 Seed Coating Application Process

1. **Seed Pre-treatment**: Before seed coating, seeds have to undergo a pre-treatment process to clean and remove dust and microbial contaminants. This step is important to achieve uniform coating and proper adhesion of the NPs or other coating materials. In general,
   i) Seeds are thoroughly cleaned to remove unwanted particles.
   ii) To ensure proper adhesion of the coating materials, moisture content in the seeds must be regulated. Seeds may be dried using controlled drying methods depending on their moisture susceptibility.
2. **Preparation of the Coating Solution**
   Preparing a coating solution of ZnO-NP is the next step in this process. This solution should be suitable to deliver the required nutrients and protective benefits to the seeds. The preparation process includes:
   - **Mixing additional components to ZnO-NP**: Depending on the application, other constituents may be added to the coating solution. These include:
     - Binders or Polymers: The role of binders is to ensure the proper adherence of the NPs to the surface of the seed. Examples of binders are natural polymers and synthetic polymers.

- **Adjusting concentration**: To balance efficacy and safety, the concentration of ZnO-NP is carefully adjusted. Since excessive amounts could be phytotoxic, precision is very important.

3. **Application of the Coating**
   There are several methods of applying the coating to seeds and the choice of method depends on the type and size of the seed, coating material, and required coating thickness. Common seed coating techniques include:
   Slurry Coating:
   - Seeds are added to the ZnO-NP slurry in a rotating mixer.
   - The continuous movement of the drum ensures even distribution of the coating solution across all seeds.
   - The viscosity of the slurry is optimized to prevent clumping on the seeds.

   Fluidized Bed Coating:
   - This technique entails suspending seeds in an airstream inside a fluidized bed chamber.
   - During fluidization of the seeds, the ZnO-NP solution is sprayed uniformly over the seeds to create a uniform layer coating.
   - Fluidized bed coating is highly efficient and controls the coating very precisely in terms of thickness and uniformity.

   Seed Tumbling and Pan Coating:
   - The coating solution is sprayed on the seeds and placed in a rotating tumbler.
   - The rotation of the pan is ensured in such a way that each seed is properly coated.
   - This method is widely used and easily adapted for large-scale seeds.
4. **Drying the Coated Seeds**
   The seeds are dried after applying the ZnO-NP coating to ensure that the coating adhered firmly and did not peel off. There are several ways in which drying was done:
   - Air Drying: These seeds are allowed to dry at room temperature in an open area. This is one of the easiest methods; however, it may be time-consuming, especially under moist conditions.
   - Controlled Drying: In industrial settings, seeds are dried using heated air in controlled environments. The temperature and humidity are regulated to prevent damage to the seeds and ensure optimal drying.

Drying is very essential as improper drying leads to poor adhesion of the coating and reduced viability of seed. Care should be taken not to overheat seeds, which can harm the seed germination and vigor.

10.7 QUALITY CONTROL AND TESTING

To ensure the effectiveness and safety of the ZnO-NP seed coating, quality control measures are implemented. This includes:

- **Uniformity Check**: The check for uniformity and coherence of the coating layer through visual and instrumental inspection.

- **Adherence Testing**: The adhesion strength is checked of the coating with the seed coat surface for durability during the process of handling and storage.
- **Seed Feasibility Tests**: In order to ensure that the coating process does not affect the sprouting ability of seeds, germination tests are conducted. Germination rates of untreated seeds are compared with the coated seeds under controlled conditions.

10.8 PACKAGING AND STORAGE

The coated seeds, after quality control checks, are carefully packaged and stored. The storage conditions must be moisture-free so that the coating is preserved. The packaging materials are chosen to effectively block humidity and pests, ensuring the seeds stay viable until it is time to plant them.

10.9 FIELD APPLICATION

The last step is the application of coated seeds into the field. The farmer applies standard agricultural methods to sow the seeds. The coating of the seeds with ZnO-NPs has several advantages, such as:

- Improved germination and seedling growth as it facilitates appropriate metabolic processes necessary for seed germination.
- Pathogen-resistance owing to ZnO-NPs conferring protection toward seeds via relatively high resistance against fungal and bacterial infections.
- Better stress tolerance because the coating itself mitigates stress on seedlings, leaving behind tougher and healthier plants.

10.9.1 Advantages of the Seed Coating Process Using ZnO-NP

1. Zn is directly provided to the seeds which boosts the nutrient uptake, reducing the dependence on fertilizers.
2. Nutrient run-off and soil contamination are decreased.
3. Better seed germination and stronger crop yields are established.

10.10 REGULATION OF PLANT HORMONES USING ZnO-NP

Phytohormones (Plant hormones) are vital chemicals to regulate plant growth and development and their response to environmental stress. The interaction between these hormones and external factors significantly influences plant health and productivity. ZnO-NPs have emerged as a powerful tool in influencing the regulation of plant hormones, thus enhancing crop growth, stress resistance, and overall plant performance. This advanced technology offers promising potential for sustainable agricultural practices [36, 37].

10.10.1 Mechanism of Regulating Plant Hormones

1. **Auxins**: ZnO-NPs improve the activity of auxins (hormones responsible for the growth of the cell and root formation) through the activation of signaling pathways and enzymatic reactions. It strengthens the root architecture that allowed for more efficient access to water and nutrients.
2. **Gibberellins (GAs)**: Gibberellins are involved in seed germination, stem elongation, and flowering. Zn is a cofactor for several enzymes involved in the biosynthesis of gibberellins. ZnO-NP stimulates gibberellin production, enhancing seed germination and increasing plant height. This effect is especially relevant to crop establishment and rapid growth in short growing seasons.
3. **Cytokinins**: These hormones promote cell division and delay senescence. Thus, it contributes to proper shoot growth and nutrient mobilization. ZnO-NP would positively impair cytokinins working favorably for the promotion of new shoots and proper chlorophyll retention. Moreover, delaying senescence has left plants actively photosynthesizing for a longer period, hence benefiting crop yield and quality. Cytokinins will also act in response to stresses occurring in plants, where ZnO-NP treatment can enhance the resilience of the plant under adverse conditions.
4. **Abscisic Acid (ABA)**: The hormone for the plant, ABA, regulates stomatal closure, seed dormancy, and response to drought and salinity. It is still unclear how ZnO-NP affects ABA regulation. In normal conditions, ZnO-NP could modulate ABA synthesis and signaling pathways in preparation for possible stress. For example, when plants are stressed from drought, treatment with ZnO-NP increases ABA levels, causing the stomata to close, leading to reduced water loss. Such adaptive mechanisms help conserve water while providing plants with a chance to survive a longer period of moisture availability.
5. **Ethylene**: Ethylene is a gaseous hormone of fruit ripening, leaf abscission, and stress response. Evidence indicates that ZnO-NP modifies ethylene biosynthesis and action. ZnO-NP can delay fruit ripening by eluding premature leaf fall through regulating ethylene production. This is of great utility in prolonging the shelf life of crops and minimizing post-harvest losses. In addition, under stress conditions, controlling ethylene levels reduces the vegetative effects of environmental challenges, such as flooding or mechanical injury.

10.10.2 Practical Applications and Benefits

1. **Improved Crop Growth and Yield**: Soil NPs of zinc oxide regulate plant hormone activity, therefore enhancing root and shoot development, nutrient uptake, and biomass accumulation. Increased yields and improved produce will constitute an answer to food security problems.

2. **Enhanced Stress Tolerance**. One of the great benefits of ZnO-NP is to make plants resistant to biotic and abiotic stresses. Hormonal modulation makes plants capable of confronting challenges inflicted by drought, salinity, extremes of temperature, and pathogen attacks. For instance, ABA raises water conservation during drought conditions, while being in the optimum range, and cytokinins support growth in nutrient-limited environments.
3. **Efficient Water and Nutrient Use**: ZnO-NP encourages root architecture development, thereby augmenting nutrient and water uptake efficiency. The use of ZnO is vital, especially in arid and semi-arid contexts, where water and nutrients are limited. The regulation of the hormones, e.g., auxins and cytokinins, enables plants to use scarce resources with utmost effectiveness toward sustainable agricultural practices.
4. **Delayed Senescence and Extended Harvest Periods**: Delaying aging in plants involves holding up high levels of cytokinins and inhibiting ethylene production by ZnO-NP. Crops can carry on performing photosynthesis for an extended time, giving yields higher than normal and extended harvest periods. Particularly for leafy vegetables and fruit crops, such prolonging of senescence equates to better marketability.

10.11 OBSTACLES AND OPPORTUNITIES

Utilizing ZnO-NP in agriculture unlocks potential for stress tolerance, nutrient uptake efficiency, and plant growth. However, despite these assurances, significant obstacles surround the wide range of application of ZnO-NP. It is important to address these issues in order to ensure the safe and sustainable application of ZnO-NP in agriculture.

10.11.1 Obstacles

1. **Environmental Impact and Persistence**: As a result of metal oxide NPs being deliberately released into the environment, their long-term environmental effects and possible fate are of concern. They may remain in soils and waters and thus phytotoxic to soil microbes and non-target organisms. In several environments, the fate, degradation, and behavior of ZnO-NP with other particulates remain obscure. Understanding particle–environment interactions, for example, their probable leaching through groundwater or effects on soil health, is necessary for the establishment of the green application.
2. **Impact on Soil Microbiome**: The soil microbiome is strongly connected with nutrient cycling, organic matter decomposition, and general soil health. While the antimicrobial properties of ZnO-NP may be beneficial for the protection of plants against pathogens, they may also have detrimental effects on beneficial soil microbes. Disruption of such microbial communities may have an extensive cascading effect on soil fertility and plant productivity. An improved understanding of the interaction between ZnO-NP and soil microorganisms is needed to avoid detrimental effects.

3. **Regulatory and Safety Concerns**: The use of NPs in agriculture is, of course, heavily regulated. The risk evaluation that one has to follow ensures that ZnO-NP is safe for human health, food security, and environmental protection – a long process indeed. Therefore, risk assessment and regulation need to be in place if we are to safely use ZnO-NP, including crop production. Other issues will pop up, like public perception or acceptance and the ways to counter them with clear communication and education.
4. **Scalability and Economic Viability**: If laboratory studies have proved the efficacy of ZnO-NP, it would be an uphill task to convert these findings to large-scale agricultural practices because of economic and logistical considerations. The synthesis and application cost of ZnO-NP must be economically viable for farmers, predominantly in developing regions. Hence, the development of more economical and large-scale methods of production and application of ZnO-NP will be vital to the technology becoming accessible and practical for wider usage.

10.11.2 Future Opportunities

1. **Precision Agriculture and Smart Delivery Systems**: Innovations in precision agricultural techniques could develop novel ZnO-NP for more targeted and effective delivery to plants. Nanocapsulation and controlled-release formulations, coupled with smart delivery systems, could minimize environmental exposure and reduce the risk of phytotoxicity.
2. **Eco-Friendly and Biodegradable Nanomaterials**: To lessen the environmental effect of using NPs in agriculture, researchers are investigating the creation of biodegradable and environmentally friendly nanomaterials. Green synthesis techniques, which create ZnO-NP using microbial or natural plant extracts, are becoming more popular. These methods can reduce the amount of dangerous chemicals used and improve the NPs' compatibility with environmentally friendly farming methods.
3. **Interdisciplinary Research for Holistic Understanding**: An interdisciplinary approach is necessary to address the issues related to ZnO-NP. To completely comprehend how NPs, plants, and the environment interact, cooperation among plant scientists, soil ecologists, toxicologists, and materials scientists is crucial. To evaluate the practical impacts of ZnO-NP on agricultural productivity, soil health, and ecosystem dynamics, long-term field research is required.
4. **Policy Development and Risk Management**: To create regulations that strike a balance between the advantages and possible concerns of nanotechnology, policymakers and regulatory agencies must collaborate with scientists. For appropriate implementation, guidelines for waste management, environmental monitoring, and nanoparticle usage must be established. Furthermore, funding studies on risk management and remedial techniques might lessen the likelihood of unfavorable results.
5. **Potential for Climate-Resilient Agriculture**: ZnO-NP' capacity to improve plant stress tolerance offers great potential as climate change continues to

jeopardize the world's food security. The development of climate-resilient crops may be greatly aided by ZnO-NP, which enhances disease, salinity, and drought resistance. Future studies may concentrate on utilizing these advantages to promote climate change-resilient agriculture.

10.12 CONCLUSION

ZnO-NP holds significant potential for enhancing agricultural and crop sustainability. By improving nutrient absorption, ZnO-NP can boost photosynthesis, support the growth of roots and shoots, and promote resilient plant growth. By protecting crops from illnesses, its antibacterial properties also reduce the need for chemical pesticides. ZnO-NP also improves a plant's tolerance to environmental challenges like salt and drought by changing how hormones are regulated in plants and how they react to stress. However, strict research and control are needed to ensure their safe and environment-friendly use.

REFERENCES

1. Nel, A., Xia, T., Madler, L., & Li, N. (2006). Toxic potential of materials at the nanolevel. *Science*, 311(5761), 622–627. https://doi.org/10.1126/science.1114397
2. Lin, D., & Xing, B. (2007). Phytotoxicity of nanoparticles: Inhibition of seed germination and root growth. *Environmental Pollution*, 150(2), 243–250. https://doi.org/10.1016/j.envpol.2007.01.016
3. Handy, R. D., Owen, R., & Valsami-Jones, E. (2008). The ecotoxicology of nanoparticles and nanomaterials: Current status, knowledge gaps, challenges, and future needs. *Ecotoxicology*, 17(4), 315–325. https://doi.org/10.1007/s10646-008-0206-0
4. Monica, R. C., & Cremonini, R. (2009). Nanoparticles and higher plants. *Caryologia*, 62(2), 161–165. https://doi.org/10.1080/00087114.2004.10589699
5. Ma, X., Geiser-Lee, J., Deng, Y., & Kolmakov, A. (2010). Interactions between engineered nanoparticles (ENPs) and plants: Phytotoxicity, uptake, and accumulation. *Science of the Total Environment*, 408(16), 3053–3061. https://doi.org/10.1016/j.scitotenv.2010.03.031
6. Bhatt, I., & Tripathi, B. N. (2011). Interaction of engineered nanoparticles with various components of the environment and possible strategies for their risk assessment. *Chemosphere*, 82(3), 308–317. https://doi.org/10.1016/j.chemosphere.2010.10.011
7. Mehta, C. M., Srivastava, R., Arora, S., & Sharma, A. K. (2016). Impact of nanoparticles on the agronomic traits of plants: A review. *Biotech*, 6(2), 254. https://doi.org/10.1007/s13205-016-0544-1
8. Beddington, J. (2010). Food security: Contributions from science to a new and greener revolution. *Philosophical Transactions of the Royal Society B: Biological Sciences*, 365(1537), 61–71. https://doi.org/10.1098/rstb.2009.0201
9. Nair, R., Varghese, S. H., Nair, B. G., Maekawa, T., Yoshida, Y., & Kumar, D. S. (2010). Nanoparticulate material delivery to plants. *Plant Science*, 179(3), 154–163. https://doi.org/10.1016/j.plantsci.2010.04.012
10. Bondarenko, O., Juganson, K., Ivask, A., Kasemets, K., Mortimer, M., & Kahru, A. (2013). Toxicity of engineered nanomaterials to aquatic invertebrates: A meta-analysis. *Archives of Toxicology*, 87(7), 1181–1200. https://doi.org/10.1007/s00204-013-1072-9
11. Connolly, M., Fernandez, M., Conde, E., Torrent, F., Navas, J. M., & Fernandez-Cruz, M. L. (2016). Nanoparticle ecotoxicology: Applications in risk assessment. *Science of the Total Environment*, 1, 334–343. https://doi.org/10.1016/j.scitotenv.2016.01.001

12. Peng, Y. H., Tsai, Y. C., Hsiung, C. E., Lin, Y. H., & Shih, Y. (2017). The environmental behavior and impacts of engineered nanomaterials. *Journal of Hazardous Materials*, 322, 348–356. https://doi.org/10.1016/j.jhazmat.2016.11.021
13. Sirelkhatim, A., Shahrom, M., Azman, S., Noor, H. M. K., Chuo, A. L., Siti, K. M. B., Habsah, H., & Dasmawati, M. (2015). Review on zinc oxide nanoparticles: Antibacterial activity and toxicity mechanism. *Micro & Nano Letters*, 7(3), 219–242. https://doi.org/10.1049/mnl.2015.0020
14. Vaseem, M., Umar, A., & Hahn, Y. B. (2010). ZnO Nanoparticles: Growth, properties, and applications. In *Nanomaterials* (4th ed., pp. 1–36). American Scientific Publishers.
15. Uikey, P., & Vishwakarma, K. (2016). A review on nanotechnology applications in food. *International Journal of Emerging Technologies in Computer Science & Electronics*, 21, 239.
16. Auld, D. S. (2001). Zinc coordination sphere in biochemical zinc sites. *Biometals*, 14, 271–313. https://doi.org/10.1023/A:1012983423042
17. Lacerda, J. S., Martinez, H. E., Pedrosa, A. W., Clemente, J. M., Santos, R. H., Oliveira, G. L., & Jifon, J. L. (2018). Nanotechnology in plant science: Enhancing agricultural productivity. *Crop Science*, 58, 1360–1370. https://doi.org/10.2135/cropsci2018.02.0110
18. Castillo, R. R., Lozano, D., Gonzalez, B., Manzano, M., Izquierdo-Barba, I., & Vallet-Regi, M. (2019). Advances in drug delivery systems using nanotechnology. *Expert Opinion on Drug Delivery*, 16, 415–439. https://doi.org/10.1080/17425247.2019.1594748
19. Farzana, et al. (2023). Advances in nanotechnology for combating microbial resistance. *Frontiers in Microbiology*, 14, 1423–1437. https://doi.org/10.3389/fmicb.2023.01423
20. Nair, et al. (2021). Recent developments in materials for nanomedicine. *Materials Today: Proceedings*, 47, 561–572. https://doi.org/10.1016/j.matpr.2020.09.240
21. Salem, et al. (2022). Antimicrobial efficacy of ZnO nanoparticles against pathogens. *Applied Nanoscience*, 12, 312–324. https://doi.org/10.1007/s13204-021-02036-7
22. Divya, et al. (2019). Zinc oxide nanoparticles: A tool for plant pathogen management. *Journal of Plant Pathology*, 101, 203–211. https://doi.org/10.1007/s42161-018-0210-y
23. Divyapriya, et al. (2020). Applications of nanomedicine in health care. *International Journal of Nanomedicine*, 15, 1129–1141. https://doi.org/10.2147/IJN.S240978
24. Ozturk, H. I., Pour-Aboughadareh, A., & Poczai, P. (2020). Applications of nanotechnology in agriculture. *Agriculture*, 10, 521. https://doi.org/10.3390/agriculture10120521
25. Niedbała, G., et al. (2024). Nanotechnology for sustainable farming practices. *Agronomy*, 14, 1462. https://doi.org/10.3390/agronomy14071462
26. Ali, M., et al. (2014). Surface interactions of nanoparticles in environmental matrices. *Applied Surface Science*, 307, 1–10. https://doi.org/10.1016/j.apsusc.2014.01.004
27. Johnson, A. M., & Kim, S. (2020). Photocatalysis mechanisms of nanomaterials. *Journal of Photochemistry and Photobiology A*, 390, 34–45. https://doi.org/10.1016/j.jphotochem.2019.112367
28. Zaman, H., et al. (2024). Catalytic applications of ZnO nanoparticles. Catalysts, 14, 420–435. https://doi.org/10.3390/catal14020420
29. Ramesh, B., & Gupta, P. (2020). Nanoparticle safety and impact assessment. *NanoImpact*, 11, 56–70. https://doi.org/10.1016/j.impact.2020.100273
30. Banerjee, P., & Roy, K. (2019). Environmental fate of engineered nanoparticles. *Chemosphere*, 218, 134–145. https://doi.org/10.1016/j.chemosphere.2018.11.072
31. Zhang, L., & Meng, Q. (2018). Advances in environmental nanotechnology. *Environmental Science & Technology*, 52, 2365–2374. https://doi.org/10.1021/acs.est.7b04364
32. Prasad, T. N. V. K. V., Sudhakar, P., Sreenivasulu, Y., Latha, P., Munaswamy, V., Raja Reddy, K., & Sreeprasad, T. S. (2012). Plant growth stimulation by nanoparticles. *Journal of Plant Nutrition*, 35, 905–927. https://doi.org/10.1080/01904167.2012.663443

33. Javed, R., Usman, M., Yücesan, B., Zia, M., & Gürel, E. (2017). Nanoparticles and their role in plant physiology. *Applied Biological Chemistry*, 110, 94–99. https://doi.org/10.1186/s13765-017-0277-8
34. Mahajan, P., Dhoke, S. K., & Khanna, A. S. (2011). Nanotechnology applications in crop protection. *Journal of Nanotechnology*, 2011, 696535. https://doi.org/10.1155/2011/696535
35. Raliya, R., & Tarafdar, J. C. (2013). Nanotechnology for enhancing crop productivity. *Agricultural Research*, 2, 48–57. https://doi.org/10.1007/s40003-012-0047-z
36. Hafizi, & Nasr. (2019). Nanoparticles in agriculture: Mechanisms and impacts. *Applied Nanoscience*, 9, 121–130. https://doi.org/10.1007/s13204-019-00998-8
37. Salem, et al. (2016). Zinc oxide nanoparticles in pathogen management. *Journal of Experimental Nanoscience*, 11, 190–199. https://doi.org/10.1080/17458080.2015.1102791
38. Wang, et al. (2023). Nanostructured materials for advanced applications. *Materials*, 16, 3097. https://doi.org/10.3390/ma16123097
39. Murali, et al. (2021). Nanomedicine applications in pharmaceuticals. *Pharmaceutics*, 13, 1662. https://doi.org/10.3390/pharmaceutics13101662
40. Afzal, et al. (2020). Agricultural nanotechnology: Opportunities and challenges. *Agriculture*, 10, 526. https://doi.org/10.3390/agriculture10120526

11 Efficacy of ZnO Adsorbents in Heavy Metal Detection and Removal

Naveen Kumar, Gurbhej Singh, Ranvir Singh Panwar, and Bastola Narayan

11.1 INTRODUCTION

Heavy metals (HMs) are naturally occurring substances that possess high density (>5 g/cm^3) and relatively high atomic weights. HMs are major polluters of water primarily because of industrial activities. These metals, mostly presented to the environment through untreated discharges of waste from industries, are hazardous for human health as well as the environment [1]. HMs such as Pb, Cd, Hg, and As are toxic and non-biodegradable. HMs can disrupt aquatic ecosystems because the metals can accumulate in the aquatic organisms that move up the food chain and get more concentrated with each trophic level, and this process can be termed bio-magnification. Such bio-magnification processes could lead to toxic effects on aquatic life through impacts on reproduction, growth, and survival rates among species like fish and amphibians [2]. These issues concerning health range from organ damage to mutagenic and carcinogenic effects. The industrial revolution in the past few decades has contaminated the environment due to the release of HMs in water through drainage, in soil due to waste, and in air due to the release of gases [3]. These sources then naturally rise by increasing human factors such as erosion. Accumulation of HMs in these water sources has major harmful effects on aquatic ecosystems as it gradually contributes to higher toxicity levels for water species as well as humans. HMs can also influence the pH and total dissolved solids (TDS) levels of water, affecting nutrient cycling and ecosystem health in general [4]. The contamination that is produced renders water unsafe, thus raising the need for effective treatment methods for water to eliminate these pollutants.

Heavy metal pollution causes major health problems and environmental consequences. In humans, it is observed that the high risks are offered by Cd, Cu, As, Hg, Pb, and others, especially at minute levels, as such low-level exposures cause very severe disorders of the body such as respiratory paralysis, renal failure, cardiac arrest, and pulmonary and intestinal hemorrhages. HMs can accumulate within the tissues and can impart diverse effects on health, which cause problems when exposure

DOI: 10.1201/9781003464907-11

occurs over extended periods [5, 6]. Zinc oxide (ZnO) is one of the extensively studied compounds, especially in the form of nanoparticles (NPs), because of their usage for cleaning the environment owing to the stability in chemical form, environment-friendly, and potent removing ability of contaminants [7]. ZnO NPs decrease HMs toxicity, especially As(III), by inhibiting the uptake of the metals in plants but simultaneously enhancing plant growth by their resistance to stress factors. For instance, in arsenic-exposed soybean plants, ZnO NPs were reported to decrease the level of arsenic in plant tissues, increase photosynthetic efficiency, and reduce oxidative stress. The protective role is partly attributed to the capability of ZnO NPs to increase the activities of antioxidant enzymes that help fight oxidative damage induced by HM stress [8].

Khezami et al. explored Cr-doped ZnO NPs for their efficiency in removing organic dyes, particularly Congo red, from wastewater. The photocatalytic properties of ZnO NPs were enhanced by Cr doping, which increases surface area and adsorption capacity. This doped ZnO shows high adsorption efficiency due to its structural modifications, which enhance pollutant binding [9]. Vasantharaj et al synthesized ZnO NPs using aqueous extract of *Ruellia tuberosa* and carried out their antibacterial activity against *Escherichia coli*. Additionally, ZnO NPs were applied to textile materials to stop *E. coli* from growing. From an environmental standpoint, ZnO NPs broke down malachite green (MG) and methylene blue (MB) by 92% and 94%, respectively, which is better compared to previously published data with other types of NPs [10]. Dhiman et al. reviewed ZnO nanostructures for eliminating toxic HMs from water, highlighting the strong potential of ZnO NPs due to their low toxicity, biocompatibility, and surface modifications that yield a negative zeta potential. Additionally, the redox capabilities of ZnO NPs were facilitated by photo-induced electron-hole pairs, enhancing its effectiveness for heavy metal remediation [11].

11.2 ZINC OXIDE (ZnO) AS A POTENTIAL CANDIDATE FOR HEAVY METAL REMOVAL

Zinc oxide (ZnO), a white crystalline solid, exists in three crystal structures: cubic zinc blende, rocksalt, and hexagonal wurtzite structure, as shown in **Figure 11.1**. At room temperature, it stabilizes as hexagonal wurtzite, featuring strong ionic bonds

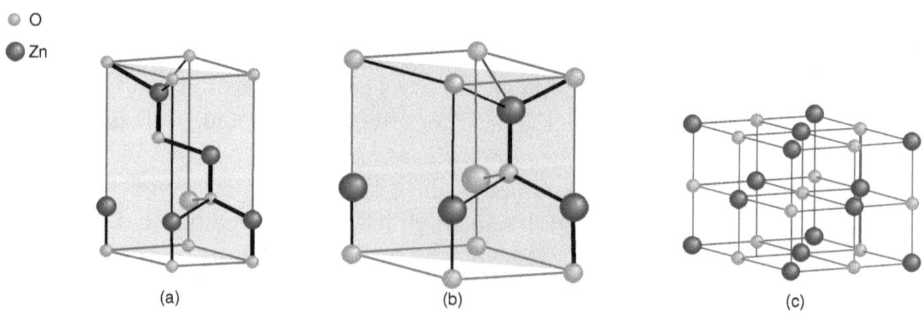

FIGURE 11.1 Crystal structure of ZnO (a) zinc blend, (b) wurtzite, and (c) rocksalt.

due to the differing electro-negativities of Zn^{2+} (1.65) and O^{2-} (3.44) ions. ZnO has sp^3 hybridized covalent bonds, with each Zn or O atom tetrahedrally surrounded by four atoms of the opposite type. ZnO in bulk form exhibits a density of 5.67 g cm^{-3} and a melting point of 1974 °C. The optical energy band gap of ZnO is 3.34 eV at room temperature [12, 13].

Zinc oxide (ZnO) is an environmentally friendly semiconducting material that possesses low cost of synthesis, biocompatibility, ease of handling, and is non-toxic, which makes it suitable for various applications, such as UV protection in skincare, antimicrobial products, piezoelectric devices, and photocatalysis [14–16]. The research has mainly focused on improving the properties of ZnO by synthesizing nanostructures such as nanowires, nano-cages, and nanosheets, which may open new avenues in industrial and environmental applications [17].

11.3 METHODS FOR SYNTHESIZING ZnO NPs

The synthesis method is an important criterion for modifying the properties of ZnO since it significantly affects factors involving crystallinity, size, morphology, and surface area, all of which directly affect the performance of ZnO in most applications. There exist several methods to prepare ZnO NPs, such as sol-gel, co-precipitation, vapor deposition, solvothermal methods, green routes, and mechano-chemical techniques, each to prepare ZnO nanostructures with precise and well-defined properties [18]. The chemical methods provide size and shape control, which further improves conductivity and photocatalytic efficiency. Instead, biosynthesis methods tend to enhance biocompatibility and decrease the environmental impact for more applicability of ZnO in pharmaceuticals and cosmetics.

11.3.1 Chemical Methods for the Synthesis of ZnO NPs

ZnO NPs are commonly synthesized through chemical methods such as sol-gel, hydrothermal, precipitation, and vapor deposition techniques, which are preferable for controlling specific features, such as structure, microstructure, and surface effects [19]. These methods are advantageous because they result in high-purity, uniform NPs, which are necessary for applications in electronics and catalysis. They also allow the tailoring of properties of ZnO for better functionalities. However, such processes are usually toxic, harmful to the environment, highly energy-consuming and costly, and require waste management to avoid ecological impact [20].

High surface-to-volume ratio, varying morphology, and strong photocatalytic properties offer great benefits for the reduction of HMs in water. ZnO NPs can thus efficiently adsorb and degrade toxic metal ions under UV or visible light, thus facilitating redox reactions to immobilize or break down metal contaminants. Additionally, the precise control over particle size and surface chemistry in chemical synthesis optimizes ZnO NPs for higher adsorption efficiency, making them particularly effective for water purification applications [21, 22].

Ahmad et al. synthesized a $ZnO/SrZnO_2$ nanocomposite and reported a 92.4% degradation efficiency for Congo red dye and carried out antibacterial activity against *E. coli* under visible light. Its photocatalytic degradation rate was 5.6 times faster than pure ZnO, attributed to enhanced charge separation and peroxide free-radical generation, making it a promising candidate for water purification [23]. Similarly, Khizami et al. developed Al-doped ZnO nanopowders, particularly with 1 wt% Al doping, which showed excellent efficiency in removing heavy metal ions from aqueous solutions [24].

11.3.2 Biogenic Synthesis Methods of ZnO NPs

Biosynthesis of ZnO NPs is achieved through biological sources such as plants, fungi, bacteria, and algae to produce NPs. Biomolecules present within neem, Aloe vera, basil plants, and *Aspergillus niger* fungi can be used to reduce zinc salts to form ZnO NPs [25]. Other important bacterial species that help in Zn2+ ion reduction and contribute toward the formation of NPs extracellularly or intracellularly are bacteria of the genus Bacillus like *B. subtilis* and algae, genus *Chlorella* and *C. vulgaris* [26]. It is cost-effective, environmentally friendly, and produces controlled-sized morphology NPs. Biogenic ZnO NPs have a plethora of applications in environmental remediation, photocatalysis, and biomedicine.

The reduction of HMs from aqueous solutions can be accomplished significantly through the biosynthesis of ZnO NPs. This biogenic route, by using plants, fungi, bacteria, and algae, provides a safe, low-cost, and environmentally benign way of preparing ZnO NPs as capable adsorbents for heavy metal removal. The biological sources contain natural biomolecules such as proteins, enzymes, and polysaccharides, which assist in reducing the zinc salts to ZnO NPs and stabilizing them as well [27, 28]. ZnO NPs exhibit high surface area, which can enhance the adsorption capacity of the heavy metal ions, for instance, Cd(II), Pb(II), Cr(VI), and Ni(II). Due to the high reactivity and surface charge, heavy metal ions are captured on the surface by effective physical adsorption, electrostatic forces, or chemical reduction in the case of enablement. These are primarily associated with electrostatic forces, coordination, or complexation sites on the surface of the NPs, which may occur at the oxygen vacancies and other active sites of the ZnO surface [29]. In addition, biosynthesized ZnO NPs have different surface functional groups from biological capping agents, which increase their metal-binding affinity and improve the quality of contaminated waters. In addition, the biogenic route facilitates control over physical properties, which maximizes the process of adsorption. The ZnO NPs synthesized using biological route, in turn, are capable of efficient removal of HMs from polluted water and, after regeneration, can be reused at an affordable cost, making this a promising means of environmental remediation [30].

ZnO NPs have also been applied to HM detection via different optical and electrochemical methods. Photoluminescence and surface plasmon resonance properties of ZnO can be used for developing sensitive assays that will change color or luminescent response in the presence of specific metal ions [31]. When combined with biosynthesis, such NPs can present selectivity for specific HMs by possibly enabling rapid in-situ testing in contaminated regions. High electron mobility and a ZnO NPs surface area will improve the sensitivity and the time of response for the electrochemical detection of trace quantities of HMs for the application of environmental monitoring [32]. Natural reducing agents present in the plant extracts are flavonoids and polyphenols, which support nanoparticle formation and are likely to induce groups, thus improving the stability and biocompatibility [33] Biosynthesized ZnO NPs are stated to have a more powerful interaction with HMs owing to these organic groups, which can develop complex formations with metal ions that eventually augment the total adsorption capacity. Abdelbaky et al. synthesized ZnO NPs using *Cymbopogon citratus* and showed the removal efficacy of Pb(II) ions from water with a high removal percentage in a short contact time [34]. Biosynthesized ZnO NPs were demonstrated to have higher adsorption capacities for Cr(VI) and Cd(II) ions with rapid kinetics and great potential for reuse after desorption treatments [35]. These results indicate the possibility of using ZnO NPs in real-world applications for wastewater treatment and environmental cleanup.

11.4 HEAVY METAL IONS REMOVAL USING ZnO NANOSTRUCTURES

According to Le et al., the mechanism behind the interaction of ZnO NPs with HMs during the adsorption process can be categorized in the following two ways [36]:
Photogenerated reduction/oxidation
Physical adsorption

11.4.1 Reduction/Oxidation via Photogenerated Electron-Hole Pairs

Exposing ZnO nanoparticles to radiation of suitable frequency facilitates the excitation of electrons from the valence band (VB) to the conduction band (CB), and thus, energized conduction band electrons and valence band holes are generated. Charge carriers initiate the formation of reactive radicals, as shown in Figure 11.2. The metal ions with an oxidation potential lower than the VB holes are oxidized, and those with redox potentials greater than the CB of ZnO are reduced [37, 38].

Le et al. reported the adsorption-based removal of HM ions, including Pb^{2+}, Cr^{6+}, Cu^{2+}, Ag^{+}, Mn^{2+}, Ni^{2+}, and Cd^{2+}, utilizing ZnO NPs by examining the adsorption

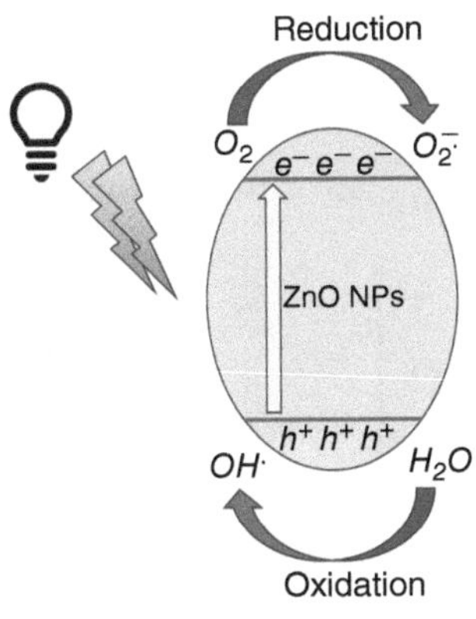

FIGURE 11.2 Photocatalytic mechanism of ZnO NPs.

efficiency under both UV and visible light, along with characterizing metal-loaded ZnO NPs post-adsorption [36]. Their findings revealed that for metal ions, like Cu^{2+}, Mn^{2+}, Ni^{2+}, and Cd^{2+}, the removal efficiency was not influenced by the light source. However, ions such as Ag^{+}, Cr^{6+}, and Pb^{2+} showed different responses under visible versus UV light.

11.4.2 Physical Adsorption

The hexagonal ZnO NPs in their as-prepared state, generally exhibit a positive zeta potential, indicating the presence of positive surface charges. However, ZnO NPs with a negative zeta potential may result from surface alterations made during the development process due to the formation of negatively charged groups on the surface, such as OH groups [39, 40]. The presence of this negative charge provides active sites that facilitate the adsorption of HM cations on the surface of ZnO NPs.

This investigation demonstrates how surface features, the growing method, the light source (visible or UV), and the particular metal ions involved affect the adsorption properties of ZnO NPs. ZnO has emerged as a prospective candidate for the elimination of various HM ions, which are known to contaminate water sources and adversely affect plant growth, as illustrated in Figure 11.3. The effectiveness of ZnO NPs in removing different HMs is explored in detail below, with findings summarized in Table 11.1.

11.5 REMOVAL EFFICACY OF ZnO NPs AGAINST VARIOUS HEAVY METALS

ZnO NPs are highly effective at removing various HMs thanks to their large surface area, strong ability to adsorb, and catalytic properties. They can efficiently adsorb and break down contaminants such as lead, cadmium, and arsenic, positioning them as a promising option for wastewater treatment and environmental remediation. The subsequent sections explain the removal efficacy of different HMs, which were detected, and the removal using ZnO NPs:

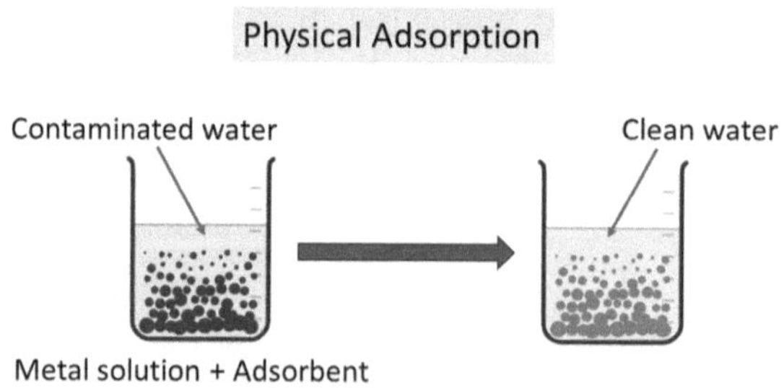

FIGURE 11.3 Illustration of physical adsorption showing the removal of contaminants from water using an adsorbent, resulting in clean water.

TABLE 11.1
Absorption Capacity of ZnO NPs Synthesized by Chemical and Green Route

Heavy Metal Removed	Type Of Synthesis	Shape/Size	Kinetic Model	Absorption Capacity (mg/g)	Reference
Cadmium ions (Cd^{2+})	Chemical	Spherical/20–30 nm	Langmuir+PSO	90.64	[71]
	Green	Beads/1.4 mm	Langmuir	18.39	[72]
Arsenic ions (As^{3+})	chemical	Nanorods/30–90 nm	Langmuir+PSO	52.63	[73]
	Green	Spherical/1–15 nm	Langmuir model	84	[74]
Mercury ions (Hg^{2+})	Chemical	Spherical/26 nm	Langmuir model	129.5	[71]
	Green	Spherical/22 nm	Langmuir model	101.01	[75]
Lead Ions (Pb^{2+})	Chemical	Spherical/5–20 μm	Langmuir model	158	[76]
	Green	Spherical/29 nm	Freundlich, Langmuir, Dubinin–Radushkevich,	22.29	[77]
Chromium ions (Cr^{3+})	Chemical	Spherical/42 nm	Langmuir & PFO	52.05	[78]
	Green	irregular rods/ Dia ~ 100 nm length-few μm	Langmuir	155.52	[9]
Cobalt ions (Co^{2+})	Chemical	Nanocones and D/~ 0.51 μm	Langmuir and PSO	76.69	[79]
	Green	Quasi-spherical/ 9 nm	Langmuir and PSO	67.93	[61]

11.5.1 Removal of Cadmium Using ZnO NPs

Cadmium (Cd) in industrial waste is highly toxic and poses significant health risks, including kidney damage, respiratory issues, and cancer. Its accumulation in the ecosystem contaminates the food chain and leads to severe ecological damage. The mobility of Cd allows it to spread easily through soil and water, contaminating agricultural lands and water sources. Additionally, it is difficult and costly to remediate, complicating environmental cleanup efforts [41].

The adsorption capacity of ZnO NPs for Cd(II) ion removal was compared by Khezami et al. to that of various other nano adsorbents reported earlier, and it observed that ZnO used in this study demonstrated a higher adsorption capacity than other nanostructured adsorbents, highlighting its potential as an effective adsorbent for Cd(II) ions in water and wastewater treatment [42]. Additionally, n-ZnO particles demonstrated a noteworthy removal effectiveness (about 92%) of Cd(II) from aqueous solutions, according to Kumar et al. The n-ZnO particles appear to be a promising material for the removal of Cd(II) from wastewater [43]. This study confirms the high efficiency of ZnO NPs in removing Cd^{2+} ions from wastewater, achieving up to 99% removal with a maximum adsorption capacity of 8.762 mg/g. Adsorption proceeds according to a pseudo-second-order model and is consistent with the Langmuir isotherm [44].

11.5.2 Removal of Arsenic Using ZnO NPs

Arsenic (As) is a toxic heavy metal commonly found in natural deposits, industrial effluents, and agricultural runoff, often contaminating groundwater and soil. Long-term exposure to arsenic, even at low levels, poses serious health risks, including skin lesions, cardiovascular diseases, and an increased risk of cancers, particularly skin, lung, and bladder cancer. It also affects plant growth and soil health, impairing agricultural productivity. The persistence of As in the environment makes it a challenging contaminant to remediate, underscoring the need for effective removal strategies to protect public health and ecosystems [45].

ZnO NPs play a key role in the photocatalytic removal of arsenic from water by converting toxic As(III) to less toxic As(V) through oxidation with hydroxyl radicals (°OH) generated under UV light in the presence of hydrogen peroxide (H_2O_2). Optimal ZnO dosage (1 g/L) achieved significant arsenic removal within 45 minutes, particularly in acidic conditions, enhancing overall treatment efficiency [46].

ZnO NPs effectively reduce As(III) accumulation in rice by enhancing growth and photosynthesis, diluting arsenic concentration in tissues. They also boost phytochelatin production in roots, sequestering more As(III) in root vacuoles and limiting its transport to shoots. An optimal concentration of ZnO NPs, i.e., 100 mg/L maximizes these protective effects [47]. ZnO NPs impregnation significantly enhances the As(V) removal capacity of biochar derived from agricultural residues, such as corn cob and coffee husk. The ZnO NPs modified with biochar, particularly the corncob-derived variant, show 25.9 mg of As(V)/g maximum adsorption capacity, making it a useful adsorbent for As removal from contaminated water [48].

11.5.3 Removal of Mercury Using ZnO NPs

Mercury is a heavy metal known for its high density and toxicity. It exists in various forms, including elemental, inorganic, and organic mercury, all of which pose severe health risks. Mercury exposure, particularly through inhalation or ingestion of contaminated fish, can damage the nervous, digestive, and immune systems and is particularly harmful to the brain and kidneys. Chronic exposure can lead to neurological and behavioral disorders, especially in children and pregnant women [49]. ZnO-based nanomaterials, including ZnO NPs, mesoporous ZnO assemblies, and Fe_3O_4-embedded ZnO nanocomposites (NCs), have shown exceptional potential for removing Hg^{2+} ions from complex wastewater. Their high adsorption abilities, alignment with isotherm and kinetic models, and multi-metal removal capability make them promising candidates for environmental remediation [50].

In addition to targeting mercury ions, Singh et al. [51] utilized mesoporous ZnO nanoadsorbents, composed of spherical assemblies of ZnO nano-crystallites, for simultaneous removal of multiple heavy metal cations, including Cd^{2+}, Pb^{2+}, Hg^{2+}, Co^{2+}, As^{3+}, Ni^{2+}, and Cu^{2+}, from wastewater. For a wastewater sample containing

47.16 mg/L Hg^{2+} along with other HMs, ZnO (200 mg) nano assemblies in 40 mL solution achieved a removal efficiency (RE) of 63.5% for Hg^{2+} [52].

11.5.4 Removal of Lead Using ZnO NPs

Lead (Pb), a toxic heavy metal, poses severe health and environmental risks due to its widespread use in industries such as batteries, paints, plumbing, and electronics. Exposure to lead, even in small amounts, can cause kidney dysfunction, neurological damage, cardiovascular issues, and growth issues, particularly in children. Lead contamination in water, soil, and air often arises from industrial effluents, mining activities, and the deterioration of lead-based materials. The removal of lead from contaminated environments is critical to protect public health and maintain ecological balance. Effective remediation requires techniques capable of addressing varying lead concentrations, diverse pH levels, and the occurrence of competing ions in complex wastewater matrices. Sustainable, efficient, and cost-effective methods are crucial to achieving regulatory standards for lead-free environments [53].

Krasovska et al. [54] highlighted the role of ZnO nanostructures in Pb^{2+} removal, demonstrating that ZnO nanotubes, with their hollow morphology and larger specific surface area, exhibited a twofold higher adsorption ability of 611 mg/g in comparison to ZnO nanorods having capacity of 256 mg/g. This highlights the importance of the structural properties of ZnO NPs in the removal of Pb. A significant increase in absorption capacity (158 mg/g) of ZnO hollow microspheres with nanostructured porous nanosheet-built was observed when compared with commercialized ZnO NPs (33 mg/g). This enhancement is considered to be because of their larger specific surface area and increased porosity, providing greater adsorption sites [55].

11.5.5 Removal of Chromium Using ZnO NPs

Chromium (Cr), widely used in industries, poses significant health and environmental risks, especially in its hexavalent form (Cr^{6+}), which is highly toxic and carcinogenic. Cr^{6+} exposure can cause respiratory disorders, organ damage, and cancer, while its mobility in water and soil leads to environmental contamination and ecosystem harm. Although trivalent chromium (Cr^{3+}) is less toxic, excessive levels can still cause oxidative stress. Removing chromium, particularly Cr^{6+}, is essential to protect human health and maintain ecological balance [56].

Yu et al. [57] demonstrated the effectiveness of ZnO NPs modified with biochar derived from water hyacinth in removing Cr^{6+} ions. At an optimized carbonization temperature of 750 °C, the ZnO-biochar composite achieved a rapid RE of ~ 87% within 2 hours, reaching 95% in 14 hours. The process followed the pseudo-second-order kinetic model and Langmuir isotherm, with a maximum adsorption capacity of 43.48 mg/g, and exhibited good recyclability, retaining ~70% efficiency after five regeneration cycles. Ahmed et al. in their report [58] mentioned that the ZnO NPs synthesized in the presence of polyethylene glycol (PEG) have increased Cr^{6+} adsorption efficiency with decreasing pH, achieving the highest RE (96.10%) at pH 2.5 within 210 minutes. They also observed that the RE increased to 58.98% from 53.10% with the rise in temperature to 60 °C starting from 25 °C, but further increasing the temperature to 100 °C reduced the efficiency to 56.70%.

11.5.6 Removal of Cobalt Using ZnO NPs

Cobalt (Co) is a heavy metal primarily utilized in various industries which include batteries, pigments, alloys, and medical implants. While cobalt is an essential trace element for humans, excessive exposure can pose serious health and environmental risks. High levels of cobalt can lead to respiratory issues, cardiomyopathy, and neurological disorders, and prolonged exposure is linked to potential carcinogenic effects. Environmental contamination from cobalt mining and industrial effluents harms ecosystems by affecting soil fertility and aquatic life. Effective cobalt removal is critical to mitigate its toxic impacts, protect public health, and prevent environmental degradation [59].

A high Co^{2+} RE (98.71%) using ZnO NPs was reported by Nalwa et al. [60] with 105 mg/10 mL of adsorbent dose and 300 mg/L of initial metal ion concentration. Equilibrium was obtained within 20 minutes at pH 3 to 5, compared to 90 minutes at pH 2. The adsorption behavior was well-described by the Freundlich isotherm and the pseudo-second-order (PSO) kinetic model. Co^{2+} RE of ZnO NPs synthesized via a semi-green method with casein as a biogenic capping and reducing agent was reported by Somu et al. [61]. At an initial metal concentration of 100 ppm, an adsorbent dose of 0.2 g/100 mL, pH 7, and 30 °C, the maximum RE was 71.23%. The Langmuir model revealed a maximal adsorption ability of 67.93 mg/g.

11.6 VARIOUS KINETIC MODELS

Several kinetic models are commonly used to describe the interactions between adsorbents and HMs, providing insights into the adsorption mechanisms and reaction pathways:

11.6.1 Pseudo-first-Order Model

This model assumes that the rate of occupation of adsorption sites is proportional to the number of unoccupied sites. The equation can be expressed as [62]:

$$\mathrm{lo}\left(q_e - q_t\right) = \log log\, q_e - \frac{k_1 t}{2.303} \tag{11.1}$$

where q_e and q_t are the amounts of adsorbate adsorbed at equilibrium and time t, respectively, and k_1 is the rate constant.

11.6.2 Pseudo-Second-Order Model

This model is based on the supposition that the adsorption follows a second-order reaction and depends on the square of the number of unoccupied sites. The expression governing the model can be represented by [63]:

$$\frac{t}{q_t} = \frac{1}{k_2 q_e^2} + \frac{t}{q_e} \tag{11.2}$$

where k_2 is the pseudo-second-order rate constant.

11.6.3 Elovich Model

Commonly used for heterogeneous adsorption systems, this model assumes a logarithmic decrease in the adsorption rate with increasing coverage. It is expressed as [64]:

$$q_t = \frac{1}{\beta} l(\alpha\beta) + \frac{1}{\beta} lt \tag{11.3}$$

where α is the initial adsorption rate and β is related to the surface coverage and activation energy.

11.6.4 Intraparticle Diffusion Model

This model describes the diffusion of adsorbate within the pores of the adsorbent as the rate-limiting step. The equation for this model can be expressed [65]:

$$q_t = k_i t^{0.5} + C \tag{11.4}$$

where k_i is the intraparticle diffusion rate constant and C represents the boundary layer effect.

11.6.5 Avrami Kinetic Model

It combines surface adsorption and diffusion processes and the governing equation can be expressed as [66]:

$$l\left[-l\left(1-F\right)\right] = nlt + lk \tag{11.5}$$

where F is the fraction of adsorption at time t, n is the Avrami constant, and k is the Avrami rate constant.

11.7 ADSORPTION ISOTHERMS

These models help predict the adsorption mechanism, evaluate the efficiency of adsorbents, and guide the optimization of adsorption systems for removing HMs from contaminated environments. Some of the well-established adsorption isotherms used to understand the adsorption behavior of heavy metal ions include:

11.7.1 Langmuir Isotherm

Assumes monolayer adsorption on a homogeneous surface with a finite number of identical adsorption sites. It is expressed as [67]:

$$q_e = \frac{q_{\max} b C_e}{1 + b C_e} \tag{11.6}$$

where q_e is the amount adsorbed, C_e is the equilibrium concentration, q_{max} is the maximum adsorption capacity, and b is the Langmuir constant.

11.7.2 Freundlich Isotherm

Describes adsorption on a heterogeneous surface with non-uniform heat of adsorption and multilayer formation. It is given by [68]:

$$q_e = K_F C_e^{\frac{1}{n}} \tag{11.7}$$

where K_F and n are Freundlich constants related to adsorption capacity and intensity, respectively.

11.7.3 Temkin Isotherm

Accounts for adsorbent-adsorbate interactions and assumes a linear decrease in adsorption energy with coverage. The equation is expressed as [69]:

$$q_e = \frac{RT}{b} \ln\left(aC_e\right) \tag{11.8}$$

where a and b are Temkin constants, R is the gas constant, and T is the temperature.

11.7.4 Elovich Isotherm

Often used for systems with heterogeneous adsorption surfaces, assuming exponential changes in adsorption energies. The expression for Elovich isotherm equation is represented as [70]:

$$q_e = \frac{1}{\beta} l\left(\alpha C_e\right) \tag{11.9}$$

where α and β are Elovich constants.

These isotherms help elucidate the adsorption mechanism, surface characteristics, and interactions between heavy metal ions and adsorbents, aiding in the design of efficient remediation systems.

11.8 CONCLUSION

The present chapter highlights that ZnO NPs possess a versatile and promising potential candidature for the detection and removal of toxic HMs from contaminated water. ZnO NPs synthesized using both chemical and biogenic methods are highly efficient, depending on the size, shape, and surface characteristics. Adsorption studies into heavy metal toxicity show that ZnO NPs can chemically eliminate HMs based on the operation of physical and photogenerated redox mechanisms. The use of various kinetic and isotherm models contributes to the optimization of the adsorption process for real applications. Moreover, the biogenic synthesis route presents an environment-friendly approach with higher sustainability and biocompatibility of ZnO NPs. Generally, ZnO NPs have unprecedented potential for environmental remediation as well as wastewater treatment in industries. Future work will focus on scaling these technologies and exploring synergistic approaches to further enhance applicability.

REFERENCES

[1] Koller, M., & Saleh, H. M. (2018). Introductory chapter: Introducing heavy metals. *Heavy Metals*, *1*, 3–11.

[2] Sharma, S. K. (Ed.). (2014). *Heavy metals in water: presence, removal and safety*. Royal Society of Chemistry.

[3] Ahmed, M. J. K., & Ahmaruzzaman, M. J. J. O. W. P. E. (2016). A review on potential usage of industrial waste materials for binding heavy metal ions from aqueous solutions. *Journal of Water Process Engineering*, *10*, 39–47.

[4] Vajargah, M. F. (2021). A review on the effects of heavy metals on aquatic animals. *J ISSN*, *2766*(2276), 854.

[5] Rehman, K., Fatima, F., Waheed, I., & Akash, M. S. H. (2018). Prevalence of exposure of heavy metals and their impact on health consequences. *Journal of Cellular Biochemistry*, *119*(1), 157–184.

[6] Engwa, G. A., Ferdinand, P. U., Nwalo, F. N., & Unachukwu, M. N. (2019). Mechanism and health effects of heavy metal toxicity in humans. *Poisoning in the Modern World-New Tricks for an Old Dog*, *10*, 70–90.

[7] Mondal, S., & Palit, D. (2022). Prospects and implementation of nanotechnology in environmental remediation and clean up. In *Natural resources conservation and advances for sustainability* (pp. 271–287). Elsevier.

[8] Ahmad, P., Alyemeni, M. N., Al-Huqail, A. A., Alqahtani, M. A., Wijaya, L., Ashraf, M.,... & Bajguz, A. (2020). Zinc oxide nanoparticles application alleviates arsenic (As) toxicity in soybean plants by restricting the uptake of as and modulating key biochemical attributes, antioxidant enzymes, ascorbate-glutathione cycle and glyoxalase system. *Plants*, *9*(7), 825.

[9] Khezami, L., Aissa, M. A. B., Modwi, A., Guesmi, A., Algethami, F. K., & Bououdina, M. (2024). Efficient removal of organic dyes by Cr-doped ZnO nanoparticles. *Biomass Conversion and Biorefinery*, *14*(3), 4177–4190.

[10] Vasantharaj, S., Sathiyavimal, S., Senthilkumar, P., Kalpana, V. N., Rajalakshmi, G., Alsehli, M.,... & Pugazhendhi, A. (2021). Enhanced photocatalytic degradation of water pollutants using bio-green synthesis of zinc oxide nanoparticles (ZnO NPs). *Journal of Environmental Chemical Engineering*, *9*(4), 105772.

[11] Dhiman, V., & Kondal, N. (2021). ZnO Nanoadsorbents: A potent material for removal of heavy metal ions from wastewater. *Colloid and Interface Science Communications*, *41*, 100380.

[12] Chen, T., & Liu, Y. (2016). *Semiconductor nanocrystals and metal nanoparticles: physical properties and device applications*. CRC Press.

[13] Singh, S., Thiyagarajan, P., Kant, K. M., Anita, D., Thirupathiah, S., Rama, N.,... & Rao, M. R. (2007). Structure, microstructure and physical properties of ZnO based materials in various forms: bulk, thin film and nano. *Journal of Physics D: Applied Physics*, *40*(20), 6312.

[14] Sharma, D. K., Shukla, S., Sharma, K. K., & Kumar, V. (2022). A review on ZnO: Fundamental properties and applications. *Materials Today: Proceedings*, *49*, 3028–3035.
[15] Coleman, V. A., & Jagadish, C. (2006). Basic properties and applications of ZnO. In *Zinc oxide bulk, thin films and nanostructures* (pp. 1–20). Elsevier Science Ltd.
[16] Sani, G. D., Yakubu, A., Saidu, A., Aati, R., Sahabi, S., & Abdullahi, S. (2023). A review on industrial applications of zinc oxide nanoparticles. *International Journal of Advances in Engineering and Management*, *5*, 1031–1041.
[17] Abdullah, F. H., Bakar, N. A., & Bakar, M. A. (2022). Current advancements on the fabrication, modification, and industrial application of zinc oxide as photocatalyst in the removal of organic and inorganic contaminants in aquatic systems. *Journal of Hazardous Materials*, *424*, 127416.
[18] Abid, N., Khan, A. M., Shujait, S., Chaudhary, K., Ikram, M., Imran, M.,… & Maqbool, M. (2022). Synthesis of nanomaterials using various top-down and bottom-up approaches, influencing factors, advantages, and disadvantages: A review. *Advances in Colloid and Interface Science*, *300*, 102597.
[19] Srujana, S., & Bhagat, D. (2022). Chemical-based synthesis of ZnO nanoparticles and their applications in agriculture. *Nanotechnology for Environmental Engineering*, *7*(1), 269–275.
[20] Madhumitha, G., Elango, G., & Roopan, S. M. (2016). Biotechnological aspects of ZnO nanoparticles: overview on synthesis and its applications. *Applied Microbiology and Biotechnology*, *100*, 571–581.
[21] Banerjee, P., Chakrabarti, S., Maitra, S., & Dutta, B. K. (2012). Zinc oxide nano-particles–sonochemical synthesis, characterization and application for photo-remediation of heavy metal. *Ultrasonics Sonochemistry*, *19*(1), 85–93.
[22] Sheikh, M., Pazirofteh, M., Dehghani, M., Asghari, M., Rezakazemi, M., Valderrama, C., & Cortina, J. L. (2020). Application of ZnO nanostructures in ceramic and polymeric membranes for water and wastewater technologies: a review. *Chemical Engineering Journal*, *391*, 123475.
[23] Ahmad, S., Aadil, M., Ejaz, S. R., Akhtar, M. U., Noor, H., Haider, S.,… & Yasmin, G. (2022). Sol-gel synthesis of nanostructured ZnO/SrZnO2 with boosted antibacterial and photocatalytic activity. *Ceramics International*, *48*(2), 2394–2405.
[24] Khezami, L., Modwi, A., Ghiloufi, I., Taha, K. K., Bououdina, M., ElJery, A., & El Mir, L. (2020). Effect of aluminum loading on structural and morphological characteristics of ZnO nanoparticles for heavy metal ion elimination. *Environmental Science and Pollution Research*, *27*, 3086–3099.
[25] Jain, A., & Bhise, K. (2025). Green Synthesis Methods for Metallic Zinc Oxide Nanoparticles: A Sustainable Angle. In *Synthesizing and Characterizing Plant-Mediated Biocompatible Metal Nanoparticles* (pp. 63–90). IGI Global.
[26] Khan, M., Ijaz, M., Chotana, G. A., Murtaza, G., Malik, A., & Shamim, S. (2022). Bacillus altitudinis MT422188: A potential agent for zinc bioremediation. *Bioremediation Journal*, *26*(3), 228–248.
[27] Ahmed, S., Chaudhry, S. A., & Ikram, S. (2017). A review on biogenic synthesis of ZnO nanoparticles using plant extracts and microbes: a prospect towards green chemistry. *Journal of Photochemistry and Photobiology B: Biology*, *166*, 272–284.
[28] Rani, S., Kumar, P., Dahiya, P., Dang, A. S., & Suneja, P. (2022). Biogenic synthesis of zinc nanoparticles, their applications, and toxicity prospects. *Frontiers in Microbiology*, *13*, 824427.
[29] Kumar, N., Jyoti, J., Aggarwal, N., Kaur, A., Patial, P., Kaur, K.,… & Raja, V. (2024). Adsorption of Pb2+ using biosynthesized ZnO nanoparticles derived using Azadirachta indica (neem) leaf extract. *Biomass Conversion and Biorefinery*, *14*(23), 1–11.

[30] Zeghoud, S., Hemmami, H., Seghir, B. B., Amor, I. B., Kouadri, I., Rebiai, A.,... & Simal-Gandara, J. (2022). A review on biogenic green synthesis of ZnO nanoparticles by plant biomass and their applications. *Materials Today Communications*, *33*, 104747.

[31] Ahmed, A., Singh, A., Padha, B., Sundramoorthy, A. K., Tomar, A., & Arya, S. (2022). UV–vis spectroscopic method for detection and removal of heavy metal ions in water using Ag doped ZnO nanoparticles. *Chemosphere*, *303*, 135208.

[32] Sulthana, S. F., Iqbal, U. M., Suseela, S. B., Anbazhagan, R., Chinthaginjala, R., Chitathuru, D.,... & Kim, T. H. (2024). Electrochemical sensors for heavy metal ion detection in aqueous medium: a systematic review. *ACS Omega*, *9*(24), 25493–25512.

[33] Hamed, R., Obeid, R. Z., & Abu-Huwaij, R. (2023). Plant mediated-green synthesis of zinc oxide nanoparticles: An insight into biomedical applications. *Nanotechnology Reviews*, *12*(1), 20230112.

[34] . Abdelbaky, A. S., Mohamed, A. M., Sharaky, M., Mohamed, N. A., & Diab, Y. M. (2023). Green approach for the synthesis of ZnO nanoparticles using Cymbopogon citratus aqueous leaf extract: characterization and evaluation of their biological activities. *Chemical and Biological Technologies in Agriculture*, *10*(1), 63.

[35] Taha, A., Da'na, E., & Hassanin, H. A. (2021). Modified activated carbon loaded with bio-synthesized Ag/ZnO nanocomposite and its application for the removal of Cr (VI) ions from aqueous solution. *Surfaces and Interfaces*, *23*, 100928.

[36] Le, A. T., Pung, S. Y., Sreekantan, S., & Matsuda, A. (2019). Mechanisms of removal of heavy metal ions by ZnO particles. *Heliyon*, *5*(4): e01440.

[37] Kim, K. M., Choi, M. H., Lee, J. K., Jeong, J., Kim, Y. R., Kim, M. K.,... & Oh, J. M. (2014). Physicochemical properties of surface charge-modified ZnO nanoparticles with different particle sizes. *International Journal of Nanomedicine*, *9*(sup2), 41–56.

[38] Litter, M. I. (2015). Mechanisms of removal of heavy metals and arsenic from water by TiO2-heterogeneous photocatalysis. *Pure and Applied Chemistry*, *87*(6), 557–567.

[39] Sharma, M., Singh, J., Hazra, S., & Basu, S. (2019). Adsorption of heavy metal ions by mesoporous ZnO and TiO2@ ZnO monoliths: adsorption and kinetic studies. *Microchemical Journal*, *145*, 105–112.

[40] Venkatachalam, P., Jayaraj, M., Manikandan, R., Geetha, N., Rene, E. R., Sharma, N. C., & Sahi, S. V. (2017). Zinc oxide nanoparticles (ZnONPs) alleviate heavy metal-induced toxicity in Leucaena leucocephala seedlings: a physiochemical analysis. *Plant Physiology and Biochemistry*, *110*, 59–69.

[41] Adedapo, A. A., & Adeoye, B. O. (2014). Cadmium, an environmental pollutant: A review. *West African Journal of Pharmacology and Drug Research*, *29*, 1–11.

[42] Khezami, L., Taha, K. K., Amami, E., Ghiloufi, I., & El Mir, L. (2017). Removal of cadmium (II) from aqueous solution by zinc oxide nanoparticles: kinetic and thermodynamic studies. *Desalination and Water Treatment*, *62*, 346–354.

[43] Kumar, R., & Chawla, J. (2014). Removal of cadmium ion from water/wastewater by nano-metal oxides: a review. *Water Quality, Exposure and Health*, *5*, 215–226.

[44] Nthwane, Y. B., Fouda-Mbanga, B. G., Thwala, M., & Pillay, K. (2022). Removal of Cd2+ by N-CNS/ZnO nanocomposite from wastewater and reuse of Cd2+-N-CNS/ZnO in blood fingerprint detection. *Results in Chemistry*, *4*, 100660.

[45] Monteiro De Oliveira, E. C., Caixeta, E. S., Santos, V. S. V., & Pereira, B. B. (2021). Arsenic exposure from groundwater: environmental contamination, human health effects, and sustainable solutions. *Journal of Toxicology and Environmental Health, Part B*, *24*(3), 119–135.

[46] Massoudinejad, M., Keramati, H., & Ghaderpoori, M. (2020). Investigation of photocatalytic removal of arsenic from aqueous solutions using UV/H_2O_2 in the presence of ZnO nanoparticles. *Chemical Engineering Communications*, *207*(11), 1605–1615.

[47] Yan, S., Wu, F., Zhou, S., Yang, J., Tang, X., & Ye, W. (2021). Zinc oxide nanoparticles alleviate the arsenic toxicity and decrease the accumulation of arsenic in rice (Oryza sativa L.). *BMC Plant Biology*, *21*, 1–11.

[48] Cruz, G. J., Mondal, D., Rimaycuna, J., Soukup, K., Gómez, M. M., Solis, J. L., & Lang, J. (2020). Agrowaste-derived biochars impregnated with ZnO for removal of arsenic and lead in water. *Journal of Environmental Chemical Engineering*, *8*(3), 103800.

[49] Zafar, A., Javed, S., Akram, N., & Naqvi, S. A. R. (2024). Health Risks of Mercury. In *Mercury Toxicity Mitigation: Sustainable Nexus Approach* (pp. 67–92). Cham: Springer Nature Switzerland.

[50] Sheela, T., Nayaka, Y. A., Viswanatha, R., Basavanna, S., & Venkatesha, T. G. (2012). Kinetics and thermodynamics studies on the adsorption of Zn (II), Cd (II) and Hg (II) from aqueous solution using zinc oxide nanoparticles. *Powder Technology*, *217*, 163–170.

[51] Singh, S., Barick, K. C., & Bahadur, D. (2011). Novel and efficient three dimensional mesoporous ZnO nanoassemblies for environmental remediation. *International Journal of Nanoscience*, *10*(04n05), 1001–1005.

[52] Singh, S., Barick, K. C., & Bahadur, D. (2013). Fe 3 O 4 embedded ZnO nanocomposites for the removal of toxic metal ions, organic dyes and bacterial pathogens. *Journal of Materials Chemistry A*, *1*(10), 3325–3333.

[53] Ara, A., & Usmani, J. A. (2015). Lead toxicity: a review. *Interdisciplinary Toxicology*, *8*(2), 55–64.

[54] Krasovska, M., Gerbreders, V., Tamanis, E., Gerbreders, S., & Bulanovs, A. (2017). The study of adsorption process of Pb ions using well-aligned arrays of ZnO nanotubes as a sorbent. *Latvian Journal of Physics and Technical Sciences*, *54*(1), 41–50.

[55] Wang, X., Cai, W., Liu, S., Wang, G., Wu, Z., & Zhao, H. (2013). ZnO hollow microspheres with exposed porous nanosheets surface: structurally enhanced adsorption towards heavy metal ions. *Colloids and Surfaces A: Physicochemical and Engineering Aspects*, *422*, 199–205.

[56] Hossini, H., Shafie, B., Niri, A. D., Nazari, M., Esfahlan, A. J., Ahmadpour, M.,... & Hoseinzadeh, E. (2022). A comprehensive review on human health effects of chromium: Insights on induced toxicity. *Environmental Science and Pollution Research*, *29*(47), 70686–70705.

[57] Yu, J., Jiang, C., Guan, Q., Ning, P., Gu, J., Chen, Q.,... & Miao, R. (2018). Enhanced removal of Cr (VI) from aqueous solution by supported ZnO nanoparticles on biochar derived from waste water hyacinth. *Chemosphere*, *195*, 632–640.

[58] Ahmed, N. M., & Yousef, N. S. (2015). Synthesis and characterization of zinc oxide nano particles for the removal of Cr (VI). *The International Journal of Scientific & Engineering Research*, *6*(7), 1235–1243.

[59] ALMisned, G., Elshami, W., Issa, S. A., Susoy, G., Zakaly, H. M., Algethami, M.,... & Tekin, H. O. (2021). Enhancement of gamma-ray shielding properties in cobalt-doped heavy metal borate glasses: the role of lanthanum oxide reinforcement. *Materials*, *14*(24), 7703.

[60] Nalwa, K., Thakur, A., & Sharma, N. (2017). Synthesis of ZnO nanoparticles and its application in adsorption. *Advanced Materials Proceedings*, *2*(11), 697–703.

[61] Somu, P., & Paul, S. (2018). Casein based biogenic-synthesized zinc oxide nanoparticles simultaneously decontaminate heavy metals, dyes, and pathogenic microbes: a rational strategy for wastewater treatment. *Journal of Chemical Technology & Biotechnology*, *93*(10), 2962–2976.

[62] Kowanga, K. D., Gatebe, E., Mauti, G. O., & Mauti, E. M. (2016). Kinetic, sorption isotherms, pseudo-first-order model and pseudo-second-order model studies of Cu (II) and Pb (II) using defatted Moringa oleifera seed powder. *The Journal of Phytopharmacology*, *5*(2), 71–78.

[63] Revellame, E. D., Fortela, D. L., Sharp, W., Hernandez, R., & Zappi, M. E. (2020). Adsorption kinetic modeling using pseudo-first order and pseudo-second order rate laws: A review. *Cleaner Engineering and Technology*, *1*, 100032.

[64] AlJaberi, F. Y. (2024). Extensive Study of Electrocoagulation-Based Adsorption Process of Real Groundwater Treatment: Isotherm Modeling, Adsorption Kinetics, and Thermodynamics. *Water*, *16*(4), 619.

[65] Wang, J., & Guo, X. (2022). Rethinking of the intraparticle diffusion adsorption kinetics model: Interpretation, solving methods and applications. *Chemosphere*, *309*, 136732.

[66] Oladoja, N. A. (2016). A critical review of the applicability of Avrami fractional kinetic equation in adsorption-based water treatment studies. *Desalination and Water Treatment*, *57*(34), 15813–15825.

[67] Can, N., Ömür, B. C., & Altındal, A. (2016). Modeling of heavy metal ion adsorption isotherms onto metallophthalocyanine film. *Sensors and Actuators B: Chemical*, *237*, 953–961.

[68] Artola, A., Martin, M., Balaguer, M., & Rigola, M. (2000). Isotherm model analysis for the adsorption of Cd (II), Cu (II), Ni (II), and Zn (II) on anaerobically digested sludge. *Journal of Colloid and Interface Science*, *232*(1), 64–70.

[69] Zhou, Y., Zhang, J., Luo, X., & Lin, X. (2014). Adsorption of Hg (II) in aqueous solutions using mercapto-functionalized alkali lignin. *Journal of Applied Polymer Science*, *131*(18), 40749.

[70] Thajeel, A. S. (2013). Isotherm, kinetic and thermodynamic of adsorption of heavy metal ions onto local activated carbon. *Aquatic Science and Technology*, *1*(2), 53–77.

[71] Sheela, T., Nayaka, Y. A., Viswanatha, R., Basavanna, S., & Venkatesha, T. G. (2012). Kinetics and thermodynamics studies on the adsorption of Zn (II), Cd (II) and Hg (II) from aqueous solution using zinc oxide nanoparticles. *Powder Technology*, *217*, 163–170.

[72] Makhlouf, M. R., Bendjaballah, M., Boukerche, I., Kouadri, I., Hamidoud, S., Benhamza, M. E. H., & Hammi, H. (2024). Eco-friendly synthesis of biosorbent based in chitosan-activated carbon/zinc oxide nanoparticle beads for efficiency reduction of cadmium ions in wastewater. *Biomass Conversion and Biorefinery*, 1–16.

[73] Yuvaraja, G., Prasad, C., Vijaya, Y., & Subbaiah, M. V. (2018). Application of ZnO nanorods as an adsorbent material for the removal of As (III) from aqueous solution: kinetics, isotherms and thermodynamic studies. *International Journal of Industrial Chemistry*, *9*, 17–25.

[74] Khamis, M., Gouda, G. A., & Nagiub, A. M. (2023). Biosynthesis approach of zinc oxide nanoparticles for aqueous phosphorous removal: Physicochemical properties and antibacterial activities. *BMC Chemistry*, *17*(1), 99.

[75] Gindaba, G. T., Demsash, H. D., & Jayakumar, M. (2023). Green synthesis, characterization, and application of metal oxide nanoparticles for mercury removal from aqueous solution. *Environmental Monitoring and Assessment*, *195*(1), 9.

[76] Wang, X., Cai, W., Liu, S., Wang, G., Wu, Z., & Zhao, H. (2013). ZnO hollow microspheres with exposed porous nanosheets surface: structurally enhanced adsorption towards heavy metal ions. *Colloids and Surfaces A: Physicochemical and Engineering Aspects*, *422*, 199–205.

[77] Al-Mur, B. A. (2023). Green zinc oxide (ZnO) nanoparticle synthesis using mangrove leaf extract from Avicenna marina: properties and application for the removal of toxic metal ions (Cd2+ and Pb2+). *Water*, 15(3), 455.

[78] Ghiloufi, I., El Ghoul, J., Modwi, A., & El Mir, L. (2016). Ga-doped ZnO for adsorption of heavy metals from aqueous solution. *Materials Science in Semiconductor Processing*, *42*, 102–106.

[79] Rahman, M. M., Khan, S. B., Marwani, H. M., & Asiri, A. M. (2014). Selective divalent cobalt ions detection using Ag2O3-ZnO nanocones by ICP-OES method for environmental remediation. *PLoS One*, *9*(12), e114084.

12 Future Prospective and Directions for ZnO Nanostructures

Kanchan Bala and Deepika Sharma

12.1 INTRODUCTION

In material science, among several metal oxide nanomaterials, ZnO is categorized as an n-type semiconductor and known as the wealthiest group of nanostructures. The remarkable features of this versatile substance make it a captivating topic for discussion, boasting exceptional compatibility, heightened sensitivity, a significant specific area, non-toxicity, distinct properties, diverse structures, a versatile nature, and an elevated isoelectric point [1]. In terms of functionality and application, each characteristic is outstanding and unique in its own way. Excellent thermal and mechanical durability at room temperature, and a significant excitation binding energy of 60 meV, have made this material appropriate for use in optoelectronics, electronics, and laser innovative technologies [2]. Due to its piezoelectric and pyroelectric properties, under visible light exposure, ZnO can be used as a photocatalyst, converter, as well as energy producer in hydrogen generation [3]. The large selectivity, sensitivity, and various physicochemical aspects, like huge surface-to-volume ratio, adsorption, photochemical durability, large electron communication biocompatibility, and reactivity of ZnO, make it an ideal material for the development of effective sensors and biosensors [4]. These materials have garnered significant attention and remain under constant analysis for their potential applications in energy storage devices (supercapacitors and lithium-ion batteries) [5, 6]. This chapter presented the recent advancements and the future perspective in ZnO-based distinctive nanostructures such as nanocomposites, nanoparticles, and functionalized materials for their applications in solar cells, conversion and storage devices, sensors and biosensors, Li-ion batteries, and environmental systems.

12.2 FUTURE PERSPECTIVE

The future prospects of ZnO nanostructures (ZnO NSs) are vast and are expected to continue playing a crucial role in numerous domains. Continued research and development efforts in ZnO NSs are anticipated to unlock the full capabilities and pave the way for transformative advancements in optoelectronics, energy storage, catalysis, and flexible electronics (Figure 12.1). The ongoing exploration of these future

 DOI: 10.1201/9781003464907-12

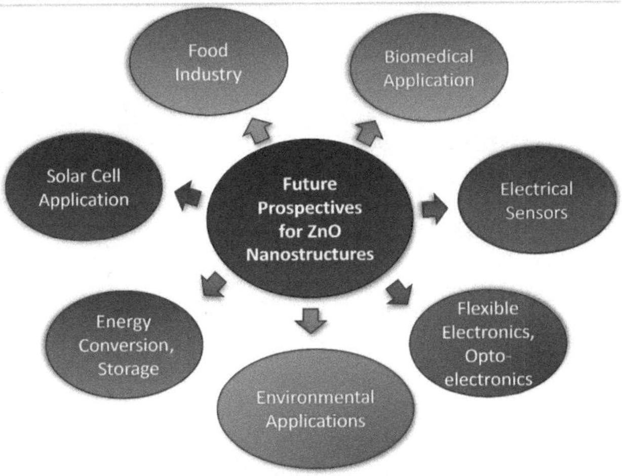

FIGURE 12.1 Future prospective applications and developments for ZnO nanostructures.

directions promises exciting opportunities for harnessing the unique properties of ZnO NSs to address diverse technological challenges and societal needs. Here are some prospective applications and developments related to ZnO NSs.

12.2.1 Food Industry

The high surface-to-volume ratio of nanomaterials admits their ability to reinforce the packaging performance, and antimicrobial activities, and elongate the standard and shelf life of food [7]. The high demand and the market for antimicrobial food packaging is expanding on a global scale, projected to grow from 6.5 to 10.09 million tons with an estimated value of 11 billion USD [8]. In comparison, other NPs such as Ag, TiO_2, and ZnO are affordable and have attracted significantly more interest in the food business. By incorporating an optimal quantity of ZnO nanoparticles (NPs), the packaging can acquire significant antimicrobial efficacy, while also enhancing essential properties like mechanical strength, barrier performance, and thermal stability. It is observed that when ZnO is included in a packaging matrix, its antibacterial efficiency is reduced. So, to make the modern and latest packaging, the dissipation of nanofiller in the matrix is an imperative feature. Enveloping ZnO NPs with polymeric nanocomposites will offer favorable prospects for the packaging due to the presence of active agents and potent antimicrobial properties. Synthetic and natural biopolymers such as polyvinyl alcohol, polyglycolic acid, poly lactic acid, polycaprolactone, polybutylene adipate terephthalate cellulose, chitosan, gelatin, starch, are environmentally friendly materials that might be employed specifically to generate great antimicrobial packaging based on ZnO NPs [9]. Nanofillers present in less than 5% of biopolymers can give outstanding results to packaging. Striking a harmonious balance between the necessary properties for packaging that contains ZnO NPs remains an ongoing challenge in the industry. Packaging disposal, as a major waste in landfills, remains a major concern in waste management;

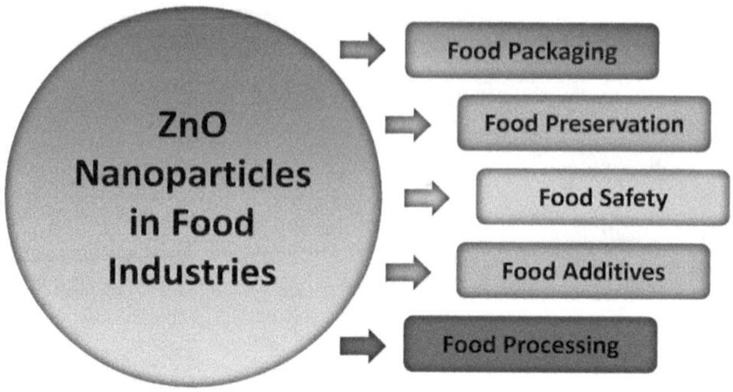

FIGURE 12.2 Applications of ZnO nanomaterials in food industries.

therefore, active packaging nanocomposite lessens the food packaging disposal and also extends their life. There is a strong recommendation to take into account six key factors in the migration of nanomaterials from food contact materials. These factors encompass the migration rate, mechanism, methods for trace analysis, prediction of migration modeling, suitable food simulants, and the associated risk of migration [10]. The following are the critical factors to consider while developing a complete regulation for nanomaterial application: (i) To attain an authentic migration detection, it is recommended to employ analytical techniques such as single particle inductively coupled plasma mass spectroscopy (sp-ICP-MS), energy-dispersive X-ray and transmission electron microscopy. These characterization methods have proven to be highly accurate in detecting nanomaterial traces. (ii) It is mandatory to use chemicals having similarity with food materials and (ii) materials used should not give harmful results to humans. The ZnO market has reached 7677 million USD from 2099 million USD in 2015 [7]. Before the practicality and usability of packaging materials, consumer acceptance is the last consideration. There are several reports on ZnO NP compatibility and safety, but caution must be exercised to limit migration under the harmful level and all customer concerns about safety. The ZnO NPs in scaling up packaging nanocomposites necessitates advanced technologies, which can potentially increase the overall cost, thereby impacting market acceptability (Figure 12.2).

12.2.2 Biomedical Applications

ZnO NSs have demonstrated potential in the biomedical field due to their antimicrobial properties, minimum-price nanomaterial, less toxicity, and biocompatibility (Figure 12.3). These melodious properties make them attractive for targeted drug delivery, anticancer, antibacterial, antioxidant activity, antidiabetic, anti-inflammatory, and imaging contrast agents [11]. ZnO NPs exhibit notable inhibitory effects against cancerous cells due to their intrinsic toxicity, positioning them as potential candidates for use in antibacterial applications. Moreover, ZnO NPs have demonstrated their effectiveness as drug carriers, enhancing the bioavailability of therapeutic drugs and improving overall therapy efficiency [12]. Additionally, ZnO NPs have demonstrated a favorable potential for managing diabetes and reducing its

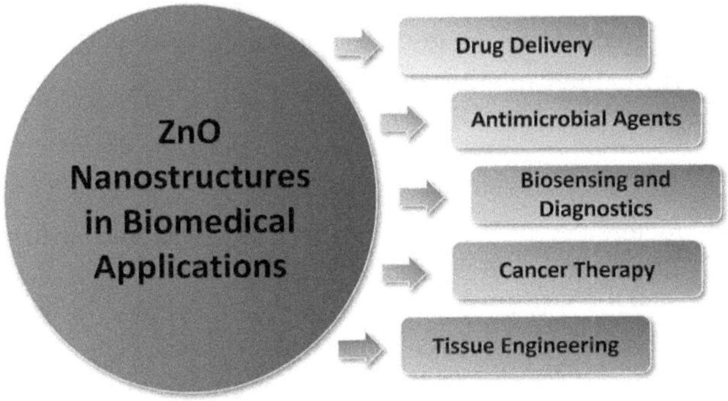

FIGURE 12.3 Biomedical applications of ZnO nanostructures.

consequences due to their capacity to raise insulin levels and lower blood glucose levels, which can be further assessed [12].

ZnO NSs may be outstanding candidates for cancer-targeted therapy. The future of nanomedicine in the realm of in vivo therapy applications lies in the development of multipurpose nanoplatforms that integrate therapeutic components with multimodality imaging capabilities [13]. SO, ZnO NPs still need the proper survey concerning their biomedical application:

a) Recent studies on the limits of ZnO NPs toxicity toward biological systems are still debatable.
b) Comparison of ZnO NSs from a biological perspective to other metal nanoparticles.
c) There is a lack of evidence-based studies examining the therapeutic functions in enhancing antibacterial and antidiabetic actions.

The study's emphasis on the above issues would bring further interpretation for the use of ZnO NPs in biomedical fields. More experimental research with tunable features such as enhancing surface area to volume proportions, diverse compositions, structures, morphology, performance, efficiency, long-term stability, biocompatibility, large-scale manufacturing, and financial viability is still expected shortly for biomedical applications. The high surface area to volume proportion will allow the system to translocate across cell membranes and will effectively bind analytes which will increase their reactivity toward biosensing applications.

Biocompatibility of ZnO nanomaterials can be increased by slumping the dissolution rate with metal doping or surface capping. To improve ZnO NP's potential for future clinical applications, their surfaces were coated with biocompatible macromolecules such as Polyethylene glycol, polylactic acid, and chitosan [14]. Clinical approval has been granted for the polymers (liposomes and dendrimers) for several medicinal uses. Therefore, it may be necessary to modify the already approved medicinal formulations with functional ligands to raise their diagnostic index. Significant efforts are required to conduct comprehensive long-term in vivo toxicology studies, which are crucial for unlocking the full potential of these fascinating nanomaterials in future biomedical applications. The versatile surface chemistry of ZnO nanomaterials enables the facile conjugation of diverse biocompatible polymers, which further facilitates their biomedical functionality.

12.2.3 Electrochemical Sensors

The attractive characteristics of ZnO NSs such as large specific surface area, powerful adsorption ability, sensitivity, and large catalytic efficiency make them suitable for detecting and sensing low concentrations of biomolecules [15]. One chapter of this book extensively covers the development of various electrochemical sensors and biosensors based on ZnO nanosheets (NSs) for applications in health diagnosis. The chapter explores the influence of multiple factors such as selectivity, sensitivity, detection limit, robustness, and different structural characteristics, providing a comprehensive analysis of their impact.

Although a wide variety of morphologies, particle sizes, and structures have been developed for engineered ZnO nano-sized structures, there is still much to learn about the effects of the synthesis parameters, morphology, pore size sensitivity, selectivity, and response time of ZnO-based sensors [16].

In the future, the activity of ZnO-based electrochemical sensors can be enhanced with surface modification, doping, various nanostructures, advanced fabrication techniques, hybridization, and integration with signal amplification techniques. Nanostructures such as nanowires, nanorods, or nanoparticles enhance the surface area which further allows for more efficient interaction with the target analytes, leading to enhanced sensitivity and detection limits. Modification of the surface of ZnO-based sensors with functional groups such as amino (-NH2), carboxyl (-COOH), hydroxyl (-OH), sulfhydryl (-SH), and phosphate (-PO3H2) groups by the chemical modification such as silanization, covalent bonding, or self-assembly methods. These groups can facilitate the immobilization and enhance the adsorption and interaction of analytes, leading to improved sensitivity and selectivity in sensing applications. The specific choice of functional group depends on the target analytes, desired sensing mechanism, and overall application requirements. Introducing dopants (metal, nonmetals, transition metals) into ZnO structures can alter the electronic properties of ZnO, leading to improved conductivity, increased charge transfer rates, and enhanced sensor response. Combining ZnO with other nanomaterials, such as carbon nanotubes, graphene, carbon nanofibers, and carbon quantum dots, can create hybrid sensor platforms with synergistic effects resulting in enhanced conductivity, increased active surface area, improved charge transfer, and amplified sensing performance. Advancements in fabrication techniques, such as micro/nanofabrication, inkjet printing, or 3D printing, can enable the precise engineering of ZnO-based sensor structures. This level of control over the sensor's geometry and architecture can optimize its performance by tailoring properties such as surface roughness, electrode configuration, and active sensing area. Integration with signal amplification techniques: Integrating ZnO-based sensors with signal amplification techniques, such as nanoparticle amplification, enzymatic amplification, or signal enhancement through catalytic reactions, can significantly boost the sensor's sensitivity and detection capabilities.

By employing these strategies, ZnO-based electrochemical sensors can be enhanced in terms of sensitivity, selectivity, stability, and overall performance, paving the way for their future advancements in various applications including environmental monitoring, healthcare, and food safety. These carbon-based materials can be directly mixed or integrated with ZnO structures through various techniques such as physical deposition, chemical vapor deposition, or surface functionalization. By incorporating these carbon groups, ZnO electrochemical sensors can benefit from enhanced conductivity, increased active surface area, improved charge transfer, and amplified sensing performance.

12.2.4 Flexible Electronics and Optoelectronics

The unique combination of transparency, flexibility, and semiconducting behavior makes ZnO NSs suitable for flexible electronics applications such as displays, wearable electronics, and electronic textiles. A tailored property like enhanced conductivity, high impact strength, low friction coefficient, and wear-resistant controlled band gap have become a prominent choice in various technological utilities [17]. Research endeavors continue to enhance the synthesis methods and mechanical properties of ZnO NSs for seamless integration into flexible electronic devices [18]. The development of ZnO NSs hybrid composites, heterostructure nanohybrids on flexible substrates, and the exploration of novel fabrication techniques are areas of interest for future advancements for aviation and industrial usage. The manufacture of large-scale nanostructured composites will be facilitated by the widespread usage of a variety of nanostructure improvised matrix products, such as nanostructured epoxy resins, prepregs, and sizings. It is currently being developed to prepare functional and structural hierarchical composites with distinct performance characteristics. The hybrid composite structures grown on the fibers, grown on epoxy hybrid, and spreading in the suitable matrix can have the potential to provide better service conditions. Additionally, nanostructured hybrid composites should be able to usher in a new era of multifunctional materials thanks to ZnO nanostructures' outstanding thermal and electrical conductivity [19].

Overall, the prospects of ZnO nanostructures are promising across various fields including electronics, energy, catalysis, biomedical applications, environmental remediation, and flexible electronics. Continued research and development efforts are expected to further enhance their properties and enable their widespread use in practical applications. The nanohybrids exhibit notable distinctions compared to devices composed solely of ZnO. They possess extraordinary high charge mobility, strong quantum confinement, and high photoconductive gain, which are pivotal factors enabling superior device performance that would otherwise be unattainable. These unique characteristics of the nanohybrids contribute significantly to their exceptional device performance.

Practical applications depend on the availability of large-area monolayers and the viability of device creation using conventional microfabrication. Recently, layer-by-layer growth of ZnO, particularly functional crystalline ZnO nanostructures with controlled morphology and orientation on graphene, has been made possible thanks to the compatibility of growth conditions of ZnO and graphene. This is important for controlling the ZnO/graphene interface, which affects the performance of ZnO/graphene nanohybrid devices, as well as scaling up for commercial applications of ZnO/graphene nanohybrid [20]. The exceptional characteristics of ZnO, including its wide bandgap, ferroelectric properties, and piezoelectric properties, along with the ambipolar gating effect of graphene, have opened up new avenues for research. The combination of these properties in ZnO/graphene nanohybrids has garnered significant attention and is driving exploration in flexible and self-powered electronic, photonic, optoelectronic, and electrochemical applications. The synergistic effects of ZnO and graphene in these nanohybrids offer great potential for groundbreaking advancements in various technological fields. Nonetheless, the ZnO/graphene interface also presents a significant "challenge" or, more precisely, a hurdle in the research and development of ZnO/graphene heterostructure nanohybrids. One crucial issue revolves around effectively controlling the ZnO/graphene interface to mitigate defects that can diminish the responsiveness and response speed of ZnO/graphene nanohybrid devices. Addressing this challenge is crucial to optimize the performance and reliability of such devices. Controlling the homogeneity of the ZnO/graphene at various scales is another problem. Future studies on ZnO/graphene nanohybrids will concentrate on this and call for the creation of fresh methods to attain atomic control on surfaces and interfaces. This suggests that in the next few years, more applications for nanostructured hybrid composites will become known.

12.2.5 Environmental Applications

Our environment is constantly under threat with the increasing industries and urbanization. The novel zinc oxide-based photocatalysts exhibited an excellent probability of the absorption of organic pollutants from wastewater, as already discussed in the previous chapter. Photocatalysis possesses numerous properties such as high stability, good probability of photocatalytic declination, and non-toxicity. The ZnO-based photocatalysts offer the best stability as well as strength [21]. The vigorous adaptive quality of ZnO allows an inclusive range of functionalization to enhance the efficiency of photocatalysis [22]. Future research can focus on enhancing the photocatalytic efficiency of ZnO nanostructures by improving their surface area, crystal structure, and charge carrier dynamics. Nevertheless, ZnO nanostructures have a large band gap of the order of 3.3 eV with a large Bohr exciton radius of 2.34 nm that is suitable only in the ultraviolet region. The photocatalysts have a lower band gap, which shows better performance than ZnO. In this regard, they find a problem of photocorrosion because of redox reaction in aqueous medium. Here, photogenerated electrons and holes combine quickly during the whole process. An overall limitation of ZnO can be avoided through their modification and doping, which decreases the band gap that allows them to absorb in the visible light region. Therefore, the ZnO can be combined with materials with a small energy band gap to increase the range of their light absorption tendency in the visible band. ZnO nanostructure increases the light absorption tendency due to an increase in exposed surface area that allows it a good dye absorption.

Hence, future studies are expected by the use of noble metals including Au, Ag, Pt, Pd, and nonmetals (S, N, and C) doping-optimized methods to improve the photocatalytic activity [23, 24]. The heterojunctions could be created with different semiconductors for better photo-degradation of wastewater. By using cheap metals like copper in bimetallic heterojunctions, the cost of photocatalysts could be reduced. The cost-effective titania and ZnO-based nanomaterials could be used in photocatalysis and wastewater treatment without affecting the quality of water during the process.

The in-situ characterizations and classifications of photocatalysts could be studied to enhance the conditions of photocatalytic process. Due to this, a better alteration of ZnO-nanomaterial could support working with different wavelengths of ultraviolet-visible light, which results in cost-effective photodegradation productivity.

12.2.6 Solar Cell Application

The growth in industry establishment rate and energy consumption has increased the demand for renewable energy resources. Solar energy earns the maximum interest among renewable energy resources. Solar cells have received great attention in terms of scientific advancement. ZnO is a semiconducting material that shows good conductivity, electron mobility, easy availability, and cheap cost [25]. These advantages make the ZnO suitable for solar cell applications. ZnO extensively used cathode interlayer material in inverted organic solar cells. Future studies are expected in terms of solar cell systems, and the interface should be more effective than the current progress and improve the light-harvesting ability and a higher surface area among the critical properties. The new solar cell technologies are considered to have low penetration in the market; therefore, progress to achieve low-cost techniques to prepare ZnO with excellent sustainability and environmental friendliness will boost the positive impacts on the social and economic aspects. Further, the fabrication of stable power conversion efficiency required the collection of a modest dispensation route for the deposition of the transport buffer layer. A suitable optimization of doping concentration and framework thickness, modifications in organic transport buffer layer which can demonstrate excellent stability under heat, light, and moisture. This can be used to make less expensive and stable solar cells [26].

ZnO nanostructures are frequently used as dynamic layers of solar cells and photoanodes, but a simple ZnO nanostructure cannot encounter extensive demand for a higher rate of photoelectric conversion. Hence, future studies can be based on the light capture efficiency of solar cells, optimization of the material growth, and structure design to recover the performance of solar cells. Refining the light capture capability of ZnO nanostructures and increasing the period of electron to synthesis of ZnO nanostructure heterojunction materials in the solar cell field are the prospects [27]. The interference area of heterojunction, the channel of carrier separation can be amplified by using the different tactics of corresponding energy band and modification on the surface.

Moreover, the transference performance of interface carriers can be extended as well as upgraded by regulating the aspect ratio of nanostructures. The density functional theory (DFT) theoretical calculation method can help the transport mechanism of heterojunction interface carriers.

12.2.7 Energy Conversion and Storage

ZnO nanostructures can be used in solar cells to convert sunlight into electricity. ZnO possesses a high surface-to-volume ratio, which increases light absorption and promotes efficient charge separation [28]. These nanostructures can convert mechanical stress or vibrations and waste heat into electrical energy. ZnO nanostructures have been also explored for battery applications such as Zn-ion and Li-ion batteries. They are used as anodes or cathodes, which can improve stability, capacity, and fast charge-discharge kinetics. Future research focuses on improving the stability as well as efficiency of ZnO nanostructures for energy conversion and storage applications. The combination of ZnO nanostructures with other semiconducting materials can lead to novel hybrid devices with improved efficiency and functionality. Semiconductor provides conduction band (CB) and valence band (VB) energy levels and the charge transfer occurring at these levels can be controlled easily. Instead of using two interfaces, one interface device such as by joining p-type and n-type semiconductors can work proficiently [29].

12.3 CONCLUSION

This chapter highlights the vast and promising prospects for ZnO nanostructures, with continued research expected to unlock new applications and address existing challenges in various fields.in food industries, ZnO nanoparticles enhance packaging performance and antimicrobial activities. In biomedical applications, the antimicrobial properties, biocompatibility, and cost-effectiveness of ZnO make it promising for drug delivery, cancer therapy, and diabetes management. In flexible electronics and optoelectronics, ZnO's transparency and semiconducting behavior support the development of display and wearable electronics. environmental applications of ZnO include photocatalysis for pollutant degradation and wastewater treatment. ZnO nanostructures are also used in solar cells for energy conversion and storage, with ongoing efforts to enhance light capture, stability, and efficiency. Hybrid devices combining ZnO with other materials hold the potential for improved performance.

REFERENCES

1. Theerthagiri, J., Salla, S., Senthil, R. A., Nithyadharseni, P., Madankumar, A., Arunachalam, P., & Kim, H. S. (2019). A review on ZnO nanostructured materials: energy, environmental and biological applications. *Nanotechnology*, *30*(39), 392001.
2. Wu, Y., Hasan, T., Li, X., Xu, P., Wang, Y., Shen, X., & Yang, Q. (2015). High efficiency single Ag nanowire/p-GaN substrate Schottky junction-based ultraviolet light emitting diodes. *Applied Physics Letters*, *106*(5), 051108.
3. Weintraub, B., Zhou, Z., Li, Y., & Deng, Y. (2010). Solution synthesis of one-dimensional ZnO nanomaterials and their applications. *Nanoscale*, *2*(9), 1573–1587.
4. Hatamie, A., Khan, A., Golabi, M., Turner, A. P., Beni, V., Mak, W. C., & Willander, M. (2015). Zinc oxide nanostructure-modified textile and its application to biosensing, photocatalysis, and as antibacterial material. *Langmuir*, *31*(39), 10913–10921.
5. Franco, M. A., Conti, P. P., Andre, R. S., & Correa, D. S. (2022). A review on chemiresistive ZnO gas sensors. *Sensors and Actuators Reports*, *4*, 100100.

6. Guo, T., Hu, P., Li, L., Wang, Z., & Guo, L. (2023). Amorphous materials emerging as prospective electrodes for electrochemical energy storage and conversion. *Chem*, *9*(5), 1080–1093.
7. Kim, I., Viswanathan, K., Kasi, G., Thanakkasaranee, S., Sadeghi, K., & Seo, J. (2022). ZnO nanostructures in active antibacterial food packaging: preparation methods, antimicrobial mechanisms, safety issues, future prospects, and challenges. *Food Reviews International*, *38*(4), 537–565.
8. Roopa, H., Panghal, A., Kumari, A., Chhikara, N., Sehgal, E., & Rawat, K. (2023). Active Packaging in the Food Industry. *Novel Technologies in Food Science*, 375–404.
9. Shankar, S., Rhim, J. W. (2018). Bionanocomposite Films for Food Packaging Applications. *Reference Module in Food Science*, *1*, 1–10.
10. Jokar, M., Pedersen, G. A., Loeschner, K. (2017). Six Open Questions about the Migration of Engineered Nano-objects from Polymer-based Food-contact Materials: A Review. *Food Additives & Contaminants: Part A 34*(3), 434–450.
11. Rasmussen, J. W., Martinez, E., Louka, P., & Wingett, D. G. (2010). Zinc oxide nanoparticles for selective destruction of tumor cells and potential for drug delivery applications. *Expert Opinion on Drug Delivery*, *7*(9), 1063–1077.
12. Jiang, J., Pi, J., & Cai, J. (2018). The advancing of zinc oxide nanoparticles for biomedical applications. *Bioinorganic Chemistry and Applications*, *2018*(1), 1062562.
13. Martínez-Carmona, M., Gun'Ko, Y., & Vallet-Regí, M. (2018). ZnO nanostructures for drug delivery and theranostic applications. *Nanomaterials*, *8*(4), 268.
14. Ledesma, F., Ozcan, B., Sun, X., Medina, S. M., & Landry, M. P. (2022). Nanomaterial strategies for delivery of therapeutic cargoes. *Advanced Functional Materials*, *32*(4), 2107174.
15. Retamal, J. R. D., Lien, W. C., & He, J. H. (2012). Biosensing applications of ZnO nanostructures. In *Handbook of Innovative Nanomaterials: From Syntheses to Applications* (pp. 665–692). Pan Stanford Publishing Pte. Ltd..
16. Napi, M. L. M., Sultan, S. M., Ismail, R., How, K. W., & Ahmad, M. K. (2019). Electrochemical-based biosensors on different zinc oxide nanostructures: A review. *Materials*, *12*(18), 2985.
17. Ravi, S., & Bajpai, V. (2021). Recent advances in ZnO nanostructures and their future perspective. *Advanced Nano Research*, *11*(1), 37–54.
18. Fortunato, E., Barquinha, P., & Martins, R. (2012). Oxide semiconductor thin-film transistors: a review of recent advances. *Advanced Materials*, *24*(22), 2945–2986.
19. Zhang, Y., Zhang, L., Cui, K., Ge, S., Cheng, X., Yan, M., … & Liu, H. (2018). Flexible electronics based on micro/nanostructured paper. *Advanced Materials*, *30*(51), 1801588.
20. Wu, J., & Gong, M. (2021). ZnO/graphene heterostructure nanohybrids for optoelectronics and sensors. *Journal of Applied Physics*, *130*(7). 070905.
21. Puri, N., & Gupta, A. (2023). Water remediation using titanium and zinc oxide nanomaterials through disinfection and photo catalysis process: a review. *Environmental Research*, *227*, 115786.
22. Abdullah, F. H., Bakar, N. A., & Bakar, M. A. (2022). Current advancements on the fabrication, modification, and industrial application of zinc oxide as photocatalyst in the removal of organic and inorganic contaminants in aquatic systems. *Journal of Hazardous Materials*, *424*, 127416.
23. Güy, N., & Özacar, M. (2016). The influence of noble metals on photocatalytic activity of ZnO for Congo red degradation. *International Journal of Hydrogen Energy*, *41*(44), 20100–20112.
24. Zhang, Q. P., Li, J., & Xu, M. (2022). Ag decorated ZnO based nanocomposites for visible light-driven photocatalytic degradation: basic understanding and outlook. *Journal of Physics D: Applied Physics*, *55*(48), 483001.

25. Wibowo, A., Marsudi, M. A., Amal, M. I., Ananda, M. B., Stephanie, R., Ardy, H., &Diguna, L. J. (2020). ZnO nanostructured materials for emerging solar cell applications. *RSC Advances*, *10*(70), 42838–42859.
26. Raj, A., Kumar, M., & Anshul, A. (2021). Recent advancement in inorganic-organic electron transport layers in perovskite solar cell: current status and future outlook. *Materials Today Chemistry*, *22*, 100595.
27. Qiao, F., Sun, K., Chu, H., Wang, J., Xie, Y., Chen, L., & Yan, T. (2022). Design strategies of ZnO heterojunction arrays towards effective photovoltaic applications. *Battery Energy*, *1*(1), 20210008.
28. Ma, W., Zhang, Y., Pan, S., Cheng, Y., Shao, Z., Xiang, H., … & Zhu, M. (2021). Smart fibers for energy conversion and storage. *Chemical Society Reviews*, *50*(12), 7009–7061.
29. Zhu, B., Fan, L., Mushtaq, N., Raza, R., Sajid, M., Wu, Y., … & Yun, S. (2021). Semiconductor electrochemistry for clean energy conversion and storage. *Electrochemical Energy Reviews*, *4*, 757–792.

Index

Pages in *italics* refer to figures and pages in **bold** refer to tables.

G

H

I

J

K

L

M

N

O

P

Q

R

FSC
www.fsc.org
MIX
Papier | Fördert
gute Waldnutzung
FSC® C083411

Kolibri 360 GmbH
Ferdinand-Jühlke-Straße 7
99095 Erfurt, Deutschland
produktsicherheit@kolibri360.de